Terahertz Antenna Technology for Imaging and Sensing Applications

Isha Malhotra • Ghanshyam Singh

Terahertz Antenna Technology for Imaging and Sensing Applications

Springer

Isha Malhotra
Electronics and Communication Engineering
Dronacharya College of Engineering
Gurugram, Haryana, India

Ghanshyam Singh
Electrical and Electronics Engineering Science
University of Johannesburg
Johannesburg, South Africa

ISBN 978-3-030-68962-9 ISBN 978-3-030-68960-5 (eBook)
https://doi.org/10.1007/978-3-030-68960-5

This Springer imprint is published by the registered company Springer Nature Switzerland AG
The registered company address is: Gewerbestrasse 11, 6330 Cham, Switzerland

Preface

Terahertz waves lies in the frequency range where the molecular resonance dominates. Earlier this relatively unexplored part of the electromagnetic spectrum has been called the terahertz gap owing to the lack of efficient sources and detectors which could generate and detects electromagnetic signals in this frequency regime of the spectrum. However, this regime of spectrum is scientifically rich with the emerging possibilities and attracting a lot of attention due to its unique properties that are favourable to various applications such as future generation communication, non-destructive testing, security scanning and process control. The unique features of terahertz waves are better resolution than a microwave, unique spectral absorption, non-ionizing radiation, and an ability to propagate through many types of dielectric materials. Moreover, the spectroscopy with high spectral resolution at frequencies in terahertz regime of the electromagnetic spectrum is a powerful analytical tool for investigating the structure and the energy levels of molecules and atoms. Using terahertz spectroscopy, it is possible to detect explosive or illicit drugs even if they are concealed because the terahertz radiation can rapidly transmit through plastics, clothing, luggage, paper products and other non-conductive materials. By comparing measured reflectivity terahertz spectra with known calibration spectra, it is easier to identify the presence of these agents and distinguish them from benign objects. Therefore, the terahertz science and technology are progressing at a tremendous speed as evidenced by the ever-increasing potential applications such as future generation communication system with imaging and sensing. The field is becoming large enough that we can begin to sub-divide it into various specializations. This technology although rapidly progressing however it is still relatively immature. Therefore, it is highly appropriate that this book focuses on emergent applications and technology in order to inspire and motivate further progress in the engineering domain and have become an important tool for the non-invasive sensing and imaging of various materials and structures, with key applications in defence, security, and healthcare industry (biomedical imaging and sensing).

For the imaging applications, the continuous wave (narrowband) or pulsed wave (broadband) terahertz systems for the generation of terahertz radiation is used and this classification is based on the laser source used to generate photocurrent. Moreover, the applications in homeland security, defence and safety in aviation industry have put large demands on the development of advanced imaging systems. While the visible and infrared imagery is providing high image quality, but they are hindered by atmospheric obscurants. Therefore, the terahertz imaging is providing an attractive method to reduce or eliminate the impact of low visibility atmospheric conditions. Moreover, in the close-range sensing and imaging applications, the pulsed terahertz sources are more favourable for acquiring depth information. The advantages of pulsed terahertz time-domain spectroscopy are that broad spectral information for 0.1–3 THz can be acquired from a single picosecond's terahertz pulse as well as the depth information from the difference in arrival times of short pulses can be obtained.

However, the challenges in present terahertz system includes size, cost, output power, signal-to-noise ratio (SNR), bandwidth, depth penetration, water sensitivity, spatial resolution, speed of data acquisition and the lack of a terahertz frequency knowledge base. Various terahertz applications would be benefited from the compact integration of terahertz devices and other types of functional devices. Further, for the terahertz imaging applications, there is a requirement of planar and compact terahertz antenna sources with on-chip fabrication and high directivity in order to achieve large depth-of-field (DoF) for better image resolution. The terahertz antennas based upon photoconduction techniques are the most common devices in terahertz systems, and the photoconductive dipole antenna (PCA), being simple in fabrication among them, is the extensively utilized terahertz source for pulsed broadband system used in terahertz imaging and spectroscopy systems. A well-designed sub-wavelength (micrometre) scale photoconductive dipole antenna structure has the potential for high output power generation and broadband terahertz pulse emission which is useful for terahertz imaging system. However, the low values of gain, directivity and optical-to-terahertz conversion efficiency are the important limitations of basic photoconductive dipole antenna. Therefore, the motivation of this book is to theoretically establish the key modalities of improving photoconductive dipole antennas performance for imaging and sensing applications and to determine methods/techniques to improve the directivity of the photoconductive dipole antenna analytically. These investigations are essential to gain a better understanding of terahertz photoconductive antenna performance.

Due to simple structure of photoconductive dipole antenna, an array implementation could be explored to enhance the radiation parameters. The photoconductive dipole array antenna improves the gain and directivity and therefore is useful to enhance the imaging capabilities to address the considerations such as limited depth-of-field (DoF), that is, the distance over which an object is considered in focus and size-weight-and-power (SWaP) of terahertz source for imaging applications. These are important considerations for applications like stand-off imaging and surveillance of moving targets where the high angular resolution as well as extended depth-of-field is the key for successful detection of hidden explosives and illicit drugs. To enhance the directivity of small-gap photoconductive dipole antenna, the frequency selective surface (FSS) as bandpass spatial filter is employed. Further, to preserve the purity of incoming THz wave on the receiving antenna side where the electromagnetic interference (EMI) is the major hindrance, the use of FSS structure with antenna geometry in THz region is beneficial to enhance the antenna performance parameters. A technique of using frequency selective surface with photoconductive array antenna is also presented to further enhance the directivity from the radiating structure to yield high image resolution. Moreover, by using this technique, radiation energy will be confined to the desired frequency band rather than spreading over a wide spectrum range. Moreover, the beam steering technique is implemented on the photoconductive dipole antenna array configuration for near-field scanning of suspicious objects for the detection of hidden explosives. The intrinsic properties of terahertz technology result in a wide scope of potential sensing and imaging applications, such as:

- Its non-ionizing nature makes it safe for use on biological and bio-medical applications.
- It can penetrate fabrics and plastics, allowing for non-destructive inspection of these materials and scanning underneath these layers.
- It cannot penetrate water or metal, enabling its use for analysis of hydration levels or detecting the presence of metallic components in samples.
- Many materials have unique spectral fingerprints in the THz range, such as different types of explosives or several compounds used in the fabrication of medicines. THz would be ideal for detection of these components in samples.

The production of terahertz waves can be divided into purely electronic sources and laser-driven/optoelectronic ones. The latter can then be split into pulsed and continuous-wave terahertz radiation, the second of which allows for frequency selectivity-which means that the source wavelength can be precisely selected and tailored to a specific task. The continuous-wave terahertz waves are produced when two laser diodes with adjacent wavelengths are combined and focused on a photomixer, producing a new laser beam in the terahertz frequency range. As the detector records the terahertz wave coherently (meaning both the amplitude and the phase of the signal are recovered), the technology allows for a more complete analysis of the sample under investigation. The production and detection of THz waves remains technically challenging and expensive, preventing its broader use in many application domains. The state-of-the-art continuous-wave THz production/detection uses a single photomixer with limited power, with imaging being performed by scanning. Further, the use of photoconductive antenna for compact THz imaging and sensing at a much lower cost with high performance is the requirement for high-end technologies. In addition to this, the emitter array generates narrow, high intensity, steerable beams like radar imaging. Moreover, the detector array allows for a fast, coherent detection of terahertz waves.

No dedicated book that discusses the use of terahertz frequency for imaging applications and the contents related to the use of photoconductive antenna technology in this frequency regime of the spectrum was available until now, and these have now been elaborated in this book. Therefore, the unique features of this book are as follows:

- Comprehensive review of terahertz source and detector for imaging and sensing.
- Examines techniques for the imaging and sensing applications.
- Discusses the photoconductive antenna technology for imaging and sensing.
- Explores applications in tomographic imaging, art conservation and the pharmaceutical and aerospace industries.
- A simple synthesis technique to find out the physical parameters of small-gap photoconductive dipole antenna is exploited.
- Certain modalities for improving the photoconductive dipole antenna performance for imaging and sensing applications are presented.
- A thorough systematic framework to assess the physical phenomenon occurring across the small-gap photoconductive dipole antenna is determined using an analytical procedure utilizing explicit mathematical expression.
- A pragmatic description of the small-gap photoconductive dipole antenna, which shows the relationship between the parameters of equivalent circuit model and dimensions of antenna electrodes, is presented.
- To enhance the directivity of small-gap photoconductive dipole antenna, the frequency selective surface (FSS) as bandpass spatial filter is employed with the antenna.

This book is intended for senior undergraduate and graduate students working in electronics/electrical and communications engineering; however, it can also be used as a reference book for engineering professionals and scientists working in academia and industry. In this book, we have tried to cover most representative achievements, with emphasis on thoughts, philosophy and methodology of terahertz wave technology for imaging and sensing application. Indeed, the existing knowledge is undoubtedly important; however, in this book, we have provided our thoughts and methodology which helps in creating new knowledge for the future imaging and sensing applications. The recent development of easy-to-use sources and detectors of terahertz radiation has enabled potential growth in the terahertz imaging and sensing technology, which is vastly adaptable and offers great potential across a wide range of applications. This book of terahertz technology for imaging and sensing applications explores the fundamental principles, recent developments and key emerging applications in this exciting field. An authoritative introduction to the fundamentals of terahertz technology for imaging and sensing applications with the generation, detection and emission of waves are discussed. All the relevant literature on the current state-of-the-art pulsed THz imaging applications and potential challenges of THz antenna technology for imaging applications has been reviewed. The topics covered in the book are concerned with the THz imaging technology and the design and development of the photoconductive antenna with its array configuration to meet the demand of imaging criteria in this regime of the spectrum. In this way, this book will attract various stages of readers as given below:

- The book will help the industry in conceptualizing the development of THz imaging and sensing systems.

- Various issues related to the THz imaging and sensing systems employing antenna technology have been dealt in this book. An additional issue of sources, detectors, interconnects and further enhancements of the gain/directivity are still open for the research, it will help to scholars and scientists to look ahead in this field.
- For the under-graduate and post-graduate students, it will assist in learning a new technology.

This book comprises 11 chapters. Chapter 1 provides a comprehensive review of the state-of-the-art terahertz sources and detectors, terahertz imaging modes, and terahertz imaging analysis. The sources include laser-based generation techniques and solid-state devices. A variety of detectors are also covered.

Chapter 2 discusses the terahertz imaging modalities in terms of transmission type and reflection type imaging. The terahertz spectrum is a rich source of material information and allows the identification of material species such as bacterial spores hidden inside optically opaque material. Therefore, the terahertz imaging with diffraction limit and tomography is presented along with its potential research challenges.

In Chap. 3, the state-of-the-art THz antennas based on integrated circuit technology for imaging and sensing applications are discussed.

In Chap. 4, a simple synthesis technique is presented to determine the physical parameters of photoconductive dipole antenna, which shows its application for THz sensing and imaging to detect the presence of hidden explosives having spectral fingerprints in the range 1–3 THz. To increase the antenna radiation efficiency, which also contributes to the total efficiency of photoconductive dipole antenna, a thin superstrate over the substrate of the dipole antenna is used to enable the antenna to withstand high biased voltage. The proposed antenna design is compact in size, which further helps to achieve the diffraction limited imaging. Moreover, the use of silicon lens with the photoconductive dipole antenna helps to focus the beam in the desired direction with increased directivity.

In Chap. 5, an analytical procedure making use of explicit mathematical expressions leading to the physical behaviour of small-gap photoconductive dipole antenna is developed, and for this, a comprehensive systematic framework to determine the physical phenomenon occurring across the small-gap photoconductive dipole antenna is presented. Moreover, an optimization of size of photoconductive gap to improve the total antenna efficiency is discussed and its consequence on radiated power is determined. As the THz antenna is one of the most important components in a THz sensing and imaging system, there is a need to have such a transmitting THz source with high directivity and optimum radiation efficiency.

Therefore, to achieve this objective to enhance the directivity of photoconductive dipole antenna, in Chap. 6, a numerical computation and design of frequency selective surface (FSS) which acts as a bandpass spatial filter is presented and is deployed with photoconductive dipole antenna. The bandpass FSS structure parameters have been computed using simple synthesis technique. The effect of slot size of FSS to improve the scattering parameter at resonating frequency of the small-gap photoconductive dipole antenna is also presented. The most basic coherent imaging can be achieved by raster scanning a sample through the THz wave focus and generating complete spectroscopic information at each pixel corresponding to the antenna structure. Therefore, to speed up the measurements, there is a necessity of making a THz system for parallel or accelerated measurements. A potential approach to parallelization is the implementation of multi-antenna setup, which enables a linear downscaling of the measurement time with the deployment of number of THz emitter and detector pairs.

However, the compact size and planar arrays have to be used to counter the increase of system cost. Looking into such necessity, in Chap. 7, we have designed an array of small-gap photoconductive dipole antenna at THz frequency by using a frequency selective surface which not only provides a planar profile for THz radiating source but also offers enhanced gain and directivity for imaging application to detect the hidden explosives. Moreover, how to control the transmission characteristics in a particular THz frequency band with the placement of array of FSS across the array of photoconductive dipole antenna is also discussed. The complete antenna array assembly with FSS is made useful to form an image by scanning a single beam in both principle planes.

In Chap. 8, the beam-steering characteristics of the small-gap photoconductive dipole phased array antenna is presented. With uniform distribution of optical source excitation and progressive phase shift in x-axis and y-axis of the antenna configuration, the beam steering phenomenon has been determined.

In Chap. 9 provides an extensive overview of state-of-the-art terahertz near-field imaging and sensing techniques and their applications. Terahertz near-field microscopy is one of the simple technique which shows the capability of beating the terahertz diffraction limit. It provides imaging with a spatial resolution much better than half- of-a wavelength imaging. The measurement techniques for near-field imaging are discussed and analytical estimate based on Kirchhoff's formalism for near-field is presented.

In Chap. 10, the terahertz biomedical imaging has become a modality of interest due to its ability to simultaneously acquire both image and spectral information. Terahertz imaging systems are being commercialized, with increasing interest in a biomedical setting. Advanced digital image processing algorithms are greatly needed to assist screening, diagnosis and treatment. Pattern recognition algorithms play a critical role in the accurate and automatic process of detecting abnormalities when applied to biomedical imaging. This goal requires classification of information-bearing physical contrast patterns and identification of information in images, for example, distinguishing between different biological tissues or materials. T-ray tomographic imaging and detection technology contributes especially to our ability for discriminating opaque objects with clear boundaries and makes possible significant potential applications in both in-vivo and ex-vivo environments.

Finally, the concept of integrated THz biosensor chip for imaging and sensing application is briefed in Chap. 11. It includes the future directions in terms of certain design techniques which can be incorporated in the basic structure of small-gap photoconductive dipole antenna to further improve its performance parameters for its use in THz imaging applications.

In summary, the book provides a unified view of the state-of-the-art of terahertz imaging and sensing technology, which should be accessible to a readership with basic knowledge about electromagnetic and RF microwave antenna theory and signal processing. The readership may find the rich set of references in each of the chapters very useful. We have done a good job by providing a concise summary of all the chapters in the preface of the book. The book is useful for graduate students, researchers and engineers working or intended to work in the area of imaging and sensing application or other Terahertz supported applications. Although numerous journal/conference publications, tutorials, and books on terahertz imaging and sensing applications have been published in the last few years, the vast majority of them focus on the various sources and detector technology. This book distinguishes itself from the existing prosperous literature of terahertz imaging and sensing applications and enabling technologies. The emphasis of such books is on the theoretical design and performance evaluation of terahertz imaging and sensing techniques employing antenna technologies.

We are indebted to numerous colleagues for the valuable suggestions during the entire period of manuscript preparation. Isha Malhotra is thankful to the Chairman and CEO of Dronacharya College of Engineering, Gurugram, India, for providing all the needed support to complete the work. She also expresses greatest gratitude to her mentor Prof. (Dr.) Ghanshyam Singh for providing an opportunity and motivation to explore the potential and work in the field of THz imaging and sensing. She conveys highest gratitude to her parents whose tremendous blessings and endless sacrifice always encouraged during writing of the manuscript as well as carrier growth. They have shown immense patience and supported in every possible way. Words alone can never express the gratitude to them. The first author also extends warm thanks to her brother, niece and son; their innocent smiles and chats always bring cheerfulness even in difficult times. We are especially thankful to Prof. (Dr.) B N Basu, IIT (BHU), India, for motivation. We would also like to thank publishers at Springer, in particular Charles B. Glaser, Brian P. Halm, Nicole Lowary and Arun Pandian KJ for their helpful guidance and encouragement during the creation of this book. We cannot not justify our work without showing gratitude to our family members, who have always been the source of strength that helped us work tirelessly to complete the manuscript. Last but not least, we thank the ultimate source of energy of every particle in the universe, the Almighty, for giving enough energy and strength to complete this work. All praise and gratitude belong to Him.

Gurugram, Haryana, India — Isha Malhotra
Johannesburg, South Africa — Ghanshyam Singh

Contents

About the Authors

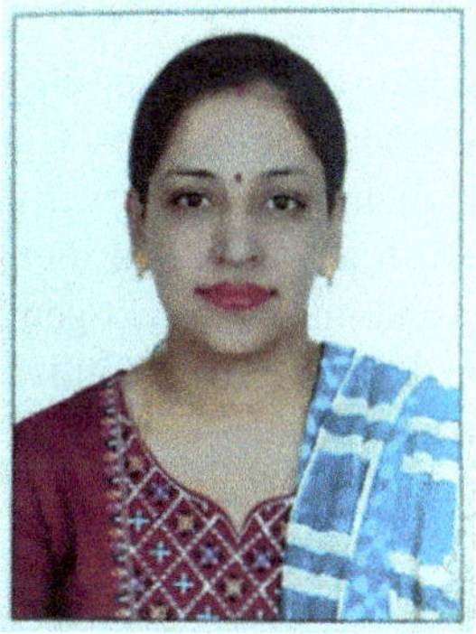

Isha Malhotra received her Ph.D. degree in electronics and communication engineering from Jaypee University of Information Technology, Waknaghat, Distt. Solan, Himachal Pradesh, India. She had received her M.Tech. degree in electronics and communication engineering from the Department of Electronics and Communication Engineering, Guru Nanak Dev Engineering College, Ludhiana, Punjab, India. She received her B.Tech. (Honours) degree from Punjab Technical University, Jalandhar, Punjab, India. She had worked as an assistant professor at the Electronics and Communication Engineering Department, Chitkara University, Rajpura, Punjab, India. Dr. Malhotra had worked as an associate professor/HoD at the Electronics and Communication Engineering Department, Dronacharya College of Engineering, Gurugram, India. At present, she is working as professor/principal at Dronacharya College of Engineering, Gurugram, India. She has published a number of peer-reviewed research articles in international journals and conferences. She has supervised multiple M.Tech. thesis. She has worked as reviewer for several reputed journals and conference. Her areas of research interest are THz imaging system, photoconductive dipole antenna, gain and directivity enhancement of conventional THz antennas for imaging application, frequency-selective surface and sensor technology.

Ghanshyam Singh received his Ph.D. degree in electronics engineering from the Indian Institute of Technology, Banaras Hindu University, Varanasi, India, in 2000. He was associated with Central Electronics Engineering Research Institute, Pilani, and Institute for Plasma Research, Gandhinagar, India, respectively, where he was research scientist. He had also worked as an assistant professor at the Electronics and Communication Engineering Department, Nirma University of Science and Technology, Ahmedabad, India. He was a visiting researcher at the Seoul National University, Seoul, S. Korea. At present, he is professor in the Department of Electronics and Communication Engineering, Jaypee University of Information Technology, Wakanaghat, Solan, India. Dr. Singh is author/co-author of more than 180 scientific papers of the refereed journasl and international conferences. His research and teaching interests include RF/microwave engineering, millimetre/THz wave antennas and their applications in communication and imaging, next generation communication systems (OFDM and cognitive radio) and nanophotonics. He has more than 14 years of teaching and research experience in the area of electromagnetic/microwave engineering, wireless communication and nanophotonics. He has supervised multiple Ph.D. and M.Tech. theses. Dr. Singh has worked as a reviewer for several reputed journals and conferences.

List of Figures

List of Tables

Chapter 1
Introduction

1.1 Introduction

Over the past two decades, the terahertz (THz) technology has been witnessing significant advancements on the scientific front. THz technology is an emerging field that has potentials to improve applications ranging from passenger scanning at airports to huge digital data transfers. THz radiation is located between the frequency bands of microwaves and infrared radiation, and is capable of easily penetrating into various materials, including biological tissue. The energy carried by THz radiation is low enough to pose no risk to any object under investigation. Technological advancements have enabled the THz frequency band to be accessible for imaging systems. Many manufacturers are currently focused on developing new devices that can send and receive radiation in the THz frequency range. The THz technology has found several applications in various fields such as spectroscopy, imaging, and communication systems, over the years. Additionally, the growing application of THz technology in a number of industries for quality check and process control monitoring is expected to enable THz components and systems to register a remarkable rise in the coming years.

Moreover, an increasing adoption of THz technology in industries, such as aerospace and defense and homeland security across regions, is thereby driving the market for THz technology over the forecast period. In addition, some of the prominent drivers for THz technology market are factors such as increasing investments in R&D activities, growing penetration of THz technology in biology and medical science sectors, and increasing applications of high-end THz devices in research laboratory and process control monitoring processes. On the other hand, factors such as the lack of awareness of THz technology can result in a slow adoption rate and concerns regarding accuracy and reliability issues may act as a major restraint for the THz technology market.

THz technology market can be segmented on the basis of component, system, application, vertical, and regions. On the basis of component, the THz technology

I. Malhotra, G. Singh, *Terahertz Antenna Technology for Imaging and Sensing Applications*, https://doi.org/10.1007/978-3-030-68960-5_1

market can be segmented into THz sources and THz detectors. On the basis of system, the THz technology market can be segmented as THz imaging, THz spectroscopy, and THz communication systems. Whereas the application segment in the THz technology market includes industrial, defense and security, biomedical, and wireless communication. Regionally, the THz technology market can be segmented into North America, Latin America, Eastern Europe, Western Europe, Asia Pacific excluding Japan, Japan, and Middle East and Africa (MEA). The THz technology market is dominated by the North America region. North American dominance is attributed to the rising adoption of process control in various industries as well as high investments in R&D activities pertaining to THz technology by key players based in the region. On the other hand, Asia Pacific excluding Japan, and Japan are expected to grow significantly, as compared to the other regions in the THz technology market. Asia Pacific excluding Japan is expected to utilize great opportunities in the THz technology market due to the wide presence of manufacturing companies in developing countries, such as China and India. Eastern Europe, Latin America, and MEA are expected to see a moderate growth rate in the THz technology market.

Among the prominent players involved in the THz technology market, companies, such as Advantest Corp, TeraView, EMCORE Corp, and Terasense Group, Inc, among others are focusing on organic as well as inorganic strategies to strengthen their position in the THz technology market. For instance, in December 2016, the Terasense Group, Inc. launched THz wave sources based on the “IMPATT” (Impact Ionization Avalanche Transit Time) technology for imaging scanners. On the other hand, in 2015, TeraView launched a continuous- wave THz system “CW Spectra 400” along with fiber-fed external devices. According to Stratistics MRC, the Global THz Technology Market accounted for $84.53 million in 2015 and is expected to reach $491.27 million by 2022 growing at a CAGR of 28.5% from 2015 to 2022. The growing demand from defense and healthcare sectors and the high adoption rate of the technology in laboratory research applications are driving the market growth. The growing recognition of the optoelectronic approaches further assists in the market growth. Moreover, the technological advancements generate more opportunities for the adoption of THz technology for a range of new applications.

1.2 Terahertz Electromagnetic Spectrum

A portion of the electromagnetic spectrum, which lies in between the optical and microwave regime, is known as the THz gap [1]. It is named so, because earlier, in comparison to a well-developed technology in the microwave regime and the optical domain of the electromagnetic spectrum, the basic research, advanced technology developments, and new initiatives in the THz band have been very limited and stayed relatively unexplored. As shown in Fig. 1.1, the electronics (millimeter waves) and photonics (infrared waves) sandwich the THz band where the

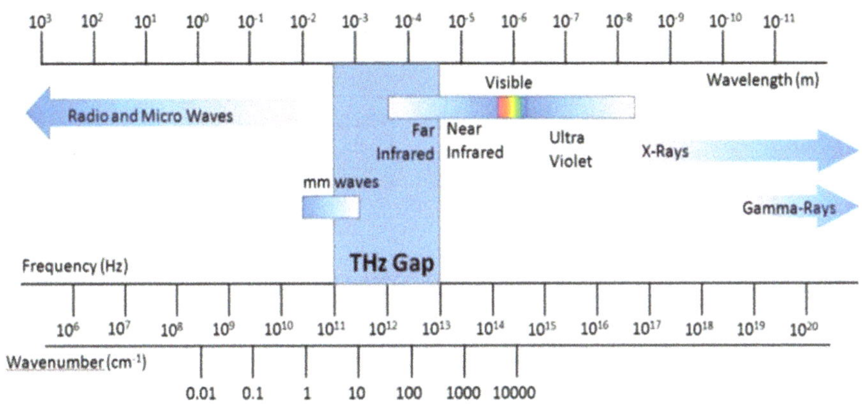

Fig. 1.1 The position of terahertz (THz) region in the electromagnetic spectrum [1]

semiconductor electronics and the optical technologies find their applications. The unavailability of reliable, compact, temperature-insensitive, and efficient power sources and detectors was the key obstacle to the popularity of the THz band of the electromagnetic spectrum. However, in the last two decades, with a significant progress in the technology, the development of solid-state mode-locked and quantum cascade lasers (QCLs), laser-based THz time-domain spectroscopy (TDS), and microelectronic fabrication of planar antennas have paved the way for the imaging technology at the THz frequency band.

Moreover, the advent of Er^+-doped femtosecond fiber lasers has replaced the costly systems made of titanium:sapphire lasers for THz systems [2]. With many more such technological advancements that came through in the field of semiconductor physics and technology, many researchers [3] are now exploring the utilization of the license-free THz band of the electromagnetic spectrum for several potential applications.

1.3 Application of Terahertz Radiations

Recently, for the THz electromagnetic spectrum, various emerging applications have been introduced in the biological imaging [4], nondestructive testing (NDT) [5], security scanning [6, 7], and process control [8] up to the next-generation wireless communication systems [9]. Within THz regime of the electromagnetic spectrum, there is a capability of the Terahertz wave for multiple interactions with the matter through which it passes away. Such interaction results into several biological, chemical, molecular as well as physical structures to show the unique spectral fingerprints in addition to the rotational modes at specific frequencies of the THz band. Moreover, the high absorption of THz radiation by oxygen molecules with polar liquids such as water is the main limitation for the wave to travel over a long

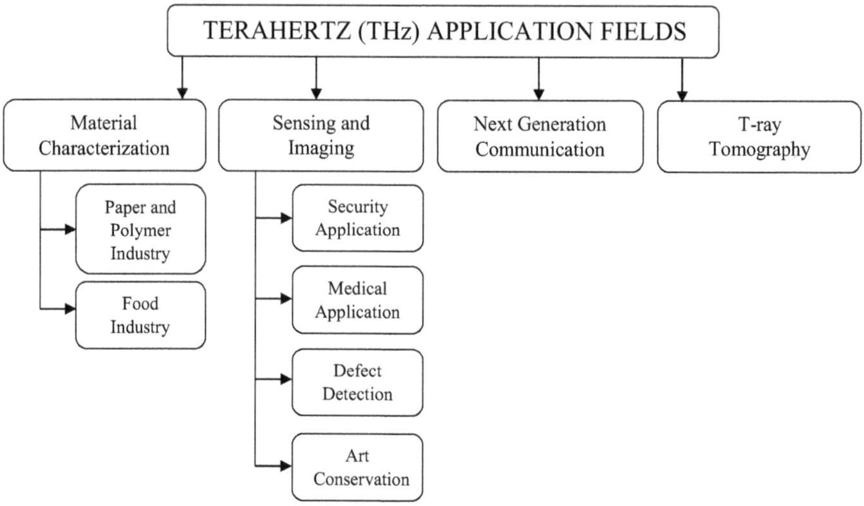

Fig. 1.2 The potential THz application areas

distance in the atmosphere [10]. Furthermore, the THz radiation penetrates the fabric, foam, and plastic [11], even as they are almost totally reflected by metals and the photon energy levels of THz rays derive an extra characteristic. Categorically, Fig. 1.2 shows that the applications of THz radiations fall within either of the material characterization, sensing and imaging, and the communication fields. However, the THz band characteristics themselves interlink these applications, which are widening gradually.

The use of THz waves in numerous application areas has become possible due to several advantages and salient features of THz waves relating to the microwave and infrared region, as mentioned below:

- In comparison to the microwave, the THz band comprises of a wider bandwidth. Moreover, the microwave band is already being utilized for different services and applications. On the other hand, the terahertz frequency range is unregulated (license-free) by regulatory authorities and a 250 GHz frequency is the maximum allocated frequency of this range, therefore, the THz band can be used for several applications.
- The THz wave is a nonionizing radiation which has an advantage over the ionized imaging systems such as positron emission tomography (PET), magnetic resonance imaging (MRI), planar X-rays, and X-ray with computed tomography (CT) scans being presently used. Therefore, the THz wave is useful for biological imaging applications. Moreover, with the frequency sensitivity feature of THz absorption, the THz waves can be used to identify differences in biological tissues without providing any harm.
- The THz wave scatters less in comparison to the infrared frequencies and light wave because scattering depends on the wavelength of the wave. Moreover, due to the longer wavelength of the THz wave, the alignment of the wave is high, and this feature helps in nondestructive testing of the materials.

- The radiating beam of the THz band of electromagnetic spectrum is highly directional, which is advantageous for the applications such as security scanning. This is because of the low diffraction of the THz beam, in contrast to the microwave range.
- Under certain atmospheric conditions such as the presence of fog, smoke, and dust, the attenuation of the THz wave is less in contrast to the infrared wave and therefore helps to utilize this license-free THz band of frequency of the electromagnetic spectrum for reliable sensing techniques.
- With an increase in the frequency of THz band, a better resolution can be achieved, which is useful for the imaging applications. The THz band offers an enhanced far-field spatial resolution of about 300 μm and reduced Rayleigh scattering (because the Rayleigh scattering intensity is $I \propto \lambda^{-4}$), in comparison to the millimeter waves and infrared rays, respectively.
- For the spread spectrum communication, the THz regime of the electromagnetic spectrum provides a large channel bandwidth. Moreover, the scintillation effect is more in the infrared communication system and the use of THz communication system can minimize this effect.

1.3.1 Material Characterization

Due to the unique signature ability, the THz band finds a number of industrial applications in material characterization. In one of the interesting applications in the paper manufacturing industries [12], the production of the finished goods is being ingeniously controlled using the THz systems where it monitors the thickness as well as the moisture content of the paper during the production process. Mousavi et al. [13] have demonstrated the experimental setup to differentiate two different paper samples using the THz wave. In addition, the THz systems also find its utility in the polymer manufacturing industry such as in the (1) online regulation of polymeric processes such as the real-time paint meter [14], (2) quality-control check of the plastic weld joints [15], (3) conductive properties of polyaniline films [16], (4) estimation of the moisture level [17], (5) fiber orientation [18], and (6) glass-transition temperature of polymers [19]. Similarly, in the food items manufacturing industry, it is desirous to detect the unwanted and harmful objects in food before its final packaging. The THz systems detect the possibility of both the metallic and nonmetallic contamination [20] in the food items. Moreover, the food items having low water content such as chocolates are transparent to the THz waves. This enabled to differentiate between the metallic and nonmetallic contamination in the chocolate bars [21]. Furthermore, for the brewery industry, the detection of corked substance is important. Therefore, the significant scattering of THz radiations due to the presence of defects or voids, in addition to the variation in the cork-cell structure, results in the emergence of the contrast images of the corked substances with imperfections [22].

1.3.2 Sensing and Imaging

Security-related applications can be subdivided into two major categories, as mentioned below.

Security Screening of Letters, Envelopes, and Small Packages

Detecting powder, liquids, explosives, and other threats in small packages and mail has recently become highly important. After discovering letters containing anthrax, such new threats as CBRE (chemical, biological, and radiological elements) are perceived as real and requiring new and effective detection approaches to counter them. Designed for inspecting flat objects (envelopes, letters, and small packages), the latest THz imaging scanner offers a brand-new approach for security screening in terms of safety, accessibility, and detection.

Security Screening of People (Body Scanner)

Unlike X-ray machines, THz waves are completely harmless to humans and have no ionizing radiation but can easily penetrate clothes and some other enclosures. These properties make THz-based people screening solutions extremely valuable for applications where human health and safety are of utmost importance. The TeraSense security body scanner operates in reflection mode and is intended for standoff detection of weapons, including cold steel and firearms, bombs and grenades, explosive belts, and various contraband items hidden under clothes.

Furthermore, among the security applications of THz imaging, the luggage and postal mail inspection [23] at airports is a major concern. The molecular crystals of the matter present specific features when they get to interact with the THz wave; therefore, the explosives [24–26] or illicit drugs [27] are well localized and explicitly recognized inside an envelope, or a parcel, or a suitcase [28]. The packages made up of a metallic sheet are opaque to the THz waves and thus, the spectrometers designed in the THz band are not probably an alternative to X-ray scanners. Though, the THz waves offer additional information about the sample under test, generally for the low-density materials and the chemical separation. For highly responsive facilities and public places like airports, railway platforms, compound areas of government offices, or open grounds during the festival celebrations, there is a requirement of full-body scanners for the security purposes. Initially, the security screening is either based on using metal detectors, X-ray backscatter, or millimeter-wave passive and/or active imaging systems. However, the deployment of security systems based on THz radiation in such areas is comparatively much safer than X-rays because of the smaller photon energy of THz waves which does not create any harm to the human tissues all the way through ionization [29]. Therefore, THz radiation shows the potential for its usage in the airport as noninvasive full-body security scanners [30, 31]. The THz system also finds its application in the liquid explosives' detection. This is due to the reason that numerous liquids exhibit a very different dielectric response in the THz band distinguishable from the other alcoholic substances [32].

The medical and biomedical imaging is also a promising field where the THz imaging technique finds several applications [33]. Due to the high degree of aligning property of the THz beam with respect to the microwave waves and the lesser Rayleigh scattering with respect to the infrared and visible beams, it finds a unique place in the biomedical imaging. An exposure to the THz wave with intensities of subhundred milliwatt per square centimeter or even greater does not show any significant variation at a cellular and molecular level of the living substance [34]. Thus, it facilitates a safer medical imaging for the human beings [35]. The THz imaging has found its potential applications in analyzing breast tumors [36], skin hydration and skin cancer [37, 38], and liver cancer [39]. Oh et al. [40] have demonstrated the diagnostic images of cancerous tumors which are acquired by using the THz molecular imaging technique (TMI). In this technique, when the near-infrared (NIR) beam is allowed to irradiate the surface of nanoparticles then a change in the THz response due to the surface plasmon resonance is measured, which helps in the detection of tumors. The authors [40] have further extended the differential measurement technique wherein the NIR beam is directly modulated and helps to eliminate the background noise which results in an improvement in the signal-to-noise ratio (SNR) values. Moreover, a high sensitivity has been achieved with this technique, which arises due to the sensitive interaction of the THz waves in the water. However, it also makes the target-specific sensing of tumors possible as well as helps to recognize the miniscule differences at a cellular level. Similarly, to show the benefits of THz radiation in the dental imaging and monitoring the tooth decay, Berry et al. [41] developed a detailed list of optical properties of the tissues that are exposed to THz pulsed radiation, experimentally. The pharmaceutical industries are also using THz radiations in the analysis of tablets and chemicals for the purpose of quality check during their production process [42, 43]. The crystalline structures of different isomers show altering spectral fingerprints in the THz range; therefore using the THz imaging technique, the polymorphic forms of the tablet coatings [44] can be detected in quality-control applications. This also helps to control the discharge of the active pharmaceutical ingredients [45]. Like microwave radiations, the THz radiation has the potential to penetrate a variety of nonconducting materials such as clothing, plastic, ceramics, wood, paper, and cardboard. However, the infiltration depth of THz waves is usually not more than that of the microwave radiation; therefore, THz waves find their applications in the nondestructive quality control with inspection of hidden defects and surfaces with nonuniformity and cracks [46]. Such a salient feature of THz waves is beneficial to evade any catastrophic failures similar to the NASA Space Shuttle Columbia in future. As the THz beams are capable of detecting the damage in the polymer foam tiles due to the induced heat and using the condition-based monitoring of the thermal protection systems (TPS), a secure operation of the space shuttle can be monitored. Zhong et al. [47, 48] have established the THz time-of-flight (TOF) tomographic imaging in the nondestructive recognition of foam insulation on space shuttle fuel tanks having prefabricated defects. The authors have determined the insulation foam defects with the help of a Gunn diode oscillator and a pyroelectric camera as a source of THz wave and detector, respectively. Rahani et al. [49] have investigated the potential of THz waves to detect heat-induced damages that occur in porous materials. The authors have analyzed the emergence of defects that are generated during the manufacturing process or in-service processes such as the formation of voids, delamination or damage due to impact/burn in the porous materials by analyzing the information on THz absorption from the transmission-type characteristics of the image. Likewise, as the THz waves can infiltrate through all composite materials the THz transmission images of composite material [50] can be obtained for security applications. Amenabar et al. [5] have reviewed the most significant feature of the THz technology as a tool for nondestructive testing (NDT) inspection of composite materials. The reflection of the THz wave (in conductive materials) and absorption of the THz wave (in polar liquids) show the way to a clear examination in the composite inspection field. However, there are certain limitations to the detection of composite matter using THz waves such as carbon fibers which is not penetrated by THz radiation and the water content in air or else in the composite itself may compromise on the correctness of the inspection system.

1.3.3 Next-Generation Communication

To cater to the requirement of high data rate communication, there is a prerequisite for using higher operating frequency in the THz range of the electromagnetic spectrum [51]. However, during the deployment of a higher operating frequency of the THz band in setting up a communication link, several design issues are required to be addressed. In a wireless communication system, an antenna plays a vital role and at THz frequencies, the design characteristics of an antenna need significant consideration. Moreover, to increase the overall length of any wireless communication link, the antenna performance parameters such as gain and directivity represent the decisive factors. Jha and Singh [52] have reviewed the technical issues of THz antennas with a particular consideration of the planar technologies that can contribute to the compact, low-profile, and inexpensive future THz wireless communication system design. As the signal attenuation and the power levels are high and low, respectively, in the THz regime of the electromagnetic spectrum, wireless communication links with a highly directive system provide point-to-point communication within some short ranges. This enables the establishment of secure short-range links to exchange the confidential information with high speed. Likewise, an active research in the field of THz communication systems [53, 54] is going on, specifically on indoor systems that are operating at point-to-point over several meters [55, 56], and are largely depending on integrated circuits so as to enable these point-to-point links for portable consumer devices [57, 58]. The THz communication systems share the characteristic of short-range communication and are primarily restricted by the strong atmospheric attenuation and scattering by building materials [59]. In case of the long-range outdoor point-to-point links, a millimeter-wave band below 300 GHz is being reported. Recently, a wireless link operating at a center frequency of 240 GHz realizes a data rate of 64 Gbit/s over a transmission distance of 850 meters by means of QPSK and 8PSK modulation in a single-channel approach without the use of a spatial diversity concept [60] and a 120 GHz link achieving 10 Gbit/s at 5.8 km [61] have been established, respectively. However, it is a myth that the millimeter-wave and the THz bands (0.1–3 THz) are unfeasible for all but short-range links due to rigorous attenuation caused by the atmospheric water vapor. Furthermore, this myth has been contradicted with the development and usage of large-aperture telescopes by the researchers in the area of submillimeter radio astronomy for the dry high-altitude locations. Suen [62] has reviewed the expertise and science essential to build up a terabit-per-second THz satellite link. The author has emphasized the benefits of developing a satellite link with ubiquitous THz communications. Moreover, several technical characteristics of a satellite link which includes atmospheric attenuation, transceiver technology, aperture, and earth station properties for the terabit-per-second satellite link are discussed in his work.

1.3.4 T-Ray Tomography

Sectional imaging using THz radiation can be achieved using T-ray tomography. The word "tomography" has been derived from the Greek words "tomos" which means slice or section and "'graphia" means "describing." The field of tomography includes methods for obtaining cross-sectional images of a target, thereby allowing to observe the internal details [63]. A tomographic imaging procedure can be used to acquire and to provide three-dimensional (3D) images in the THz frequency range. The principle of THz tomography to extract 3D information using the pulsed THz imaging system is broadly classified in terms of its application as: (1) THz computed tomography (CT), (2) tomosynthesis (TS), (3) synthetic aperture processing (SAP), and (4) time-of-flight (TOF) THz tomography. In THz CT, Maxwell's equations are used to describe the relationship between the THz wave distribution and object's refractive index as a function of position. The first demonstration of THz CT reported in [64] imaged a simple polyethylene cylinder and used a reconstruction method based on the Born linearization of the wave equation. This helped to measure amplitude as well as spectral phase information which was not available with X-ray imaging. Tomosynthesis (TS) is another method for reconstructing tomographic images using few slices of a 3D object from a limited number of projections [65]. Such a technique is useful to retrieve a layer of interest against out-of-focus layers. It involves the use of absorption images of an object and different projections from various views at selected angles which is achieved with a moving THz source along a linear trajectory. A series of radiographs of an object are obtained according to different points of view. THz pulsed imaging carries a unique feature of taking 3D map of the object with the use of time-of-flight (TOF) of the reflected signal. The temporal situation of the reflected THz pulses indicates the presence of the interfaces along the propagation direction of the beam. Therefore, using the difference of TOF from pixel to pixel, the depth information of 3D profile of the target can be determined [66]. Moreover, TOF THz resolution depends on the width of incident THz pulse and shows good results with short THz pulse. TOF tomography is useful for layered targets having well-defined boundaries. However, a sample having complicated constitution, the multiple reflections, and refractions of THz radiation superpose and result in scattering. Synthetic aperture processing (SAP) of tomography uses a diverging or unfocused beam to collect data. In this, a target positioned in a measurement scene is illuminated by the sensor from several adjacent scanning positions using a wide beam. A coherent integration of the received signals is applied, which results in an increase in the signal-to-noise ratio of the resulting image of the target [67]. Such a technique provides a constant, range-independent lateral or cross-range resolution, which is useful to measure thick samples [68]. The data processing and reconstruction using T-ray tomography is generally based on several direct methods such as random transfer, Fourier slice theorem, and acquisition properties.

1.4 Overview of Continuous and Pulsed Terahertz Imaging Systems

The THz radiations provide a higher spatial resolution, in comparison to the microwave radiations because of the shorter wavelength. Moreover, the THz radiations do not cause any recognized harm to the living organisms, which makes THz imaging a powerful and most likely safe imaging technology. Furthermore, the applications, such as homeland security, defense, and safety in aviation industry, have put a huge demand on the expansion of advanced imaging systems. The types of THz sources used in imaging system are briefly discussed to have an idea about the availability of types of THz sources for THz continuous and pulsed wave imaging systems.

1.4.1 THz Sources for Imaging Systems

Generally, the electronic solid-state sources have a limited operating bandwidth due to the transit time of carriers through a semiconductor specimen, which causes a high-frequency roll-off. This makes a limitation in the use of solid-state devices for THz frequencies. Recently, the development of a variety of THz sources is gradually filling the THz gap, providing complementary characteristics in terms of various parameters such as operating frequency, an average and peak power. The classification of THz sources used in the THz imaging system is shown in Fig. 1.3. Numerous systems have been developed to generate the THz radiations, with each

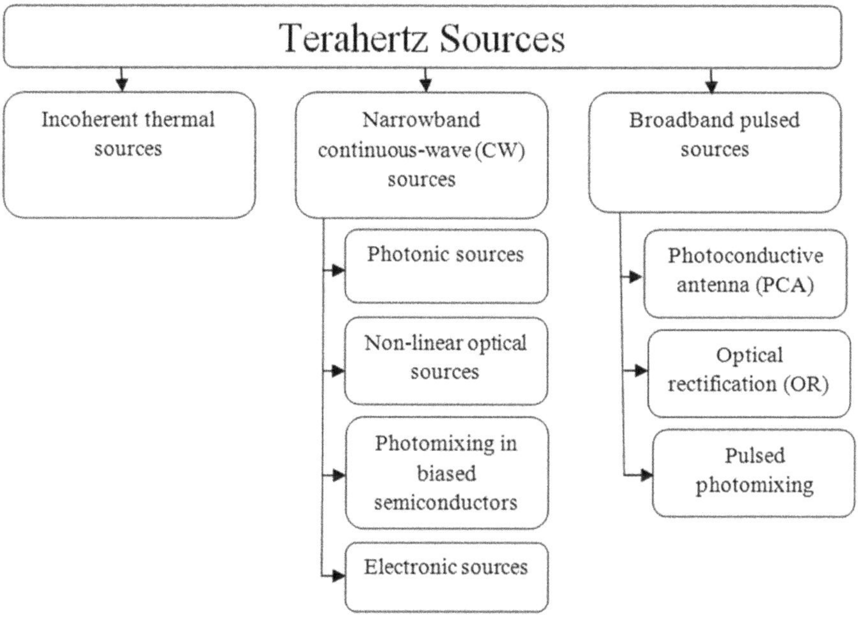

Fig. 1.3 Classification of THz sources

system offering different output powers, sensitivities, and bandwidths. On the basis of spectroscopic techniques used in sensing and imaging applications, the THz sources have been broadly classified into three categories such as:

1. Incoherent thermal THz sources
2. Continuous-Wave (CW) THz sources (also known as Narrowband THz source)
3. Pulsed THz sources (also known as Broadband THz source)

The narrowband continuous-wave sources are further classified as photonic sources, nonlinear optical sources, photomixing in biased semiconductors, and electronic sources. Similarly, the broadband pulsed sources are classified as photoconductive antenna (PCA), optical rectification, and the pulsed photomixing. The incoherent thermal sources generate THz radiations through mercury arc lamp or SiC rod (Globar) in an optical interferometer. In the interferometer, the characteristic interference pattern of the sample is considered through scanning the arm length. Using the fast Fourier transform (FFT) and the numerical FFT (NFFT), the interferometer responds to the sample spectrum with zero-padding and aliasing [69]. Among the models for THz imaging, numerous noncoherent techniques like microbolometer arrays are presented by researchers, whereas to increase the measurement speeds such approaches provide only limited information due to lack of information about the phase. However, due to the rapid development of coherent THz sources, the trend shows the advancement in THz imaging systems and it presents an opportunity for high-resolution, potentially noninvasive imaging suitable for security or quality-control applications [70, 71].

The photonic CW THz source operation depends on the type of the laser used. Generally, the main hindrance in developing a THz laser is to find an inexpensive and suitable gain medium which can pump more resourcefully with high gain in addition to high output power. The nonlinear optical materials of susceptibility $\chi^{(2)}$ with nonlinear optical coefficient act as a medium for the generation of THz radiation at the beat frequency of two pump sources. In case of pump sources which are having wavelengths offset by THz frequencies, the nonlinear optical material can act as a source of THz radiation. In this technique, a material with second-order optical nonlinearity such as 4-dimethylamino-N-methyl-4-stilbazolium tosylate (DAST), two continuous-wave lasers of offset resonance frequencies are combined resulting in difference frequency generation (DFG). With photomixing in biased semiconductors' generation technique for CW THz radiation, the mutual interference of two laser frequencies with semiconductor material results in output oscillations that occur at the sum as well as the difference of the laser frequencies. The systems are fabricated in such a way that the difference term of laser frequencies lies in the THz range. For the low-power CW operation, the high-speed electronic devices can be used such as in the case of a noncontact system developed using complex impedance bridges to precisely characterize the dielectric constant of thin films between the range of 30 GHz and 1 THz.

The advancement in the generation of pulsed THz radiation with the help of ultrafast optical lasers offers a new way for accessing the THz frequency range. However, there are several mechanisms to produce pulsed THz radiation, which includes the techniques such as (1) photocarriers' acceleration in photoconducting antennas, (2) second-order nonlinear effects in electro-optic crystals, (3) plasma oscillations, and (4) electronic nonlinear transmission lines. Initially, the pulsed THz techniques were originally developed for the waveguide and circuit characterization. In Table 1.1, the generation techniques of both continuous-wave THz radiations and pulsed-wave THz radiations are summarized for a quick and easy comparison of each technique.

The photoconductive approach to generate pulsed THz radiation uses the formation of the transient current due to high-speed photoconductors across the radiating antenna. The optical rectification is another method for the generation of pulsed THz radiation, which is an opposite process of the electro-optic effect. The optical rectification (OR) is a second-order consequence, which occurs in materials with a nonzero $\chi^{(2)}$ coefficient and is also known as the Pockels effect. The pulsed photomixing is a technique used to generate THz signals in the photoconductive antenna, CW photomixing, and CW nonlinear DFG.

1.4.2 Performance Comparison Between Continuous and Pulsed THz Imaging System

The narrowband/CW THz sources have a variety of applications in high-resolution spectroscopy, telecommunications, and high bandwidth intersatellite links. Figure 1.4 shows the CW imaging system wherein the THz continuous-wave

Table 1.1 Summary of generation techniques of THz radiation

	Continuous-wave THz radiation		Pulsed THz radiation	
S.No.	Generation technique	Generated material/ medium	Generation technique	Generated material/ medium
1	Photomixing [72]	Power cable (PC) switch	Transient photoconductive switching [73]	Photoconductive antenna (PCA)
2	Difference frequency generation using parametric oscillation [74]	Nonlinear crystal	Optical rectification [75]	Dielectrics, semiconductors, organic materials
3	Rotational transitions [76]	Far-infrared gas lasers	Emission from a periodically undulated electron beam [77]	Electron accelerators
4	Streaming motion and population inversion [78]	Semiconductor laser	Surge current (super depletion field) [79]	Semiconductor surface
5	Frequency multiplication of microwaves [80]	Schottky barrier diode	Tunneling of electron wave packet [81]	Quantum semiconductor structures
6	Transitions in superlattice [82]	Quantum cascade lasers (QCLs)	Coherent longitudinal optical phonons [83]	Semiconductors, semimetals, superconductors
7	Electron interactive with a traveling electromagnetic wave [84]	Backward wave oscillator (BWO)	Optically short-circuiting the switch [85]	High-temperature superconductor bridge
8	Relativistic electron interaction with transverse magnetic field [86]	Free electron lasers (FELs)	Nonlinear transmission line (NLTL) [87]	Electronic circuits consisting of NLTL

generation is achieved using the Gunn diode assembly. Initially, a Gunn diode oscillator generates a beam of 0.2 THz with 12 mW power. With the help of a parabolic mirror, the beam is allowed to focus on to a 4 mm spot and is then modulated with the help of a chopper at 1.2 kHz. A polyethylene Fresnel lens placed at a distance of 408 mm from the parabolic mirror focused the modulated beam with a focal length of 204 mm. The modulated beam is allowed to pass through the sample that is positioned at this focus. Later than passing through the sample, the beam once again gets focused through a small polyethylene spherical lens and recollects at the horn of the detector, which is an unbiased GaAs Schottky barrier diode. Then, the signal received by the detector is read using a lock-in amplifier.

However, the CW imaging system, presented in Fig. 1.4, only provides intensity data and does not give any depth as well as the frequency-domain or the time-domain information of the subject, while a fixed-frequency source along with a single detector is used. For many imaging applications [88], only the plots of transmitted energy are adequate to detect the sample under test. Moreover, the CW imaging system provides a compact, simple, fast, and comparatively low-cost system at the cost of loss of depth information of the sample under test and

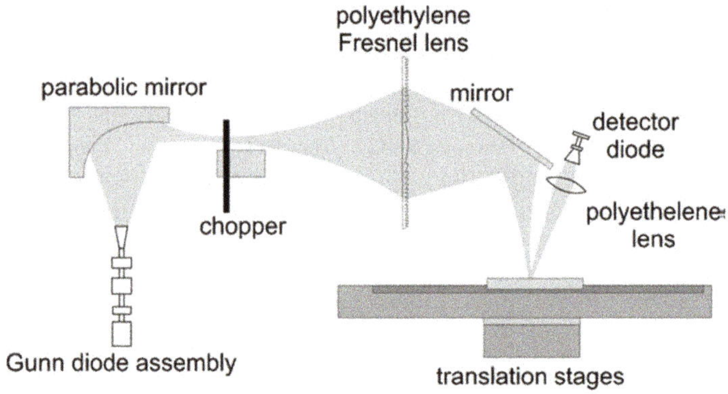

Fig. 1.4 Schematic illustration of a CW THz system [90]

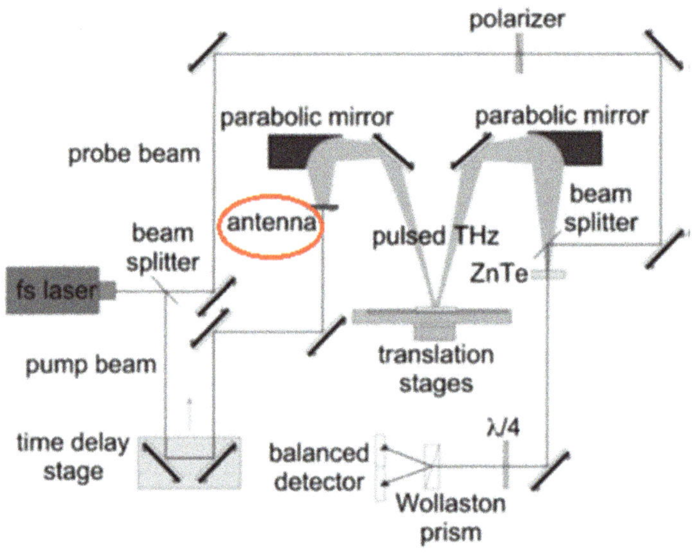

Fig. 1.5 Schematic illustration of a pulsed THz imaging system [90]

frequency-domain or time-domain information of the subject. The simplicity of the optics involved in the CW imaging system is due to the absence of a pump-probe system [89]. Therefore, there is no requirement of a time-delay scan as is needed in a pulsed imaging system and hence the image formation takes place more quickly in the CW imaging system. In the pulsed terahertz imaging system, a photoconductive antenna [91] is illuminated with the help of a femtosecond laser pulse acting as a source and the generated THz electromagnetic pulse is detected by a nonlinear crystal made up of ZnTe by means of an electro-optic sampling technique based on the optical pump-probe setup [92]. Figure 1.5 shows the pulsed imaging system wherein the THz pulsed wave generation is achieved using a photoconductive antenna.

Table 1.2 Comparison of CW and pulsed THz imaging systems [90]

S.no.	Parameter	CW imaging system	Pulsed imaging system
1.	Approximate total system cost	\$50 000–\$150 000	\$300 000–\$1 000 000
2.	System complexity	Low	High
3.	Approximate weight of imaging components	Approx. 2 Kg	Approx. 300 Kg
4.	Information provided	Transmitted energy	(1) Magnitude of electric field. (2)Total transmission time (optical path length). (3) Absorption spectrum. (4) Depth of scattering centers (5) Information about the phase. (6) Shape of pulse.
5.	Scanning speed	0.005 s per point 1 mm step size (which is limited by translation stage maximum speed of 0.2 m s−1)	20–0.05 s per waveform
6.	Data complexity	Low	High

The Coherent RegA laser used in the setup has the characteristic features such as (1) a repetition rate of 250 kHz, (2) a pulse duration of 200 fs, and (3) an average power of 400 mW. Initially, the THz pulses are generated using a large-aperture photoconductive antenna with a peak frequency of 0.4 THz. Then, the generated THz beam is allowed to get focused onto the target using a parabolic mirror, which gets reflected through the sample layers as well as the aluminum substrate. The reflected beam is collected using another parabolic mirror, which it then sends to a ZnTe crystal acting as a THz detector where it overlies with the probe beam. Inside the ZnTe crystal, the probe beam gets modulated by the THz field. Such an imaging technique records the waveform of the reflected THz pulses and therefore preserves the phase and amplitude information.

With the detailed working principle of both the CW imaging system and the pulsed imaging system as discussed above, the strengths and limitations of both imaging systems with respect to each other are now summarized in Table 1.2 for a quick and easy reference. Using a more complicated arrangement for the CW imaging system, additional information can be extracted. Likewise, the overall weight issue of the pulsed imaging system can be abridged appreciably with the utilization of a more compact laser such as a fiber laser for the pump-probe setup.

1.5 Terahertz Pulsed Imaging System for Detection of Hidden Explosives

In the imaging application, with visible light, only a photograph can be created and with X-rays, a view within the human body through penetration at a time is performed; however, it is beneficial to use THz waves, as they can create the image as well as transmit the information about the spectral response of the material. Moreover, in THz imaging at low values of THz frequency, the dielectric response of several organic and inorganic materials is responsible for their spectral response and at high frequencies, the spectral response is subjugated by specific intramolecular or intermolecular vibrations as well as the rotations which are similar to those that occur with infrared radiation in case of infrared spectroscopy. The salient features of THz radiation, which makes it a potentially influential technique in THz pulsed imaging for security applications [93] in detection of hidden explosives, are as follows:

- *Penetration:* THz radiation has the ability to pass through the nonmetallic and low-absorption materials like clothing and packaging materials. Moreover, these radiations can partially reflect from the interfaces between materials having a different refractive index, which enables to detect the plastic and ceramic objects containing powdered explosives hidden beneath the clothes.
- *High-resolution 3D imaging*: The wavelengths, which are used in pulsed imaging, provide images with submillimeter resolution. Like a radar system, the use of extremely short pulses in pulsed THz techniques helps to take 3D imaging. Moreover in a pulsed imaging system, the range gating technique increases the contrast and the discrimination ability of the imaging system.
- *Spectroscopy:* Many explosives and explosive-related compounds (ERCs) show spectral features in the broad frequency range of 0.1–4 THz and thus provide unique spectroscopic information which helps in their detection using THz radiations, even when they are sealed inside an object [6].
- *Safety:* The THz radiation is nonionizing in nature and uses very low-power levels in the nanowatt range because of the availability of highly sensitive coherent detection schemes.
- *Low scattering:* The longer wavelength of the signal from the pulsed THz imaging system allows for much lower scattering from the object under detection.
- *Intensity:* In comparison to radio waves, the THz signals are much easier to focus and collimate, which further helps to realize high depth-of-field (DOF) for pulsed imaging system.

In the THz pulsed imaging system, there is a use of THz electromagnetic waves to spectroscopically detect and identify concealed explosives such as the research department explosive RDX (having chemical name 1,3,5-trinitroperhydro-1,3,5-triazine) and high melting explosive HMX (having chemical name octahydro-1,3,5,7-tetranitro-1,3,5,7-tetrazocine) through their characteristic transmission or reflectivity spectra which lies in the THz range [28]. Moreover, in the THz regime of the electromagnetic spectrum, spectroscopy with high spectral resolution is an effective analytical tool to determine the structure as well as the energy levels of the molecules and atoms. The THz pulsed spectrometers are essentially broadband systems because of the use of ultrafast optical pulses to create the THz radiation. Furthermore, the THz spectroscopy has the ability to determine the far-infrared optical properties of the material as a function of frequency which provides the insight into material characteristics for a broad range of security applications.

Related to the security threats to contemporary society, there is a requirement of scanning systems for screening of persons, so as to identify the concealed perilous objects. However, on the other side, the use of full-body scanning techniques based on X-rays hoists problems about the human health issues, therefore limiting their acceptance by the public. Another significant security need such as recognizing a suicide bomber from a safe distance is not convened at all by the existing scanning technologies. However, the most constructive property of imagers, which are operating in the THz range, has an ability to detect the small variation in temperature on the object's surface and has the ability to see even through clothing. Besides the background of the radiating human body, the reflecting and the absorbing objects become noticeable with THz radiations. Therefore, the objects such as metals (which are generally highly reflective), ceramic materials, and explosives, which show characteristic absorption spectra in the submillimeter range [94] and even if they are concealed beneath the clothing, can be detected easily using THz radiations. Therefore, due to the potential features of THz radiations in imaging applications, the THz pulsed imaging systems on the basis of the type of imagers used in the THz pulsed imaging system are classified as:

- Passive imaging
- Active imaging

The THz pulsed imagers offer salient features such as the low signal-to-noise ratio, low spatial resolution, and limited distance of imaging. However, the performance of THz pulsed imagers in terms of noise-equivalent temperature difference (NETD) value is lower and is in the range of $0.5 - 5^{\circ}K$ (degree Kelvin) [95]. The passive THz imagers record the contrast in radiometric temperature within an object under scene and the active THz imagers record the contrast in the scattered radiance within an object when it is illuminated with THz source. In active imaging, the imager makes the active image of the object wherein all the radiations are confined with all of its illumination to a single mode and the receiver observes the same mode on the other side of the object. However, the passive THz imaging systems which are inherently multimode has a small dynamic range, in comparison to the active THz imaging system [96].

With the advances in THz monolithic and array compatible integrated circuits (TMICs) [97, 98] operating at room temperature, a fully passive approach is implemented with the use of heterodyne receivers. However, the active imaging system offers an advantage of reducing the sensitivity requirement on the THz receivers and in such a case the receiver can improve the acquisition speed and the number of image pixels (format). Friederich et al. [99] have presented active imaging systems to determine the pulsed imaging system's potential in real-time imaging, which includes (a) active electronic imaging system, (b) optoelectronic THz imaging system, and (c) THz focal plane arrays.

1.5.1 Potential Challenges of THz Pulsed Imaging System

The THz pulsed imaging is an emerging technology that finds diverse applications as discussed in the preceding sections. Despite being an influential technique based on salient features of THz radiations, there are many obstacles that stand in the way of large-scale industrial induction of THz pulsed imaging system. However, there are some key research areas which invite the researchers to work. Current efforts in the hardware are vital to transform the THz imaging systems from laboratory to the industry. The progresses in imaging architectures as well as algorithms are equally important for the exact and quick data processing. The THz imaging era has begun after Hu and Nuss developed the first scanning system in 1995 wherein the sample under test was scanned in x–y translation [100]. They have demonstrated a practical THz imaging system wherein the THz transients are focused on to the diffraction-limited spot localized on the sample. The transmitted THz waveforms are then acquired as well as processed in real time at each point of the sample. The authors have used THz pulses to stare through the packaging of a semiconductor chip and also to find the water content of tree leaves. In case of a conventional THz imaging system, a sample is scanned to take an image, which poses a severe limitation on the available acquisition speed. However, a two-dimensional (2D) electro-optic sampling is used with a charge-coupled device (CCD) camera to enhance the imaging speed. Moreover, with the necessity of accelerating the image acquisition speed and high absorption of many materials, there is a requirement of significant advancement in terms of compactness of size, cost-effective feature, and portability of THz systems to bring about better feasibility of real-time imaging. Along with this, other major issues and challenges, such as (1) output power, (2) signal-to-noise ratio (SNR), (3) water sensitivity, (4) depth of penetration, (5) bandwidth, (6) spatial resolution, and (7) lack of a THz frequency knowledge base, need to be considered for pulsed THz imaging systems. The spatial resolution in THz pulsed imaging is principally limited by the diffraction limit that is the function of wavelength and numerical aperture of the optical system.

Various applications in THz imaging rely on either single plano-convex lens of a spherical shape or off-axial parabolic mirrors but an increase of the numerical aperture for the former makes it inefficient leading to a significant rise of aberrations; however, the latter have good aberration correction and high resolution although they suffer from the overlapping of incident and focused beams [101]. The THz near-field microscopy is an alternative approach to overcome the diffraction limit. However, it offers numerous disadvantages such as:

1. Detection of light scattered on very small diaphragms or confined at a tip apex placed at the object plane requires powerful emitters and highly sensitive detectors.
2. A near-field imaging requires a very short object distance; thus, the scanning probe may interact with the sample and even perturb its structure.
3. Requires a long scanning time which further strongly limits its reliability for certain specific applications [102].

The signal-to-noise ratio (SNR) improvement is another challenge being faced by pulsed THz imaging systems. This is inherently tied to the average power of the THz emitter. In THz time-domain spectroscopy systems, a high SNR can be achieved. In imaging applications, there are certain factors that result in the reduction of SNR to a level where it becomes a matter of concern. However, the use of a THz source with high power can improve the SNR as well as the dynamic range of sensing and imaging systems by increasing the penetration depth in the scattering or absorbing materials. The polar molecules interact strongly with pulsed THz wave and due to this property the water molecules absorb the THz waves and therefore, the depth of penetration of the THz wave is limited in moist substances. Moreover, in the THz imaging applications to yield enhanced depth-of-field (DoF) for imaging purpose, a radiating source with highly directive antenna is required.

Several state-of-the-art THz spectroscopy systems rely on the ultrafast laser-based systems which are bulky and therefore nonportable. In addition, such spectroscopy systems are complicated because of the essential mechanical parts such as a raster scan with single detector to obtain a two-dimensional image and are therefore somewhat expensive to develop as well as to operate. Moreover, the scan process generally takes tens of minutes to produce a high-resolution THz picture of a scene. Thus, all the electronic-based THz systems are required toward reducing the space, weight, and power, and thus facilitate future sensing and imaging applications. Furthermore, the THz imaging technology faces some inherent problems which arise from the (1) specular reflections, (2) unwanted clothing reflections (clutter), (3) interference effects (speckles), and (4) angular orientation effects, that result in degradation in the image quality and resolution. Moreover, in case of active imaging techniques there are certain threat scenarios also such as the presence of nonreflecting objects carried directly on the human skin. However, in case of active imaging, the degradation in THz images due to speckles can be minimized by adding angular diversity or multimode mixing to the illumination THz source [103]. As the noise-equivalent power (NEP) of an imaging array depends on the antenna efficiency, therefore, it is essential that the high-efficiency on-chip antennas are needed for a low system NEP. Figure 1.6 shows a broadband focal plane array camera with intrinsic cut-off frequency of 2.5 THz with antimonide-based heterostructure backward diodes, which are monolithically integrated by means of planar THz antennas for each sensor pixel [104] for the real-time THz imaging application.

Therefore, considering the potential challenges of pulsed THz imaging system for security applications, the advancements in the area of THz antenna technology for imaging application will open a new avenue in the biomedical and security applications in the THz regime of the electromagnetic spectrum. Moreover, both wavelength-scaled feed systems that integrate directly with electronic components and large-scale beam forming systems are required. Several groups of researchers are pursuing the concept of small THz antenna, which is especially targeted for shorter wavelengths to be useful for imagers, interferometers, and broad-bandwidth spectrometers in THz imaging systems. As at shorter wavelengths, there is a difficulty in coupling together the multifunction elements which are fabricated individually such as diodes, transistors, or passive transmission lines; therefore, more

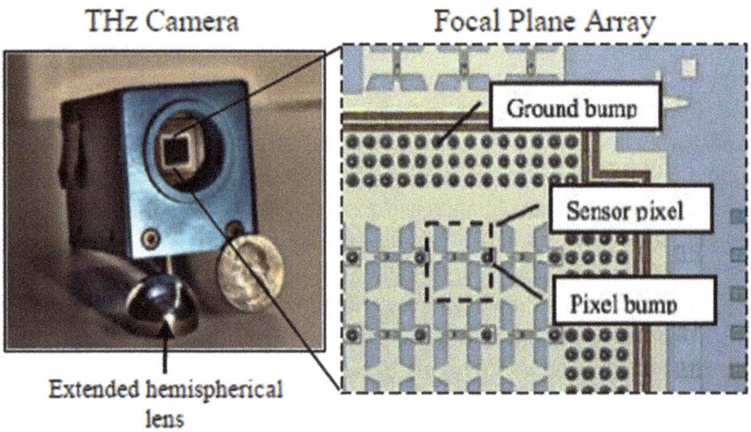

Fig. 1.6 The THz camera comprising of focal plane array pixels consisting of broadband slot-type photoconductive antenna with integrated sensors [104]

emphasis is being laid down on realizing these antennas [105] with matching fabrication process as of the sensing and power converting devices which are used at the antenna terminals. Such technique results in a reduction in the losses associated with excessive mode, beam mismatch, and absorption. However, still there are some other challenges such as recognizing specific impedance matching, tuning, or phase scanning, high gain, broad bandwidth, which need to be considered simultaneously. Moreover, to realize a diffraction-limited image using focal plane arrays, novel compact antenna designs with areas of one-half wavelengths squared or less need to be designed. In Table 1.3, the performance evaluation of the characteristic parameters related to the THz imaging system for the feasibility of real-time imaging system is briefed. To make realizable pulsed THz imaging system, the techniques, which can be deployed in THz antennas to satisfy the demand of imaging system, are mentioned with the respective potential challenges to work upon.

1.6 Related Work

A well-designed subwavelength (micrometer/nanometer) radiating structure shows the potential for high output power generation and broadband THz pulse emission, which is useful for THz imaging system. An antenna is a vital component in the THz imaging system, as it plays a significant role in both impedances' matching in addition to a power radiating source. Researchers have designed several types of THz antennas having different structures. Some of them are (1) THz dipole antenna [106], (2) Yagi–Uda antenna [107], (3) spiral-type antenna [108], and (4) butterfly-shaped antenna [109] with lens. Moreover, the diffraction limit in imaging system can also be overcome by the use of metamaterial biochips and nanotechnology-based contrast agents. The THz antennas are broadly divided into two distinct areas:

Table 1.3 The performance evaluation of various characteristic parameters associated with the feasibility of real-time imaging system

Characteristic parameter	Performance evaluation	Technique to achieve	Limitation
Acquisition speed	High	Line detection using microlens array	High system volume with portability challenge
Spatial resolution	High	Focal plane arrays	Diffraction limit and rise in aberrations
Signal-to-noise ratio (SNR)	High	High-power THz source	Unavailability of a highly efficient single unit of THz source
Dynamic range	Large	Beam-steering technique and frequency scanning system	Propagation losses and nonspecular reflection on the object under test
Depth-of-field	High	Highly directive THz source	High atmospheric attenuation
Noise-equivalent power	Low	PCA which has low optical and thermal noises in comparison to the electro-optic rectification method	Low antenna efficiency
Size-weight-and-power (SWaP)	Small	High-efficiency on-chip antennas	Critical fabrication issue

- One is the use of THz antenna as wavelength scale beam forming or feed elements. In this form, the THz antenna couples the energy acquired from the free space into or out of subwavelength generators/receivers such as diodes, bolometers, transistors, or photoconductive elements.
- Second is the use of a much larger aperture THz antenna to accumulate the signals and to focus the beam. This form is being used in a variety of applications such as high-resolution imaging or scanning and the light gathering power, which is required by large radio telescopes.

For the emission or detection of THz waves using optical and electrical methods, a THz photoconductive antenna (PCA) is used. However, the THz PCA lacks in harnessing the modern technological advancement for the high-power THz emission. Moreover, for the advanced imaging and sensing applications, there is a rising attention in developing high-power THz sources and sensitive detectors. The accessibility of high-power, tunable wavelength, as well as compact optical sources by means of pulsed and continuous-wave operation becomes the major driving force of the photoconductive THz sources/detector applications. However, the inherent trade-off amidst high quantum efficiency and ultrafast operation is the main obstacle in developing high-power source and sensitive detectors using conventional photoconductors. This trade-off is because of the restricted carrier transport velocities inside the semiconductor substrates that are bounded with the carrier scattering occurring within the semiconductor lattice. Moreover, for the efficient generation/ detection of THz radiation in photoconductive device there is a requirement of small carrier lifetime of the photogenerated carriers to move toward the contact electrodes of the antenna. To achieve this, many researchers have used the nanoplasmonic light concentrators as well as the photoconductive sources/detectors designed from optical nanoantennas. Jarrahi [110] has shown and discussed in detail the impact of using optical nanoantennas in the active area of photoconductive THz sources. Likewise, Berry et al. [111] have reported the performance comparison between a plasmonic photoconductive emitter and a photoconductive emitter having nonplasmonic contact electrodes. In their research work, they have claimed for 50 times higher THz-radiated power from a plasmonic photoconductive emitter. Similarly, a photoconductive detector with plasmonic effect offers 30 times higher THz detection sensitivities, in comparison to a similar photoconductive detector with nonplasmonic contact electrodes. They have also shared the future prospects of their prototype device that can be further improved with the deployment of the resonant cavities and antennas having higher radiation resistance and bandwidth. A considerable enhancement in THz radiation power and the detection sensitivity can be achieved by using plasmonic contact electrodes with high aspect ratio. Such electrodes are embedded within a photoabsorbing semiconductor material which results in the generation of a large number of carriers in close proximity to the photoconductor contact electrodes. Therefore, the nano- and micro-fabrication can open a way to numerous opportunities for enhancing the emission power of THz photoconductive antennas.

Park et al. [112] have demonstrated an increase in the THz emission power using the optical nanoantennas by tuning plasmon resonance with antenna geometry. They have designed an antenna for the high-power THz emission using large-area excitation with multiple interdigitized microelectrodes underneath the restricted optical power. Moreover, the plasmon resonance gets precisely controlled through variation in the width of nanoantenna. The designed antenna configuration features nanorod arrays placed in between two microelectrodes on the photoconductive substrate [112]. The optical nanoantennas enhance the light concentration near the nanorod arrays by the side of localized surface plasmon resonance. Moreover, the light confinement within the photoconductive region also results in the generation of more photocarriers, which further contribute to THz emission power enhancement. The authors of [112] have integrated the optical nanoantennas within the interstitial microgap of a bowtie-type photoconductive antenna on SI-GaAs photoconductive substrate by integrating multiple techniques such as the conventional photolithography, electron-beam lithography, and metal lift-off technology. Zhu et al. [113] have discussed three linearly polarized photoconductive THz antennas designed for the THz imaging system. A bow-tie antenna with a finite ground plane and DC biasing lines has been used as the reference design. In [113], the authors compare the performance of three linearly polarized photoconductive THz antennas designed using artificial magnetic conductors (AMCs). The use of AMC enhances the gain of the antennas. Nguyen et al. [114] have designed a full-wavelength THz dipole antenna holdup by means of a GaAs membrane structure with high input impedance. The radiation efficiency also gets improved, which further improves the overall efficiency of a THz photomixer. Furthermore, the authors have also shown the effects of membrane thickness as well as the diameter of a hole inside the ground plane in a back-excitation configuration of THz dipole antenna.

Through the optimization process, they have observed two interesting properties. One is that a relatively thin cavity is appropriate for the antenna designs, which demands high input impedance as in THz photomixer antennas. Another is the use of bulk GaAs substrates, which results in enhancement in the antenna performance and also possesses high input impedance along with high radiation efficiency. Maraghechi et al. [115] have developed a correlation between the lengths of bow-tie antenna (type of photoconductive antenna) with its THz spectral emission response. Moreover, the capability to fine-tune the center frequency of THz photoconductive switch improves the accuracy of detectors/emitters of the spectroscopy, imaging, and sensing systems. In [115], for computing the effective antenna length, three approximation methods, such as (1) quasi-static, (2) high-frequency, and (3) Brown and Woodward, have been used. Accordingly, a number of bow-tie antennas with different lengths have been fabricated and are analyzed [116, 117]. The authors have also concluded that a simple quasi-static approach to approximate the effective permittivity of the substrate can be applicable for frequencies up to 1.5 THz.

Beck et al. [118] have reported an impulsive THz radiator with characteristics such as (1) 36 kV/cm vacuum electric field, (2) 250 kHz repetition rate, (3) 1.5 mW average thermal powers, and (4) 2×10^{-3} NIR-to-THz conversion efficiency. This has been obtained using photoexcitation of a biased large-area photoconductive emitter with near-infrared (NIR) femtosecond pulses having μJ energy. They investigated the THz emission due to the acceleration of photoinduced carriers within GaAs photoconductive substrate at high excitation densities. However, to achieve this specific metal-semiconductor-metal (MSM) structure photoconductive antenna is utilized. With the use of a specific MSM structure along with a large active photoconductive area of approximately 1×1 mm^2, the saturation effects including bias screening have been successfully reduced. Moreover, this also allows ensuring a suitable scattering strength to keep up the eigenmodes with the desired field distribution. Furthermore, a peculiar scattering response can be achieved by making use of the excitation of the resonant eigenmodes in an aptly shaped metallic antenna. Tani et al. [119] have reported two different forms of photoconductive antennas: (a) Schottky photoconductive antenna and (b) multicontact photoconductive antenna, respectively. The Schottky photoconductive antenna has the ability to sense THz radiation intensity without the time-delay scans; therefore, it is helpful for applications wherein spectroscopic information is not significant like THz intensity imaging. Moreover, the multicontact photoconductive antenna finds applications in the polarization-sensitive THz spectroscopy like the THz ellipsometry. The authors have studied the characteristic features of these photoconductive antennas with the help of a THz time-domain spectroscopy system. By means of a cautious design of the contacts such as a point contact with a metal tip on n-type GaAs, the Schottky barrier diode has the ability to detect THz radiation. However, an ordinary Schottky diode is not appropriate for the detection of pulsed THz radiation (~ 1 ps) generated by the excitation of a photoconductive antenna or some other THz emitting devices with femtosecond laser since the authors have also detected the continuous thermal background radiation whose power is comparable to or even higher than the average power of the pulsed THz waves. However, if the Schottky barrier diodes are also photoactivated using the same laser pulses which are used to pump the emitter then the problem can be solved. In other words, if a photoconductive antenna is activated using the laser pulses which can set right the THz signal within a limited time window while rejecting the majority of the thermal radiation then they can detect the pulsed THz radiation intensity without time-delay scans. In multicontact photoconductive antenna design [119], the authors have placed a cross-shaped photoconductive antenna on LT-GaAs photoconductive substrate using a standard photolithography and the method of chemical etching. On applying a bias voltage to the two adjacent electrodes while the other two electrodes grounded, then the bias electric field in the photoconductive gap is focused to ±45° from the horizontal axis. When a short pulse light is irradiated to the biased photoconductive gap, it results in the generation of a linearly polarized THz radiation from the transient photocurrent directed toward the bias field. By providing a phase shift of 90° to the input source, the antenna's electric field polarizes in the orthogonal direction. This kind of polarization modulator is helpful for polarization-sensitive THz spectroscopy like the THz ellipsometry and THz-vibrational circular dichroism (VCD) spectroscopy.

Hara et al. [120] have demonstrated an enhanced THz detection using photoconductive antennas based on self-assembled ErAs:GaAs nanoisland superlattices. The authors have compared three detectors, each fabricated on low-temperature grown GaAs (LT-GaAs), radiation-damaged silicon-on-sapphire (SI-GaAs), and an ErAs:GaAs superlattice, respectively. The ErAs:GaAs-based detector shows a strong enhancement in the THz detection efficiency with respect to the incident optical power. Moreover, the results show improved THz bandwidth and signal-to-noise ratios. To reduce the image acquisition time, an array arrangement supported by electronic beam steering is useful in imaging application. The photoconductive antenna is fairly stable against optical and thermal noises in comparison to the electro-optic rectification. However, the total antenna efficiency, which depends on the multiplication of optical laser to THz conversion efficiency, impedance matching efficiency, and radiation efficiency, is low. To increase the impedance matching efficiency as well as the radiation efficiency, it is required to consider the physical phenomena's contribution to the enhancement of efficiency in photoconductive antenna. In [121], the study of equivalent circuit model of photoconductive dipole antenna and its use for photomixer are described in detail. The equivalent circuit is determined on the basis of source conductance developed in the photoconductive dipole antenna with laser excitation on the photoconductive gap. The authors have emphasized over modified expression for source conductance which occurs as a physical phenomenon across a photoconductive gap in photoconductive dipole antenna. The outcomes of variation of antenna parameters on the conductance are observed, which facilitates to optimize the antenna impedance matching efficiency. Three designs of small-gap photoconductive antennas as emitters are analyzed in [122] to increase the directivity and radiation efficiency of small-gap photoconductive dipole antenna which has a major limitation of showing low directivity values. A simple synthesis technique is presented to resolve the antenna parameters corresponding to 1.5 THz to detect the powdered hidden explosives such as RDX,

HMX, and PETN. The simulation results shown in the paper recommend the use of thin-layer superstrate with silicon lens for diffraction-limited imaging using THz radiation in the low THz frequency region. The choice of specific photoconductive materials with essential electro-optic characteristics is a major factor which enables the photoconductive antenna for its effective performance in terms of its radiation parameters. Burford et al. [123] have reviewed the use of different photoconductive materials for antenna fabrication. Even though LT-GaAs remains the standard for photoconductive antenna, however the potential for all fiber-based spectroscopy systems motivates to develop other photoconductive material systems. In imaging system, there is a requirement of high acquisition speed, which can be achieved using array configuration of THz antenna and is discussed by Pradarutti et al. [124], wherein a THz line detection with 16 channels using microlens array coupled photoconductive antenna (PCA) array is implemented to improve the acquisition speed.

Moreover, the THz antennas based upon the photoconduction method have numerous advantages that they offer in comparison to the other THz sources. These antennas work in the room-temperature environment, are compact in size, and can operate both at the emission and detection sides. Therefore, from the detailed explanation about the THz antennas for imaging applications, it is apparent that there are many avenues to improve the performance of THz photoconductive antenna technology for imaging application [125]. Moreover, the key challenge is to design an optimized photoconductive antenna in terms of the radiation pattern, bandwidth, and radiation efficiency, so as to counter the problem of low radiated THz power for the pulsed THz imaging application. In Table 1.4, the recently developed THz photoconductive antenna design with respect to the imaging applications is summarized for quick reference.

1.7 Problem Formulation

The photoconductive antennas (PCAs) form the basis of many THz imaging and spectroscopy systems and find promising applications in various scientific fields. On the basis of the architecture of photoconductive antennas for THz pulsed systems, they are classified as aperture antennas (large and small compared to the wavelength), spiral antennas, bowtie antennas, and dipole antennas. One advantage of using the photoconductive antenna in pulsed imaging system is that the photoconductive antenna can be used on both transmitter and receiver sides. Only the difference lies in the use of biased voltage. On the transmitter side, the photoconductive antenna is biased with a fixed external voltage and the optical source, however, on the receiver side no biased voltage is required. For the imaging applications, there is a requirement of planar and compact THz antenna sources amidst on-chip fabrication along with high directivity in order to achieve large depth-of-field (DoF) for better image resolution. A small-gap photoconductive dipole antenna being simple in fabrication and planar in design shows its easy deployment as THz source for pulsed broadband system utilized in THz imaging and spectroscopy systems. Moreover, a femtosecond oscillator-based THz photoconductive dipole antenna

Table 1.4 Recent developments in THz photoconductive dipole antenna design with respect to the imaging applications

Type of THz antenna	Main characteristic	Application area	Reference
Photoconductive nanoantennas with plasmonic contact electrode gratings	Effective in enhancing the radiation power and detection sensitivity of photoconductive THz sources and detectors	Time-domain and Frequency-domain THz imaging and spectroscopy systems	Jarrahi [110]
Photoconductive antenna with plasmonic contact electrodes	Enhances the optical-to-THz conversion efficiency of photoconductive THz emitters and the detection sensitivity of photoconductive THz detectors	Material characterization, biological sensing, and medical imaging	Berry et al. [111]
Nanoplasmonic photoconductive antenna (NP-PCA)	Enhanced THz emission power by tuning the plasmon resonance	Emitters for spectroscopy, imaging, and sensing systems	Park et al. [112]
Bowtie-shaped photoconductive dipole antenna: a) With silicon-based lens and artificial magnetic conductor (AMC), b) capacitively loaded dipole, c) grid antenna	Enhanced directivity and front-to-back ratio of photoconductive antenna	High-resolution THz spectral imaging system	Zhu et al. [113]
Dipole antenna with GaAs membrane structure	Thin cavity provides high input resistance and the use of bulk GaAs substrates provides high radiation efficiency	Supports applications that demand large coverage, easy alignment, and high scanning speed rather than high resolution	Nguyen et al. [114]
Bow-tie THz photoconductive antenna	Possesses the optimum radiation bandwidth	Emitters for spectroscopy, imaging, and sensing systems	Maraghechi et al. [115]
Amplifier-driven large-area photoconductive antenna	Impulsive the THz radiation with high electric fields	Material Characterization	Beck et al. [118]
Schottky photoconductive antenna	Detection of THz radiation intensity without the time-delay scan required for the ordinary PC antenna	THz sensing applications	Tani et al. [119]
Four contact photoconductive antenna	To generate orthogonally polarized THz radiation	Polarization-sensitive THz spectroscopy	
Photoconductive antennas based on self-assembled ErAs:GaAs nanoisland superlattices	Enhanced THz detection	THz sensing and imaging	Hara et al. [120]

(continued)

Table 1.4 (continued)

Type of THz antenna	Main characteristic	Application area	Reference
THz photoconductive dipole antenna	Develop an equivalent circuit model based on source conductance occurring as a physical phenomenon across a photoconductive gap which is analyzed to optimize the antenna in terms of output power, impedance matching efficiency, and SNR.	Emitters for spectroscopy, imaging, and sensing systems	Khiabani et al. [121]
Photoconductive dipole antenna using LT-GaAs superstrate and silicon lens for the detection of powdered hidden explosives such as RDX, HMX, and PETN.	For directivity enhancement, antenna structure parameters are computed using the synthesis technique. For increasing the radiation efficiency, thin superstrate over the substrate of dipole antenna is used. For diffraction-limited imaging, a shorter wavelength antenna with silicon lens is proposed.	Sensing and imaging applications	Malhotra et al. [122]

system can generate high signal-to-noise ratio (SNR) broadband THz waves and detect them with high sensitivity. However, one of the major restrictive factors of THz photoconductive dipole antenna technology is saturation at high optical pump powers. To overcome the saturation limits of PCAs, two different approaches can be used.

- Implementation of large device apertures
- Implementation of interdigitated electrodes

Moreover, to increase the optical-to-THz conversion efficiency, the use of large-area THz photoconductive antenna emitters is proposed by the researchers. However, large-area emitters have complex fabrication as well as packaging considerations. Similarly, a plasmonic nanostructure THz photoconductive dipole antenna also offers high device quantum efficiency and the antenna design comprises of nanostructured electrodes with large-area emitters. However, because of the lack of maturity in nanoscale lithography technology, the fabrication of such ordered metallic nanostructures remains more difficult. Furthermore, to increase the THz photoconductive antenna bandwidth the use of broadband antennas is required. Therefore, the use of log spiral and log periodic antenna patterns to simple face-to-face dipole THz photoconductive antenna can yield broadband response. To enhance the radiation efficiency of THz photoconductive antenna, a log spiral antenna with nanostructured electrodes can be used. However, the issues, such as varying polarization and resonant/nonresonant regions across the desired operating band, need to be considered while implementing these complex antenna geometries into practice.

The performance limitation of conventional small-gap THz photoconductive dipole antenna in terms of low gain and low directivity values with small optical-to-THz conversion efficiency are also the major concerns in its use in security application of THz pulsed imaging systems for the detection of hidden explosives and illicit drugs. This indicates that the directivity enhancement mechanism of THz photoconducting dipole antenna needs a fair dealing. The radiation efficiency is also an important concern due to the reduced conductivity of the metal at the THz frequencies. Therefore, the key modalities of improving small-gap photoconductive dipole antennas' performance are identified for imaging applications and ways to improve the directivity of the photoconductive dipole antenna are discussed. Considering the need for an effective measure to combat the problem of low radiated THz power, a detailed synthesis technique to obtain the physical parameters of the small-gap photoconductive dipole antenna is presented. The analysis also includes the key challenges such as how to design an optimized small-gap photoconductive dipole antenna in terms of radiation pattern, bandwidth, and radiation efficiency. The simulation results show the potential for further improvement of THz-TDS systems used in security applications of imaging systems.

Various researchers have reported different directivity and gain enhancement techniques such as the use of Fabry-Perot cavity resonator, electromagnetic band gap, and left-handed metamaterial with antenna. Moreover, these materials have attracted significant research interest because of their special electromagnetic properties, which are applicable to a wide range of electromagnetic devices. Therefore, a novel idea based on using bandpass frequency selective surface (FSS) as a superstrate with photoconductive dipole antenna for the prospective improvement of gain and directivity of the THz antenna is investigated [125, 126]. The proposed antenna configuration having planar and compact assembly is ideal for its use as an efficient THz source for the detection of powdered hidden explosives in security applications of pulsed THz imaging system. In one such application, when an individual carrying hidden explosives is moving through a corridor toward an imaging system then their hidden explosives will be visible only for the brief moment while they are in the DoF then in such situation with the use of a highly directive THz antenna with short-range imaging system having a narrow depth-of-field (DoF) we can detect the explosive easily. However, such scanning over an extended volume could provide security such as in a public marketplace where the security is important, but a visible display is not so much important. Therefore, for such application of imaging, a compact array structure of THz antenna with pulsed optical beam from femtosecond laser pulse is proposed with highly directive ability of the radiating THz source. A technique of using FSS with photoconductive array antenna is also presented to further enhance the directivity from the radiating structure to yield high image resolution [126, 127]. Moreover, by using this technique, radiation energy will be confined to the desired frequency band rather than spreading over a wide spectrum range. We believe that the proposed advancements in the area of THz antenna technology for imaging application will open a new avenue in the biomedical and security applications in the THz regime of the spectrum.

1.8 Organization of Book

The remainder of this book is organized as follows. Chapter 2 includes the THz imaging modalities in terms of transmission-type and reflection-type imaging. The THz spectrum is a rich source of material information and allows the identification of material species such as bacterial spores hidden inside an optically opaque material. Therefore, the THz imaging with diffraction limit and tomography is presented along with its potential research challenges. In Chap. 3, the state-of-the-art THz antennas based on integrated circuit (IC) technology for imaging and sensing applications are discussed. In Chap. 4, a simple synthesis technique is presented to determine the physical parameters of photoconductive dipole antenna, which shows its application for THz sensing and imaging to detect the presence of hidden explosives which have spectral fingerprints in the range 1–3 THz. To increase the antenna radiation efficiency which also contributes to the total efficiency of photoconductive dipole antenna, a thin superstrate over the substrate of the dipole antenna is used to enable the antenna to withstand high biased voltage. The proposed antenna design is compact in size which further helps to achieve the diffraction-limited imaging. Moreover, the use of silicon lens with the photoconductive dipole antenna helps to focus the beam in the desired direction with increased directivity. In Chap. 5, an analytical procedure making use of explicit mathematical expressions leading to the physical behavior of small-gap photoconductive dipole antenna is developed and for this, a comprehensive systematic framework to determine the physical phenomenon occurring across the small-gap photoconductive dipole antenna is presented. Moreover, an optimization of the size of photoconductive gap to improve the total antenna efficiency is discussed and its consequence on radiated power is determined. As the THz antenna is one of the most important components in a THz sensing and imaging system, there is a need to have such a transmitting THz source with high directivity and optimum radiation efficiency. Therefore, to achieve this objective to enhance the directivity of photoconductive dipole antenna, in Chap. 6, a numerical calculation and design of frequency selective surface (FSS), which acts as a bandpass spatial filter, is presented and is deployed with photoconductive dipole antenna. The bandpass FSS structure parameters have been computed using a simple synthesis technique. The effect of the slot size of FSS to improve the scattering parameter at resonating frequency of the small-gap photoconductive dipole antenna is also presented. The most basic coherent imaging can be achieved by raster scanning a sample through the THz wave focus and generating complete spectroscopic information at each pixel corresponding to the antenna structure. Therefore, to speed up the measurements, there is a necessity of making a THz system for parallel or accelerated measurements. A potential approach to parallelization is the implementation of multiantenna setup which enables a linear downscaling of the measurement time with the deployment of a number of THz emitter and detector pairs. However, the compact size and planar arrays have to be used to counter the increase of system cost. Looking into such necessity, in Chap. 7, we have designed an array of small-gap photoconductive dipole antennas at THz frequency by using a frequency selective surface which not only provides a planar profile for THz radiating source but also offers enhanced gain and directivity for imaging application to detect the hidden explosives. Moreover, to control the transmission characteristics in a particular THz frequency band with the placement of array of FSS across the array of photoconductive dipole antennas is also discussed. The complete antenna array assembly with FSS is made useful to form an image by scanning a single beam in both principal planes. In Chap. 8, the beam-steering characteristics of the small-gap photoconductive dipole phased array antenna are presented. With uniform distribution of optical source excitation and progressive phase shift in x-axis and y-axis of the antenna configuration, the beam-steering phenomenon has been determined. THz near-field imaging and sensing is explored in Chap. 9. The measurement techniques for near-field imaging are discussed and an analytical estimate based on Kirchhoff's formalism for near-field is presented. In Chap. 10, an important feature of THz imaging/spectroscopy arises from its low-energy photons which are well below the ionization energy of biomolecules and have been explored for biomedical application. Therefore, a molecular system can be studied without being disturbed by a probing tool. In the medical imagery, the radiation damage from ionization—as induced, for example, by X-ray—can also be avoided using THz waves, due to their low energies. Furthermore, it also provides a reasonably good spatial resolution of less than a few hundred microns. This diffraction-limited resolution is maintained even in biological cells and tissues because the dimensions of cells are comparable to or smaller than the wavelength of THz waves. Finally, the concept of integrated THz technology for imaging and sensing applications is briefed in Chap. 11. It starts with silicon-based terahertz integrated circuit (THz-IC) technology followed by antenna arrays for imaging and sensing applications. Reasonably, the silicon technologies that are the core of the vast majority of commercial and high-end electronics products seem to be a tempting solution to bridge this terahertz gap. It includes the future directions in terms of certain design techniques which can be incorporated in the basic structure of small-gap photoconductive dipole antenna to further improve its performance parameters for its use in THz imaging applications.

1.9 Summary

This chapter presents a comprehensive review of the terahertz electromagnetic spectrum with its potential applications. Among the most compelling and scientifically stimulating regimes of the electromagnetic spectrum is the terahertz region. The beauty of electromagnetic spectrum is that it conveys various transverse characteristics across its breath, and based on interactions and behaviors within the media in which the wave propagates. This variety of characteristics with distinct and diverse interest leading to multiple applications from communication to imaging and sensing as well as fundamental scientific research has been presented extensively. Current terahertz products for imaging and sensing technology have been limited to single or few pixels only with raster scanning techniques to produce single terahertz image frame. Therefore, contrary to the current state of the art, developing such applications with commercial viability will require portability and high integration level, video rate, low-power consumption as well as room-temperature operations.

References

1. X. Zhang, Terahertz wave imaging: Horizons and hurdles. Phys. Med. Biol. **47**(21), 3667–3677 (2002)
2. M. Nnagai, K. Tanaka, H. Ohtake, T. Bessho, T. Sugiura, T. Hirosumi, M. Yoshida, Generation and detection of terahertz radiation by electro-optical process in GaAs using 1.56μm fiber laser pulses. Appl. Phys. Lett. **85**(18), 3974–3976 (2004)
3. S.S. Dhillon, M.S. Vitiello, E.H. Linfield, A.G. Davies, M.C. Hoffmann, J. Booske, C. Paoloni, M. Gensch, P. Weightman, G.P. Williams, E.C. Camus, D.R.S. Cumming, F. Simoens, I.E. Carranza, J. Grant, S. Lucyszyn, M.K. Gonokami, K. Konishi, M. Koch, C.A. Schmuttenmaer, T.L. Cocker, R. Huber, A.G. Markelz, Z.D. Taylor, V.P. Wallace, J.A. Zeitler, J. Sibik, T.M. Korter, B. Ellison, S. Rea, P. Goldsmith, K.B. Cooper, R. Appleby, D. Pardo, P.G. Huggard, V. Krozer, H. Shams, M. Fice, C. Renaud, A. Seeds, A. Stohr, M. Naftaly, N. Ridler, R. Clarke, J.E. Cunningham, M.B. Johnston, The 2017 terahertz science and technology roadmap. J. Phys. D. Appl. Phys. **50**(4), 043001/1-49 (2017)
4. C. Yu, S. Fan, Y. Sun, E. Pickwell-MacPherson, The potential of terahertz imaging for cancer diagnosis: A review of investigations to date. Quant. Imaging Med. Surg. **2**(1), 33–45 (2012)
5. I. Amenabar, F. Lopez, A. Mendikute, In introductory review to THz non-destructive testing of composite mater. J. Infrared, Millimeter Terahertz Waves **34**(2), 152–169 (2013)
6. L.A. Skvortsov, Standoff detection of hidden explosives and cold and firearms by terahertz time-domain spectroscopy and active spectral imaging. J. Appl. Spectrosc. **81**(5), 725–749 (2014)
7. C. Zhang, C. Deng, Y. Zhao, X. Zhang, Passive terahertz imager with an opto-mechanical scanner, Proc. 36th international conference on infrared, millimeter and terahertz waves (IRMMW-THz), pp. 1–2, 2–7, Houston, TX, USA, Oct 2011
8. E.V. Yakovlev, K.I. Zaytsev, I.N. Dolganova, S.O. Yurchenko, Non-destructive evaluation of polymer composite materials at the manufacturing stage using terahertz pulsed spectroscopy. IEEE Trans. Terahertz Sci. Technol **5**(5), 810–816 (2015)
9. J. Federici, L. Moeller, Review of terahertz and subterahertz wireless communications. J. Appl. Phys. **107**(11), 111101/ 1-23 (2010)
10. S. Sy, S. Huang, Y.-X.J. Wang, J. Yu, A.T. Ahuja, Y.-t. Zhang, E. Pickwell-MacPherson, Terahertz spectroscopy of liver cirrhosis: Investigating the origin of contrast. Phys. Med. Biol. **55**(24), 7587–7596 (2010)
11. K.B. Cooper, R.J. Dengler, N. Llombart, T. Bryllert, G. Chattopadhyay, E. Schlecht, J. Gill, C. Lee, A. Skalare, I. Mehdi, P.H. Siegel, Penetrating 3-D imaging at 4-and 25-m range using a submillimeter-wave radar. IEEE Trans. Microwave Theory Tech **56**(12), 2771–2778 (2008)
12. E.K. Rahani, T. Kundu, Z. Wu, H. Xin, Mechanical damage detection in polymer tiles by THz radiation. IEEE Sensors J. **11**(8), 1720–1725 (2011)
13. P. Mousavi, F. Haran, D. Jez, F. Santosa, J.S. Dodge, Simultaneous composition and thickness measurement of paper using terahertz time-domain spectroscopy. Appl. Opt. **48**(33), 6541–6546 (2009)
14. T. Yasui, T. Yasuda, K.-i. Sawanaka, T. Araki, Terahertz paintmeter for noncontact monitoring of thickness and drying progress in paint film. Appl. Opt. **44**(32), 6849–6856 (2005)

15. S. Wietzke, C. Jördens, N. Krumbholz, B. Baudrit, M. Bastian, M. Koch, Terahertz imaging: A new non-destructive technique for the quality control of plastic weld joints. J. Eur. Opt. Society-Rapid Publications **2**, 07013/1–07013/5 (2007)
16. E. Nguema, V. Vigneras, J.-L. Miane, P. Mounaix, Dielectric properties of conducting polyaniline films by THz time-domain spectroscopy. Eur. Polym. J. **44**(1), 124–129 (2008)
17. D. Banerjee, W. Von Spiegel, M. Thomson, S. Schabel, H. Roskos, Diagnosing water content in paper by terahertz radiation. Opt. Express **16**(12), 9060–9066 (2008)
18. J.-W. Park, K.-H. Im, D.K. Hsu, J.-A. Jung, I.-Y. Yang, Terahertz spectroscopy approach of the fiber orientation influence on CFRP composite solid laminates. J. Mech. Sci. Technol. **26**(7), 2051–2054 (2012)
19. K. Kawase, T. Shibuya, S.I. Hayashi, K. Suizu, THz imaging techniques for nondestructive inspections. Comptes Rendus Physique **11**(7–8), 510–518 (2010)
20. S. -T. Han, W.K. Park, Y.-H. Ahn, W.-J. Lee, H.S. Chun, Development of a compact subterahertz gyrotron and its application to t-ray real-time imaging for food inspection. Proc. 37th International conference on infrared, millimeter, and terahertz waves (IRMMW-THz), Wollongong, NSW, Australia, pp. 1–2, 23–28 September, 2012.
21. C. Jördens, M. Koch, Detection of foreign bodies in chocolate with pulsed terahertz spectroscopy. Opt. Eng. **47**(3), 037003/1–037003/5 (2008)
22. Y.L. Hor, J.F. Federici, R.L. Wample, Nondestructive evaluation of cork enclosures using terahertz/millimeter wave spectroscopy and imaging. Appl. Opt. **47**(1), 72–78 (2008)
23. H. Hoshina, Y. Sasaki, A. Hayashi, C. Otani, K. Kawase, Noninvasive mail inspection system with terahertz radiation. Appl. Spectrosc. **63**(1), 81–86 (2009)
24. M.C. Kemp, Explosives detection by terahertz spectroscopy—A bridge too far? IEEE Trans. Terahertz Sci. Technol **1**(1), 282–292 (2011)
25. H. Zhong, A. Redo-Sanchez, X.-C. Zhang, Standoff sensing and imaging of explosive related chemical and bio-chemical materials using THz-TDS. Int. J. High Speed Electron. Syst **17**(02), 239–249 (2007)
26. Y. Kurmi, V. Chaurasia, Hidden explosive detection systems for vehicle. Int. J. Comput. Appl **130**(10), 16–19 (2015)
27. M. Lu, J. Shen, N. Li, Y. Zhang, C. Zhang, L. Liang, X. Xu, Detection and identification of illicit drugs using terahertz imaging. J. Appl. Phys. **100**(10), 103104/1–103104/6 (2006)
28. J.F. Federici, B. Schulkin, F. Huang, D. Gary, R. Barat, F. Oliveira, D. Zimdars, THz imaging and sensing for security applications—Explosives, weapons and drugs. Semicond. Sci. Technol. **20**(7), S266–S280 (2005)
29. G.J. Wilmink, J.E. Grundt, Invited review article: Current state of research on biological effects of terahertz radiation. J. Infrared, Millimeter Terahertz Waves **32**(10), 1074–1122 (2011)
30. A. Amicelle, C. Aradau, J. Jeandesboz, G. Valkenburg, I. van der Ploeg, Materialities between security and privacy: A constructivist account of airport security scanners. Secur. Dialogue **46**(4), 326–344 (2015)
31. E. Heinz, T. May, D. Born, G. Zieger, S. Anders, G. Thorwirth, V. Zakosarenko, M. Schubert, T. Krause, M. Starkloff, A. Kruger, Passive submillimeter-wave stand-off video camera for security applications. J. Infrared, Millimeter Terahertz Waves **31**(11), 1355–1369 (2010)
32. P.U. Jepsen, U. Møller, H. Merbold, Investigation of aqueous alcohol and sugar solutions with reflection terahertz time-domain spectroscopy. Opt. Express **15**(22), 14717–14737 (2007)
33. J.-H. Son, *Terahertz Biomedical Science and Technology* (CRC Press, Boca Raton, 2014)
34. G.J. Wilmink, B.D. Rivest, C.C. Roth, B.L. Ibey, J.A. Payne, L.X. Cundin, J.E. Grundt, X. Peralta, D.G. Mixon, W.P. Roach, In vitro investigation of the biological effects associated with human dermal fibroblasts exposed to 2.52 THz radiation. Lasers Surg. Med. **43**(2), 152–163 (2011)
35. J.-H. Son, Terahertz electromagnetic interactions with biological matter and their applications. J. Appl. Phys. **105**(10), 102033/1–11 (2009)

36. T.C. Bowman, M. El-Shenawee, L.K. Campbell, Terahertz imaging of excised breast tumor tissue on paraffin sections. IEEE Trans. Antennas Propag. **63**(5), 2088–2097 (2015)
37. Z.D. Taylor, R.S. Singh, D.B. Bennett, P. Tewari, C.P. Kealey, N. Bajwa, M.O. Culjat, A. Stojadinovic, H. Lee, J.P. Hubschman, E.R. Brown, THz medical imaging : In vivo hydration sensing. IEEE Trans. Terahertz Sci. Technol **1**(1), 201–219 (2011)
38. R. Woodward, V. Wallace, D. Arnone, E. Linfield, M. Pepper, Terahertz pulsed imaging of skin cancer in the time and frequency domain. J. Biol. Phys. **29**(2–3), 257–259 (2003)
39. C. Zhang, G. Zhao, B. Jin, Y. Hou, H. Jia, J. Chen, P.H. Wu, Terahertz imaging on subcutaneous tissues and liver inflamed by liver cancer cells. Terahertz Sci. Technol **5**(3), 114–123 (2012)
40. S.J. Oh, Y.-M. Huh, J.-S. Suh, J. Choi, S. Haam, J.-H. Son, Cancer diagnosis by terahertz molecular imaging technique. J. Infrared, Millimeter Terahertz Waves **33**(1), 74–81 (2012)
41. E. Berry, A.J. Fitzgerald, N.N. Zinov'ev, G.C. Walker, S. Homer-Vanniasinkam, C.D. Sudworth, R.E. Miles, J.M. Chamberlain, M.A. Smith, Optical properties of tissue measured using terahertz-pulsed imaging, Proceedings of SPIE: Medical Imaging 2003: Physics of Medical Imaging, San Diego, California, United States, vol. 5030, pp. 459–470, 15–20 Feb 2003
42. D.M. Charron, K. Ajito, J.-Y. Kim, Y. Ueno, Chemical mapping of pharmaceutical cocrystals using terahertz spectroscopic imaging. Anal. Chem. **85**(4), 1980–1984 (2013)
43. Y.-C. Shen, Terahertz pulsed spectroscopy and imaging for pharmaceutical applications: A review. Int. J. Pharm. **417**(1), 48–60 (2011)
44. Y.-C. Shen, P.F. Taday, Development and application of terahertz pulsed imaging for nondestructive inspection of pharmaceutical tablet. IEEE J. Select. Top. Quantum Electron **14**(2), 407–415 (2008)
45. L. Maurer, H. Leuenberger, Terahertz pulsed imaging and near infrared imaging to monitor the coating process of pharmaceutical tablets. Int. J. Pharm. **370**(1), 8–16 (2009)
46. J.P. Guillet, B. Recur, L. Frederique, B. Bousquet, L. Canioni, I. Manek-Hönninger, P. Desbarats, P. Mounaix, Review of terahertz tomography techniques. J. Infrared, Millimeter Terahertz Waves **35**(4), 382–411 (2014)
47. H. Zhong, J. Xu, X. Xie, T. Yuan, R. Reightler, E. Madaras, X.C. Zhang, Nondestructive defect identification with terahertz time-of-flight tomography. IEEE Sensors J. **5**(2), 203–208 (2005)
48. H. Zhong, N. Karpowicz, J. Xu, Y. Deng, W. Ussery, M. Shur, X.C. Zhang, Detection of space shuttle insulation foam defects by using a 0.2 THz Gunn diode oscillator and pyroelectric detector, In Frontiers in optics, Optical Society of America, p. FTuG28, 2004
49. E.K. Rahani, T. Kundu, Z. Wu, H. Xin, Heat induced damage detection by terahertz (THz) radiation. J. Infrared, Millimeter Terahertz Waves **32**(6), 848–856 (2011)
50. P.D. Cunningham, N.N. Valdes, F.A. Vallejo, L.M. Hayden, B. Polishak, X.-H. Zhou, J. Luo, A.K.-Y. Jen, J.C. Williams, R.J. Twieg, Broadband terahertz characterization of the refractive index and absorption of some important polymeric and organic electro-optic materials. J. Appl. Phys. **109**(4), 043505/1–043505/5 (2011)
51. T. Nagatsuma, G. Ducournau, C.C. Renaud, Advances in terahertz communications accelerated by photonics. Nat. Photonics **10**(6), 371–379 (2016)
52. K.R. Jha, G. Singh, Terahertz planar antennas for future wireless communication: A technical review. Infrared Phys. Technol. **60**, 71–80 (2013)
53. T. Kürner, S. Priebe, Towards THz communications-status in research, standardization and regulation. J. Infrared, Millimeter Terahertz Waves **35**(1), 53–62 (2014)
54. Z. Chen, Z. Tan, Y. Han, R. Zhang, X. Guo, H. Li, J. Cao, H.C. Liu, Wireless communication demonstration at 4.1 THz using quantum cascade laser and quantum well photodetector. Electron. Lett. **47**(17), 1002–1004 (2011)
55. G. Ducournau, P. Szriftgiser, F. Pavanello, E. Peytavit, M. Zaknoune, D. Bacquet, A. Beck, T. Akalin, J.F. Lampin, THz communications using photonics and electronic devices: The race to data-rate. J. Infrared, Millimeter Terahertz Waves **36**(2), 198–220 (2015)

56. X. Li, Z. Dong, J. Yu, N. Chi, Y. Shao, G. Chang, Fiber-wireless transmission system of 108 Gb/sdata over 80 km fiber and 2× 2multiple-input multiple-output wireless links at 100 GHz W-band frequency. Opt. Lett. **37**(24), 5106–5108 (2012)
57. H.-J. Song, J.-Y. Kim, K. Ajito, M. Yaita, N. Kukutsu, Fully integrated ASK receiver MMIC for terahertz communications at 300 GHz. IEEE Trans. Terahertz Sci. Technol **3**(4), 445–452 (2013)
58. K.R. Jha, G. Singh, "*Terahertz Planar Antennas for Next Generation Communication*" Springer: Switzerland, Jan. (2014)
59. C. Jansen, S. Priebe, C. Moller, M. Jacob, H. Dierke, M. Koch, T. Kurner, Diffuse scattering from rough surfaces in THz communication channels. IEEE Trans. Terahertz Sci. Technol **1**(2), 462–472 (2011)
60. I. Kallfass, F. Boes, T. Messinger, J. Antes, A. Inam, U. Lewark, A. Tessmann, R. Henneberger, 64 gbit/s transmission over 850 m fixed wireless link at 240 GHz carrier frequency. J. Infrared, Millimeter Terahertz Waves **36**(2), 221–233 (2015)
61. A. Hirata, T. Kosugi, H. Takahashi, J. Takeuchi, K. Murata, N. Kukutsu, Y. Kado, S. Okabe, T. Ikeda, F. Suginosita, K. Shogen, 5.8-km 10-Gbps data transmission over a 120-GHz-band wireless link, IEEE international conference on wireless information technology and systems (ICWITS), Honolulu, HI, USA, pp. 1–4, 28 August-3 September, 2010
62. J.Y. Suen, Terabit-per-second satellite links: A path toward ubiquitous terahertz communication. J. Infrared, Millimeter Terahertz Waves **37**(7), 615–639 (2016)
63. I. Malhotra, K.R. Jha, G. Singh, Terahertz antenna technology for imaging applications: A technical review. Int. J. Microw. Wirel. Technol. **10**(3), 271–290 (2018)
64. B. Ferguson, D. Gray, D. Abbot, X.C. Zhang, T-ray computed tomography. Opt. Lett. **27**(15), 1312–1314 (2002)
65. N. Sunaguchi, Y. Sasaki, N. Maikusa, M. Kawai, T. Yuasa, C. Otani, Depth-resolving THz imaging with tomosynthesis. Opt. Express **17**(12), 9558–9570 (2009)
66. D.M. Mittleman, S. Hunsche, L. Boivin, M.C. Nuss, T-ray tomography. Opt. Lett. **22**(12), 904–906 (1997)
67. A. Bandyopadhyay, A. Stepanov, B. Schulkin, M.D. Federici, A. Sengupta, D. Gary, J.F. Federici, R. Barat, Z.H. Michalopoulou, D. Zimdars, Terahertz interferometric and synthetic aperture imaging. J. Opt. Soc. Am. A **23**(5), 1168–1178 (2006)
68. J. O'Hara, D. Grischkowsky, Synthetic phased-arrayterahertz imaging. Opt. Lett. **27**(12), 1070–1072 (2002)
69. G. Grüner, Millimeter and submillimeter wave spectroscopy of solids, in *Millimeter and Submillimeter Wave Spectroscopy of Solids*, vol. 1, (Springer, Berlin, 1998)
70. D.L. Woolard, J.O. Jensen, R.J. Hwu, *Terahertz Science and Technology for Military and Security Applications*, vol 46 (World Scientific, Singapore, 2007)
71. J.A. Zeitler, L.F. Gladden, In-vitro tomography and non-destructive imaging at depth of pharmaceutical solid dosage forms. Eur. J. Pharm. Biopharm. **71**(1), 2–22 (2009)
72. T. Kleine-Ostmann, P. Knobloch, M. Koch, S. Hoffmann, M. Breede, M. Hofmann, G. Hein, K. Pierz, M. Sperling, K. Donhuijsen, Continuous-wave THz imaging. Electron. Lett. **37**(24), 1461–1463 (2001)
73. N.T. Yardimci, S.-H. Yang, C.W. Berry, M. Jarrahi, High-power terahertz generation using large-area plasmonic photoconductive emitters. IEEE Trans. Terahertz Sci. Technol **5**(2), 223–229 (2015)
74. T. Satoh, Y. Toya, S. Yamamoto, T. Shimura, K. Kuroda, Y. Takahashi, M. Yoshimura, Y. Mori, T. Sasaki, S. Ashihara, Generation of mid-to far-infrared ultrashort pulses in 4-dim ethylamino-N-methyl-4-stilbazolium tosylate crystal. J. Opt. Soc. America B **27**(12), 2507–2511 (2010)
75. H. Hirori, A. Doi, F. Blanchard, K. Tanaka, Single-cycle terahertz pulses with amplitudes exceeding 1 MV/cm generated by optical rectification in LiNbO3. Appl. Phys. Lett. **98**(9), 091106/1–091106/4 (2011)

76. M. Ghoranneviss, M.A. Kashani, A. Hogabri, A. Kohiyan, A. Anvari, Design and modification of the FIR HCN laser, SPIE's international symposium on optical science, engineering, and instrumentation, pp. 85–87, San Diego, CA, USA, 19–24 July 1998
77. M.Y. Frankel, J.F. Whitaker, G.A. Mourou, Optoelectronic transient characterization of ultrafast devices. IEEE J. Quantum Electron. **28**(10), 2313–2324 (1992)
78. S. Pavlov, H.-W. Hübers, J.N. Hovenier, T.O. Klaassen, D.A. Carder, P.J. Phillips, B. Redlich, H. Riemann, R.K. Zhukavin, V.N. Shastin, Stimulated terahertz stokes emission of silicon crystals doped with antimony donors. Phys. Rev. Lett. **96**(3), 037404/1–037404/4 (2006)
79. J. Hebling, A. Stepanov, G. Almási, B. Bartal, J. Kuhl, Tunable THz pulse generation by optical rectification of ultrashort laser pulses with tilted pulse fronts. Appl. Phys. B Lasers Opt. **78**(5), 593–599 (2004)
80. H. Eisele, A. Rydberg, G.I. Haddad, Recent advances in the performance of InP Gunn devices and GaAs TUNNETT diodes for the 100–300-GHz frequency range and above. IEEE Trans. Microwave Theory Tech **48**(4), 626–631 (2000)
81. M. Tani, O. Morikawa, S. Matsuura, M. Hangyo, Generation of terahertz radiation by photomixing with dual-and multiple-mode lasers. Semicond. Sci. Technol. **20**(7), S151–S163 (2005)
82. F. Capasso, C. Gmachl, R. Paiella, A. Tredicucci, A.L. Hutchinson, D.L. Sivco, J.N. Baillargeon, A.Y. Cho, H.C. Liu, New frontiers in quantum cascade lasers and applications. IEEE J. Select. Top. Quantum Electron **6**(6), 931–947 (2000)
83. A. Leitenstorfer, S. Hunsche, J. Shah, M. Nuss, W. Knox, Detectors and sources for ultrabroadband electro-optic sampling: Experiment and theory. Appl. Phys. Lett. **74**(11), 1516–1518 (1999)
84. D. Fast, W. Hurlbut, V.G. Kozlov, Extending spectral coverage of BWOs combined with frequency multipliers to 2.6 THz, International Society for Optics and Photonics, SPIE OPTO, San Diego, California, USA, pp. 82610L/1–82610L11, 12–16 Aug 2012
85. N. Large, M. Abb, J. Aizpurua, O.L. Muskens, Photoconductively loaded plasmonic nanoantenna as building block for ultracompact optical switches. Nano Lett. **10**(5), 1741–1746 (2010)
86. P. Piot, D. Douglas, G. Krafft, Longitudinal phase space manipulation in energy recovering linac-driven free-electron lasers. Phys. Rev. Spec. Top. Accel Beams **6**(3), 030702/1–11 (2003)
87. M.J. Rodwell, M. Kamegawa, R. Yu, M. Case, E. Carman, K.S. Giboney, GaAs nonlinear transmission lines for picosecond pulse generation and millimeter-wave sampling. IEEE Trans. Microwave Theory Tech **39**(7), 1194–1204 (1991)
88. T. S. Hartwick, Far infrared imaging techniques for law enforcement applications, SPIE/SPSE technical symposium, East-Reston, pp. 139–140, 18 April 1977
89. K.J. Siebert, T. Löffler, H. Quast, M. Thomson, T. Bauer, R. Leonhardt, S. Czasch, H.G. Roskos, All-optoelectronic continuous wave THz imaging for biomedical applications. Phys. Med. Biol. **47**, 3743–3748 (2002)
90. N. Karpowicz, H. Zhong, J. Xu, K.-I. Lin, J.-S. Hwang, X. Zhang, Comparison between pulsed terahertz time-domain imaging and continuous wave terahertz imaging. Semicond. Sci. Technol. **20**(7), S293–S299 (2005)
91. D. Auston, X. Zhang, Large-aperture photoconducting antennas. Proc. Opt. Soc. America OSA **9**, 2–8 (1991)
92. Q. Wu, X.C. Zhang, Free-space electro-optic sampling of terahertz beams. Appl. Phys. Lett. **67**(24), 3523–3525 (1995)
93. V.P. Wallace, E. MacPherson, J.A. Zeitler, C. Reid, Three-dimensional imaging of optically opaque materials using nonionizing terahertz radiation. J. Opt. Soc. America A **25**(12), 3120–3133 (2008)
94. D.M. Sheen, T.E. Hall, R.H. Severtsen, D.L. McMakin, B.K. Hatchell, P.L. Valdez, Standoff concealed weapon detection using a 350-GHz radar imaging system, SPIE defense, security, and sensing, Orlando, USA, pp. 767008/1–76700812, 5–9 April 2010

95. M. Kowalski, M. Kastek, Comparative studies of passive imaging in terahertz and mid-wavelength infrared ranges for object detection. IEEE Trans. Inf. Forensics Secur **11**(9), 2028–2035 (2016)
96. E. Grossman, J. Gordon, D. Novotny, R. Chamberlin, Terahertz active and passive imaging, Proceedings of 8th European conference on antennas and propagation (EuCAP), Netherlands, pp. 2221–2225, 6–11 April 2014
97. W. Deal, X. Mei, K.M. Leong, V. Radisic, S. Sarkozy, R. Lai, THz monolithic integrated circuits using InP high electron mobility transistors. IEEE Trans. Terahertz Sci. Technol **1**(1), 25–32 (2011)
98. V. Radisic, K.M. Leong, X. Mei, S. Sarkozy, W. Yoshida, W.R. Deal, Power amplification at 0.65 THz using INP HEMTs. IEEE Trans. Microwave Theory Techniq **60**(3), 724–729 (2012)
99. F. Friederich, W. Von Spiegel, M. Bauer, F. Meng, M.D. Thomson, S. Boppel, A. Lisauskas, B. Hils, V. Krozer, A. Keil, T. Loffler, THz active imaging systems with real-time capabilities. IEEE Trans. Terahertz Sci. Technol **1**(1), 183–200 (2011)
100. B. Hu, M. Nuss, Imaging with terahertz waves. Opt. Lett. **20**(16), 1716–1718 (1995)
101. Y.H. Lo, R. Leonhardt, Aspheric lenses for terahertz imaging. Opt. Express **16**(20), 15991–15998 (2008)
102. A.J. Huber, F. Keilmann, J. Wittborn, J. Aizpurua, R. Hillenbrand, Terahertz near-field nanoscopy of mobile carriers in single semiconductor nanodevices. Nano Lett. **8**(11), 3766–3770 (2008)
103. M.A. Patrick, J.A. Holt, C.D. Joye, F.C. De Lucia, Elimination of speckle and target orientation requirements in millimeter-wave active imaging by modulated multimode mixing illumination. J. Opt. Soc. America A **29**(12), 2643–2656 (2012)
104. G.C. Trichopoulos, H.L. Mosbacker, D. Burdette, K. Sertel, A broadband focal plane array camera for real-time THz imaging applications. IEEE Trans. Antennas Propag. **61**(4), 1733–1740 (2013)
105. K.R. Jha, G. Singh, Analysis and design of terahertz microstrip antenna on photonic bandgap material. Journal of Computational Electronics. **11**(4), 364–373 (2012)
106. I.C. Mayorga, A. Schmitz, T. Klein, C. Leinz, R. Gusten, First in-field application of a full photonic local oscillator to terahertz astronomy. IEEE Trans. Terahertz Sci. Technol **2**(4), 393–399 (2012)
107. K. Han, T.K. Nguyen, I. Park, H. Han, Terahertz Yagi-Uda antenna for high input resistance. J. Infrared, Millimeter Terahertz Waves **31**(4), 441–454 (2010)
108. R. Singh, C. Rockstuhl, C. Menzel, T.P. Meyrath, M. He, H. Giessen, F. Lederer, W. Zhang, Spiral-type terahertz antennas and the manifestation of the Mushiake principle. Opt. Express **17**(12), 9971–9980 (2009)
109. K. Topalli, G.C. Trichopoulos, K. Sertel, An indirect impedance characterization method for monolithic THz antennas using coplanar probe measurements. IEEE Antennas Wirel. Propag. Lett **11**, 3–5 (2012)
110. M. Jarrahi, Advanced photoconductive terahertz optoelectronics based on nano-antennas and nano-plasmonic light concentrators. IEEE Trans. Terahertz Sci. Technol **5**(3), 391–397 (2015)
111. C.W. Berry, N. Wang, M.R. Hashemi, M. Unlu, M. Jarrahi, Significant performance enhancement in photoconductive terahertz optoelectronics by incorporating plasmonic contact electrodes. Nat. Commun. **4**, 1622/1–10 (2013)
112. S.-G. Park, K.H. Jin, M. Yi, J.C. Ye, J. Ahn, K.-H. Jeong, Enhancement of terahertz pulse emission by optical nanoantenna. ACS Nano **6**(3), 2026–2031 (2012)
113. N. Zhu, R.W. Ziolkowski, Photoconductive THz antenna designs with high radiation efficiency, high directivity , and high aperture efficiency. IEEE Trans. Terahertz Sci. Technol **3**(6), 721–730 (2013)
114. T.K. Nguyen, H. Han, I. Park, Numerical study of a full-wavelength dipole antenna on a GaAs membrane structure at terahertz frequency. J. Infrared, Millimeter Terahertz Waves **32**(6), 763–777 (2011)

115. P. Maraghechi, A.Y. Elezzabi, Experimental confirmation of design techniques for effective bow-tie antenna lengths at THz frequencies. J. Infrared, Millimeter Terahertz Waves **32**(7), 897–901 (2011)
116. D. Rutledge, D. Neikirk, D. Kasilingam, Integrated-circuit antennas. Infrared Millimeter Waves **10**(2), 1–90 (1983)
117. M.Y. Frankel, S. Gupta, J.A. Valdmanis, G.A. Mourou, Terahertz attenuation and dispersion characteristics of coplanar transmission lines. IEEE Trans. Microwave Theory Tech **39**(6), 910–916 (1991)
118. M. Beck, H. Schäfer, G. Klatt, J. Demsar, S. Winnerl, M. Helm, T. Dekorsy, Impulsive terahertz radiation with high electric fields from an amplifier-driven large-area photoconductive antenna. Opt. Express **18**(9), 9251–9257 (2010)
119. M. Tani, Y. Hirota, C.T. Que, S. Tanaka, R. Hattori, M. Yamaguchi, S. Nishizawa, M. Hangyo, Novel terahertz photoconductive antennas. Int. J. Infrared Millimeter Waves **27**(4), 531–546 (2006)
120. J.F. O'Hara, J. Zide, A. Gossard, A. Taylor, R. Averitt, Enhanced terahertz detection via ErAs: GaAs nanoisland superlattices. Appl. Phys. Lett. **88**(25), 251119/1–251119/4 (2006)
121. N. Khiabani, Y. Huang, Y.-C. Shen, S. Boyes, Theoretical modeling of a photoconductive antenna in a terahertz pulsed system. IEEE Trans. Antennas Propag. **61**(4), 1538–1546 (2013)
122. I. Malhotra, K.R. Jha, G. Singh, Analysis of highly directive photoconductive dipole antenna at terahertz frequency for sensing and imaging applications. Opt. Commun. **397**, 129–139 (2017)
123. N.M. Burford, M.O. El-Shenawee, Review of terahertz photoconductive antenna technology. Opt. Eng. **56**(1), 010901–010901 (2017)
124. B. Pradarutti, R. Müller, W. Freese, G. Matthäus, S. Riehemann, G. Notni, S. Nolte, A. Tunnermann, Terahertz line detection by a microlens array coupled photoconductive antenna array. Opt. Express **16**(22), 18443–18450 (2008)
125. I. Malhotra, K.R. Jha, G. Singh, Design of highly directive terahertz photoconductive dipole antenna using frequency selective surface for sensing and imaging application. Journal of Computational Electronics, **17**(4), 1721–1740 (2018)
126. I. Malhotra, K.R. Jha, G. Singh, Design of highly directive lens-less photoconductive dipole antenna with frequency selective surface for terahertz imaging applications. OPTIK: International Journal of Electron Optics **173**, 206–219 (2018)
127. I. Malhotra, K.R Jha, G. Singh, Beam steering characteristics of highly directive photoconductive dipole phase array antenna for terahertz imaging applications. Optical and Quantum Electronics, **51**(1), 27/1–19 (2019)

Chapter 2
Terahertz Imaging Modalities: State-of-the Art and Open Challenges

2.1 Introduction

Imaging in the THz regime of the spectrum is eye-catching for the reason that wavelengths in the range 100 m to 0.5 mm (i.e., for the frequency range 3 MHz to 0.6 THz) are short enough to offer high resolution with modest apertures thus far long enough to penetrate materials such as cloth or cardboard. Therefore, the utilization of THz radiation to form images has several advantages. Moreover, as compared to infrared and higher frequencies, many common materials are relatively transparent for THz imaging, including common packaging materials such as paper, as well as many plastic and composites [1]. The image contrasts can also be obtained through polarization-resolved measurements, especially in the samples having anisotropic conductivity due to structural properties [2, 3]. Likewise, materials having large electrical conductivity such as metals and having large static dipoles such as water are the strong absorbers of THz radiations. Such absorption results in the formation of contrast in the resultant image. Since THz radiation is nonionizing and its sensitivity to water contents is high, there is no health risk to cells except for heating. This feature has motivated many researchers to make studies of THz imaging for agriculture as well as biomedical sensing [4–8]. The penetration depth of THz radiation into living tissue is small almost 100 microns; therefore, several diagnostic applications are being explored. Similarly, due to the vibrational absorption, bands possessed by the molecular crystals in the THz band act as spectroscopic fingerprints for material identification [9–12].

Among the concepts for THz imaging, several noncoherent techniques like micro-bolometer arrays are presented by researchers; whereas to increase the measurement speeds, these approaches provide only limited information due to lack of information about the phase. However, due to the rapid development of coherent THz sources, the trend shows the advancement in THz imaging systems and it presents an opportunity for high-resolution, potentially noninvasive imaging suitable for security or quality-control applications [13, 14]. The low energy of THz photons

I. Malhotra, G. Singh, *Terahertz Antenna Technology for Imaging and Sensing Applications*, https://doi.org/10.1007/978-3-030-68960-5_2

and extreme sensitivity to water absorption helps in characterization and imaging of biological systems [15]. Since the structure of biomolecules is closely related to their functionality, therefore, the pulsed THz sensing has its various applications in the field of biomedicine, such as the DNA sensing [16] and skin cancer detection. However, the pulsed THz radiations are able to penetrate a few millimeters of tissue; 85% of all cancers lie in the epithelium, allowing the THz to sense very small cancers such as basal cell carcinomas (BCCs). As the breathing modes of DNA caused by the stretching of hydrogen bonds between the two DNA strands such as vibrations, twisting and global stretching modes are excited in the 0.1–10 THz band, and the pulsed THz spectroscopy provides information about DNA structure and dynamics. The THz-based imaging technologies find a unique place in healthcare savings [17]. The low-cost, portable systems are helping in detecting skin cancer, reducing the need of costly and traumatic surgery.

Moreover, the screening of persons to detect concealed hazardous objects is a common task arising from various security threats to contemporary societies. However, the full-body scanning techniques based on X-rays on the other side raise questions about the health issues, thus limiting their acceptance by the public. Another important security need like recognizing a suicide bomber from a safe distance is not met at all by existing technologies. However, the most useful property of imagers operating in THz range has an ability to detect small temperature differences on the object's surface as well as the ability to see through clothing. Against the background of the radiating human body, reflecting and absorbing objects are visible. Hence, the objects concealed under the clothing can also be detected. This applies to metals, which are highly reflective, ceramic materials, and even explosives which show typical absorption spectra in the submillimeter range [18]. Therefore, due to the potential features of THz radiations in imaging applications, the THz imaging systems on the basis of type of imagers used in the THz imaging system are classified as (1) passive imaging and (2) active imaging. The THz imagers provide images with low signal-to-noise ratio and low spatial resolution and offer limited distance of imaging. The performance of THz imagers in terms of noise-equivalent temperature difference (NETD) value is lower and is in the range of 0.5–5 k [19]. The passive THz imagers record the contrast in radiometric temperature within an object under scene, and active THz imagers record the contrast in the scattered radiance within an object when it is illuminated with THz source. In active imaging, the imager is used to make the active image which confines all of its illumination to a single mode and the receiver observes the same mode. However, the passive THz imaging systems which are inherently multimode have a small dynamic range in comparison to active THz imaging system [20].

Considering one simple example of THz imaging system [21] as shown in Fig. 2.1, a picture is divided into several pixels and each individual pixel's information is convoluted to a time-domain signal using a transmitting terahertz wave. Initially to compress the pixel's information, the THz beam is divided into multiple subbeams having variant time delays in ascending order. For this to achieve, a two-dimensional echelon having 16 rungs with continuous increased thickness from left toward right and from up toward down is used made from THz transparent material.

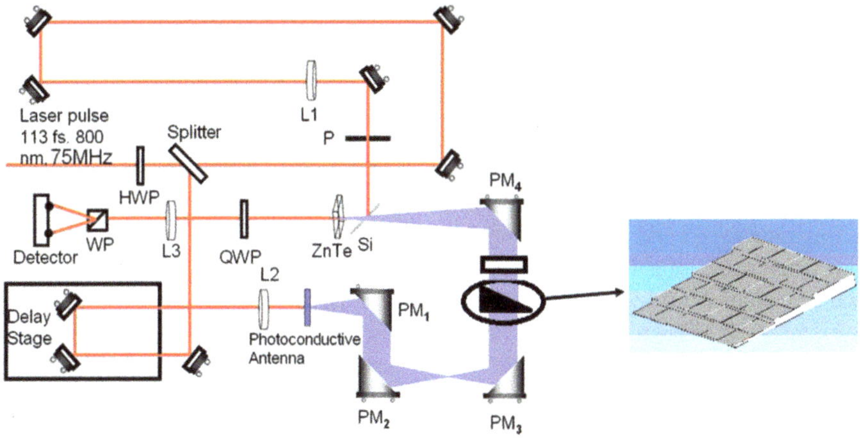

Fig. 2.1 Schematic setup of a THz-TDS system with a square-shaped 2D echelon [21]

Table 2.1 Comparison of typical specifications of four types of commercially available focal plane cameras used to visualize THz beam [22]

	CMOS	GaAs heterostructure	Microbolometer	Pyroelectric
Number of pixels	32 × 32	64 × 64	320 × 240	320 × 320
Pixel size (μm)	80	1500	50	75
Operating range (THz)	0.3–1.3	0.05–0.7	0.6–4.0	0.1 and up
NEP ($nWHz^{-1/2}$)	0.4	1	0.03	13
Responsivity (kV/W)	140	50	18 (at peak)	N/A
Frame rate (fps)	30	24	25	50

When a parallel THz beam transmits through the echelon, then each subbeam passing through each rung will get delayed differently and in order in time scale. In this setup, each subbeam with specific time delay takes along the specific pixel's information. Then, one scanned time-domain signal is coupled by 16 subsignals. Using deconvolution, the pixel's information is extracted and then THz image of the picture is obtained by reconstruction.

In general, there are four distinguished technology platforms which are in use and form the basis for pixel array in commercial THz cameras, such as (1) III–V high-electron-mobility transistors (HEMTs), (2) silicon CMOS circuits, (3) microbolometers, and (4) pyroelectric devices. Table 2.1 shows the comparison of typical specifications of these four types of cameras to visualize THz beams, each operating at room temperature. The experimental comparison of some of these cameras is discussed in [23].

Furthermore, the THz imaging technology faces a number of inherent problems arising from the specular reflections, angular orientation effects, interference effects ("speckles"), and unwanted clothing reflections ("clutter"), which degrades the image quality and resolution. Moreover, certain threat scenarios, such as nonreflecting objects carried directly on the human skin, are in principle a challenge to active imaging techniques. The degradation in THz images caused by speckle as in case of active imaging can be minimized by adding angular diversity or multimode mixing to the illumination THz source [24]. With the advances in THz monolithic and array compatible integrated circuits (TMICs) [25, 26] operating at room temperature, a fully passive approach is implemented with the use of heterodyne receivers. However, the active imaging system offers an advantage of reducing the sensitivity requirement on the THz receivers and in such case the receiver can improves the acquisition speed and number of image pixels (format). Friederich et al. [27] have presented active imaging systems to determine their potential in real-time imaging, which include (a) active electronic imaging system, (b) optoelectronic THz imaging system, and (c) THz focal plane arrays. Therefore, both the active and passive imaging systems have their respective advantages and their use depends on the application in use. Moreover, the THz imaging modalities developed over the decade are broadly classified as imaging with THz time-domain spectroscopy (THz-TDS), transmission-type imaging, reflection-type imaging, THz imaging based on conductivity, pulsed THz imaging, THz computed tomography, THz near-field imaging, etc. All the mentioned methods are discussed in detail in this chapter.

2.2 Transmission-Type and Reflection-Type Terahertz Imaging

Using THz-TDS measurements, the amplitude and the phase of a THz waveform can be obtained. Moreover, with the help of numerical treatment in the imaging system, the complex refractive index can be obtained using Fresnel coefficients, which are expressed in terms of refractive indices of the relevant material. Figure 2.2 shows the setup with transmission imaging module and the same can be replaced by reflection imaging module using the same setup.

For an interface between media 1 and 2, each media is characterized with its own complex refractive index. According to the setup shown in Fig. 2.2 when a THz parallel beam is transmitted from media 1 into media 2, then the transmission and reflection coefficients are obtained as follows:

$$\text{Transmission coefficient}, t_{12} = \frac{2n_1}{n_1 + n_2},$$

$$\text{Reflection coefficient}, r_{12} = \frac{n_1 - n_2}{n_1 + n_2}$$

where n_1 and n_2 are the refractive index of media 1 and 2, respectively. Using experimentally measured transmission and refraction coefficients with the corresponding analytical expressions mentioned in above equations, the material complex refractive indices can be obtained. This will provide a direct imaging of the material

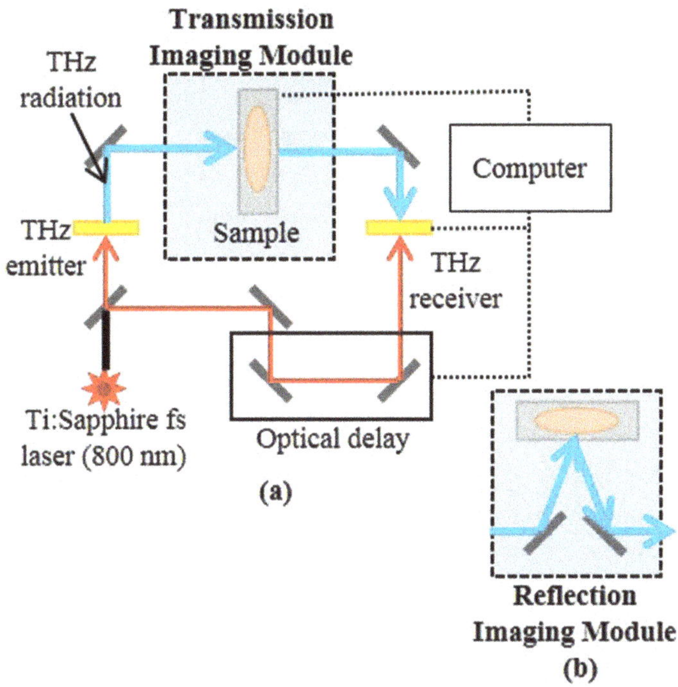

Fig. 2.2 General THz imaging setup: (**a**) transmission imaging module and (**b**) reflection imaging module

refractive index distribution across the sample. Such method is useful in a case when collimated THz beam is used in the imaging setup. However, for nonplanar wave front of the THz probing beam [28], the expressions for the Fresnel coefficients need to be modified [29].

The first demonstration of THz imaging of a circuit and a leaf was realized by using optically gated THz transmitter and the corresponding detector [30]. Afterward, the research on imaging with THz radiation had been rapidly developed in the past few decades [31]. To understand transmission-type THz imaging, let us consider a setup as shown in Fig. 2.3 in which a sample is considered a plate of thickness "L." Assume that the complex refractive index of the sample is n_2, and media across the sample before and after have complex refractive indices n_1 and n_3, respectively.

In this setup, GaAs/AlGaAs-based THz laser light source quantum cascade laser (QCL) is used, which has three-well modules emitting at 3.90THz and operates at 11 K [32]. THz QCL is fixed to a copper heat sink which is further mounted onto the cold finger of a closed-cycle helium cryostat (no. 1). The light receiver is a GaAs/AlGaAs-based THz QWP, which is cooled down by a closed-cycle helium cryostat (no. 2). A sample to be detected is placed at the focal plane (X-Y plane) of two ERMs and is mounted on a computer-controlled X-Y translation stage. The

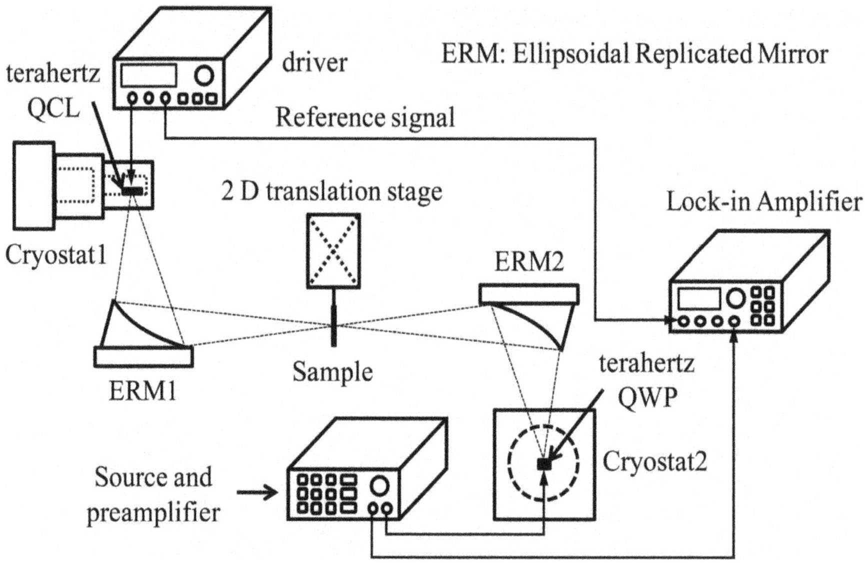

Fig. 2.3 Transmission-type THz imaging system [32]

QCL is driven by a high-power pulse generator. The THz wave emitted from QCL is collected by the ERMs and is focused onto QWP. Then, the signal current generated by the detector is extracted as a voltage by a low noise current preamplifier. The lock-in amplifier which is controlled by a computer for synchronization with X-Y translation stage will read the amplified voltage. The read-out value is displayed on the oscilloscope simultaneously. Mathematically, to obtain the normalized transmission function, $S_t(\omega)$ for a collimated THz beam, there is a need to perform two measurements: (a) THz transmission through the sample $E_s(\omega)$ and (b) THz transmission through an empty system, that is, when sample is removed and is replaced by media 1, this is referred as reference measurement $E_{ref}(\omega)$. The normalized transmission function at the angular frequency "ω" is determined as follows:

$$S_t(\omega) = \frac{E_s(\omega)}{E_{ref}(\omega)} = \frac{t_{12}\, t_{23}\, \exp\left(\frac{i n_2\, \omega L}{c}\right)}{t_{13}\, \exp\left(\frac{i n_1\, \omega L}{c}\right)} * \mathrm{FP}(L,\omega)$$

$$= \frac{2n_2\left(n_1 + n_3\right)}{\left(n_1 + n_2\right)\left(n_2 + n_3\right)} * \exp\left[i\frac{\left(n_2 + n_1\right)\omega L}{2}\right] * \mathrm{FP}(L,\omega)$$

where t_{12} is the transmission coefficient determined from refractive index of media 1 and 2, t_{23} is the transmission coefficient determined from refractive index of media 2 and 3, and t_{13} is the transmission coefficient determined from refractive index of media 1 and 3. FP (L, ω) is the Fabry–Perot term which accounts for multiple reflections inside the sample [33]. This term has the significance while characterizing thin films [34]; however, this term can be neglected if the sample is optically thick. In THz-TDS technique, a sample is said to be optically thick if the echoes, that is, multiple reflections, are well separated in time and only the directly transmitted wave term is sampled during time-domain measurement [35]. On the other hand, a sample can be considered optically thick when the sample material loss is high enough such that multiple reflected waves are so small in magnitude that they can be ignored. Moreover, several theories are used to solve the above equations [36, 37], one such says if the sample is suspended in air such that $n_2 = n - ik$ and $n_1 = n_2 = n_{air} = 1$ with ignoring FP (L, ω) term, then the overall expression of normalized transmission function, $S_t(\omega)$, get reduced to

$$S_t(\omega) = \frac{4n}{(n+1)^2} * \exp\left[-k\frac{\omega L}{c}\right] * \exp\left[i(n-1)\frac{\omega L}{c}\right]$$

and can be rearranged to

$$n(\omega) = \frac{c}{\omega L} * \left\{\arg\left[\frac{(n+1)^2}{4n} S_t(\omega)\right]\right\} + 1$$

$$k(\omega) = \frac{-c}{\omega L} * \ln\left|\frac{(n+1)^2}{4n} S_t(\omega)\right|$$

where arg (z) is the phase of the complex number z. Then, a fixed-point algorithm can be used to generate $n(\omega)$ and $k(\omega)$. Using this algorithm, the initial values of $n0(\omega)$ and $k0(\omega)$ are put in the above equations to generate new values of $n(\omega)$ and $k(\omega)$. On repeating the iteration of the process, a fixed point is found for which the values converge. The assumptions for $n0(\omega)$ and $k0(\omega)$ can be made referring [38]. As discussed earlier, a standard transmission imaging involves sample standing free in a sample space or sample placed between plates of low-absorption material. The signal which is generated by the THz system travels at a normal incident to the sample surface and the transmitted signal is received on the other side. However, to obtain a much shorter imaging time, a faster mechanical scanning mechanism may be used.

For THz imaging applications, other common orientation of the sample under test is using reflection-type imaging. In THz reflection imaging system, the reflected signal is measured on the same side as the incident signal. To achieve this in an imaging setup, there are two ways: (a) setting the incident and reflection signals at an oblique angle or (b) using a beam splitter separate the incident and reflected signals when both occur at normal incidence [39]. Figure 2.3 shows a common experimental setup for reflection-type imaging system. The sample to be analyzed present in media 3 is placed behind a window with media 2 of thickness L. The whole unit of sample under test is placed in air of media 1. In reflection-type THz imaging, the THz beam incident the sample normally and both THz emitter and THz detector are kept on one side of the sample for imaging. If the window (within which the sample is hold) is sufficiently thick such that it is easy to separate the traces on the basis of time domain which are coming from the air/window reflection $E_{ref}(w)$ and the window/sample reflection $E_s(w)$, then the normalized reflection function is determined as follows:

$$S_r(w) = \frac{E_s(w)}{E_{ref}(w)} = \frac{t_{12} r_{23} t_{21}}{r_{12}} \exp\left(\frac{-i n_2 w L}{c}\right)$$

$$= \frac{4 n_1 n_2}{n_1^2 - n_2^2} \exp\left(\frac{-i n_2 w L}{c}\right) \frac{n_2 - n_3}{n_2 + n_3}$$

Suppose media 1 is air, so $n_1 = 1$, and media 2 is a window of a known refractive index $n_2 = n_w$, then

$$C = S_r(w) = \frac{E_s(w)}{E_{ref}(w)} = \frac{4 n_W}{1 - n_W^2} \exp\left(\frac{-i n_W w L}{c}\right) \frac{n_W - n_3}{n_W + n_3}$$

$$C = S_r(w) = \frac{E_s(w)}{E_{ref}(w)} \frac{1 - n_W^2}{4 n_W} \exp\left(\frac{i n_W w L}{c}\right) = \frac{n_W - n_3}{n_W + n_3}$$

Therefore, n_3 can be determined analytically as follows:

$$n_3(w) = \left(\frac{1 - C}{1 + C}\right) n_w(w)$$

In the above equations, the Fabry–Perot effect in the window is ignored [33]. Moreover, when the sample is strongly absorbing, then the reflection geometry provides superior results than in comparison to transmission-type imaging system. In general, the transmission imaging system is limited by the maximal dynamic range of the imaging system which is defined as the frequency-dependent maximal signal amplitude relative to the noise floor. However, when the sample is strongly absorbing, then the noise floor is quickly reached for a short length of the sample [33]. In contrast, the reflection-type imaging system does not rely on transmission through the sample; however, the evaluation is based on amplitude and phase accuracy of the reflected signal. Consequently, the maximal absorbing coefficient depends on the signal-to-noise ratio (SNR), which is defined as the average signal divided by its standard deviation [40, 41]. Our body comprises of large amount of water, and since liquid water is strongly absorbing in the THz spectral range, many biomedical applications of THz waves are performed in reflection-type imaging [42]. Moreover, in biomedical applications, the transmission-type THz imaging shows less image resolution because of fewer focusing elements in the THz beam path and the calculation of tissue properties has less sensitivity to tissue adhesion which further causes vibration in phase. On the other hand, for such applications, the reflection-type THz imaging shows higher imaging resolution and promises more suitable mode for imaging fresh tissues of body. Similar type of study is well performed in skin cancer imaging by the authors in [43]. In Fig. 2.4, a reflection-type THz imaging system is shown. In the setup, an impulse THz wave is generated and is detected with the help of photoconductive dipole antenna (PCA) [44] having InGaAs substrate and dipole-type electrode structure. The transmitting and receiving antennas are coupled with optical fiber and are located at an arbitrary position without optical alignment, respectively.

An Er-doped fiber laser is used to provide pulse duration of about 100 fs with a repetition rate of 100 MHz and has an average power of 50 mW. The output of the fiber laser is divided for the transmitter and the receiver beam paths in free space after precompensation of the group velocity dispersion [45]. The optical path length for the receiver is controlled using a linear delay stage. The sample whose image is to be obtained is set in an imaging area centered at the pivot point of the rotation of the receiver. Moreover, the position of the sample is required to be aligned in the x-y plane using linear stages. In the setup, the transmitter and the receiver are arranged in alignment with facing one another. For imaging, the beam width of the THz wave is measured using the pulse peaks of direct waves by arranging the identical distance between the transmitter and the receiver but keeping different angles between their central axes.

2.3 Terahertz Imaging Based on Conductivity

The THz pulse has the capability to provide an insight into the charge-carrier dynamics at the picosecond timescale. A technique known as time-resolved THz spectroscopy (TRTS), which is also popularly known as optical-pump-THz-probe

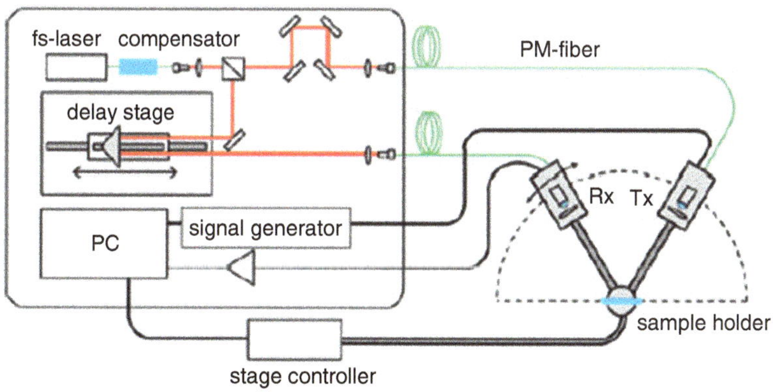

Fig. 2.4 Reflection-type THz imaging system [45]

spectroscopy and is derived from THz time-domain spectroscopy (THz-TDS), uses an optical pump to photo excite the charge carrier in a material under observation. With the help of optical delay line, the transient conductivity is then probed with the THz pulse [46–49], and such setup helps to study conductive materials and is generally used in THz near-field imaging. Moreover, to extract the electrical properties of a material from the complex refractive index, there is a need to determine the complex permittivity to electrically characterize a material accurately. One such method is to use THz-TDS, which provides a direct experimental contactless measurement of the complex permittivity without resorting to the Kramers–Kronig relations as is used in Fourier transform infrared spectroscopy (FTIR). In this, the complex permittivity $\varepsilon\,(w)$ [46, 47] is obtained as follows:

$$\begin{aligned}\varepsilon(w) &= \varepsilon_1(w) + i\varepsilon_2(w)\\ &= 1 + i\frac{\sigma}{\varepsilon_0 w}\end{aligned}$$

where $\varepsilon_1(w) = n\,(w)^2 - k\,(w)^2$ and $\varepsilon_2(w) = 2\,n\,(w) * k(w)$. The complex permittivity is related to the complex conductivity $\sigma(w)$ through $\varepsilon(w) = 1 + i\frac{\sigma(w)}{\varepsilon_0 w}$ as mentioned earlier, where ε_0 is the vacuum permittivity. Therefore, $\sigma(w) = \sigma_1(w) + i\,\sigma_2(w)$ and $\sigma_1(w) = \varepsilon_0\varepsilon_2(w)w$, $\sigma_2(w) = -[\varepsilon_1(w) - 1]\varepsilon_0\,w$. The nature of conductivity can be obtained by fitting the frequency-dependent conductivity with a conductivity model such as Drude model.

That is,

$$\sigma(w) = \frac{\sigma_{\mathrm{DC}}}{1 - iw\tau}, \text{ and } \sigma_{\mathrm{DC}} = ne\mu = \frac{ne^2\tau}{m} = \varepsilon_0 w_{\mathrm{P}}^2\tau.$$

where n is the charge-carrier density, e is the elementary charge, μ is the mobility, m is the mass of the carrier, τ is the Drude scattering time, and w_{P} is the plasma frequency.

2.4 Classification of Terahertz Imaging with Diffraction Limit

The spatial resolution in THz imaging is predominantly limited by the diffraction limit. In general, a diffraction limit means an imaging lens is unable to resolve two adjacent objects which are located closer than $\lambda/2NA$, where λ represents the wavelength of light and NA is the numerical aperture of the lens. Moreover, the resolution of optical imaging instruments is primarily limited by the diffraction of light. Consider a case in which if an image is made through a small aperture, then there is a point at which the resolution of an image is restricted by the aperture diffraction.

A small aperture gives greater depth of field (DoF) and a sharper image [50]. However, if the aperture is kept too small, then the effect of diffraction will increase, which results in reduction in the sharpness of an image. Such situation results in diffraction-limited imaging. Many applications in THz imaging rely on either single plano-convex lens of spherical shape or off-axial parabolic mirrors [51]. If a single spherical lens is having NA of 0.45 approximately, then it provides a spatial resolution (it is a measure of the smallest object that can be resolved by the sensor, or the ground area imaged for the instantaneous field of view (IFOV) of the sensor, or the linear dimension on the ground represented by each pixel) of about 1.22λ. On increasing the NA of such lens will result in the occurrence of aberration which is a property of an imaging system that causes light to be spread out over some region of space rather than focused to a point. Such image is usually blurred or distorted depending on the type of aberration. Therefore, wide aperture off-axis parabolic mirrors can be used in THz imaging systems which have good aberrational correction and high resolution; however, they suffer from overlapping of incident and focused beams [52]. Further, owing to the broadband nature of the THz radiation, the diffraction-limited focal spot in the center of the THz beam path is generally having a complicated nature which depends on the detailed design of the optical systems. The focal spot diameter is in general frequency dependent, and so the care must be exercised in defining the spatial resolution of an image. To construct an image, the received THz waveforms (either through transmission-type imaging or through reflection-type imaging) are converted to the Fourier domain using numerical Fourier transform. Therefore, by choosing a specific frequency within the bandwidth of THz, the spot size of the THz beam can be selected, which varies in the proportion to the wavelength (λ). Moreover, it is observed from the study in [53] that images created using high-frequency component show less blurring due to the small spot size of the THz beam at the selected frequency of operation. However, images formed using a low-frequency component of THz beam show more blurring.

Since THz spectroscopy, imaging, time-of-flight, and computed tomography are mostly applied in the applications of biology and medicine, nondestructive evaluation technology, pharmaceutical industry, and security tasks, it is important to discuss THz imaging modalities with reference to diffraction limit in detail for such applications. Further sections are provided for the concerned readers who are interested to learn more about the detailed operation of the following THz systems as mentioned next.

2.4.1 Terahertz Imaging Below Diffraction Limit

In diffraction-limited imaging system, the spot size of the focused THz beam is nearly equal to the wavelength multiplied by the f – the number of focusing optics [53]. It means that the smallest resolvable features in an image are not smaller than the wavelength. Moreover, for 1 THz radiation, the resolution is nearly 0.5 mm, which is comparable to the resolution of the human eye. In other words, THz images are visually reminiscent because such radiations typically show features which our eyes could see. However, such resolution limits the applicability of THz imaging because smaller objects cannot be resolved in detail. Therefore, similar to optical imaging [54], there is a need to push the resolving power below the wavelength limit. Further, to image a small object, that is, small wavelength object, there is a need to consider not only the propagating waves but also the evanescent waves which exist in a region near to the object itself. As these evanescent waves do not propagate, so measuring must be accomplished closer to the object. One such technique known as near-field imaging [55] has been developed, which involves keeping the detecting probe very close to the object within the wavelength of the object. The near-field imaging is further discussed in detail in the next section of this chapter.

The term diffraction [56] was elaborated by the scientist named Ernst Abbe who had found experimentally that when a light having wavelength λ traveled in a medium with refractive index n and converged to a spot with half-angle θ results in a spot with radius, $d = \lambda * 2n\sin\theta$, where $n\sin\theta$ = numerical aperture, (NA). For vacuum, NA = 1 the lowest spot size, that is, resolution becomes half the wavelength, that is, $\lambda/2$. After the study, many-a-times the diffraction limit is also named as Abbe diffraction limit. In [57], the authors have demonstrated an approach in which the image formation mechanism becomes independent of the wavelength of the radiation used and so have achieved to avoid Abbe diffraction limit. They have achieved to obtain high-resolution THz multispectral reconstructive imaging of nanometer sized metal lines, which are fabricated on a silicon wafer by overcoming the Abbe diffraction limit. Moreover, there are some diffraction-free beams (DFBs) which are nontrivial in nature. Such extraordinary monochromatic light beams in free space have been successfully generated since the Durin's pioneer work in 1987 [58]. Bessel, Mathieu, parabolic, and Weber optical beam are some of the examples of DFBs [59–66]. In general, a diffraction is a fundamental feature for an electromagnetic field which propagates in free space. However, in case of above-mentioned DFBs, the transverse spatial distribution and the size of beam spot do not change in finite distance duration of propagation. Most of the DFBs have two-dimensional invariant transverse profiles in free space and are known as 2D-DFBs. However, to get monochromatic diffraction-free solutions with one-dimensional transverse profile in free space, the authors of [64] have investigated two such beams (with one-dimensional monochromatic diffraction-free solution) to propagate in free space named cosine and airy beam. Due to the lack of spatial localization, the cosine beam is not useful in practical work; however, an accelerating airy beam having its transverse spatial profile with one-dimensional localization does not propagate along a straight line but follows a parabolic trajectory [65]. Such beam is just like a piece of bended light paper. The researchers [67–71] have shown that in several applications, the use of 2D-DFBs also extends the desired depth of field (DoF) of an imaging system. In [72], the authors have used a three-dimensional printed element named ridge prism to generate diffraction-free structured THz beam. Likewise, in [73], the authors presented an experimental realization of diffraction-free space-time light sheet with the use of spatial light modulator. The resulted polychromatic beam propagates along nearly 100 mm straight line without any change in the transverse profile of one-dimensional beam. Similarly, in [74], the authors have an easy method to implement the concept of wave front manipulation so as to achieve ultrabroadband THz spectroscopy system with diffraction-limited approach.

2.4.2 *Terahertz Time-of-Flight Imaging*

Time-of-flight imaging technique resolved the distance between the THz transmitting beam and the object or sample under observation for each point of the image by measuring the round-trip time of the signal. Such reflection-type imaging technique is useful for obtaining an image of opaque substances and the materials which are highly reflective. In biomedical THz imaging, the living tissues which are opaque due to the presence of high liquid water content can be imaged using reflection-type imaging. Moreover, the broad bandwidth of the THz radiation is also an advantageous factor which supports the reflection mode of THz imaging because it allows not only obtaining the depth information of the object but also helps to construct a complete three-dimensional representation of the object. Initially, a three-dimensional THz imaging was implemented in 1997, wherein the researchers [75] used an optical setup at the location of the sample such that the THz beam reflected off the sample under observation instead of getting transmitted through the sample. When the sample under observation for imaging is having smooth dielectric, the layers are illuminated through single-cycle THz pulse kept at normal incidence, and then every layer generates a reflected THz pulse. Each interface of the sample under consideration is being located at different distance from the receiver; therefore, each reflected wave is arrived with unique time delay at the receiver. In reflection mode, the signal measured hence consists of a train of THz pulses. The amplitude of each reflected pulse in the train provides the information about the magnitude of the dielectric change across the interface from which it was originated. The time delay between successive pulses in the train provides the information about the thickness of the intervening layer of the sample under observation. Such measurements performed on the sample having parallel transparent layers helped to obtain the spatial variation among the refractive index profile along the direction of propagation of the THz beam. Since the work presented in [75] was originally developed as tomographic measurement, it was more appropriately estimated as time-of-flight imaging. The mode of imaging is also comparable with ultrasound B-scan in which an oscillating sound beam is emitted from a probe placed on the eye. The sound waves strike the intraocular structures and get reflected back to the probe and get converted into an electrical signal from which various ocular and orbital diseases are inspected. Further, as in case of B-scan, in THz imaging, the length of the temporal data acquisition window establishes the upper limit on the depth up to which an image data can be acquired. Moreover, THz imaging also provides detailed spectroscopic analysis from the specific interactions of THz radiation with certain materials [76–78]. However, in time-of-flight imaging measurements due to multiple discrete reflections involved in the measurement procedure, the achievable spectral resolution is difficult to determine. Therefore, in THz time-of-flight imaging, special numerical deconvolution procedures are generally developed to extract spectroscopic data from a waveform which has information of multiple reflections. Initially, the reflectivity and the distance to a single reflecting surface are to be determined, and if the reflection coefficient is complex in nature, then it becomes difficult to distinguish out the phase changes which are associated with the reflectivity from the linear phase shift due to small displacement of the object under consideration along the direction of propagation. In [79], the authors have suggested a phase retrieval algorithm to separate out these effects and had obtained a measurement of the object height profile. Moreover, in THz time-of-flight imaging system, depth resolution plays a significant role and it is important to determine how effectively the system can resolve two closely spaced reflecting interfaces. For this, one effective method is to resolve the temporal duration of the THz pulses. An equivalent formula is framed in terms of spectral bandwidth of the THz pulse in [52] for time-of-flight imaging. The resolution is determined from half of the coherence length of the radiation, that is, $L_C = c/\Delta\omega$, where c is the speed of light in the intervening medium and $\Delta\omega$ is the spectral bandwidth. Moreover, to enhance the penetration depth of the beam in THz time-of-flight imaging, the authors [80] used sequential glycerol delivery images to display the location of artificial tumor which was located under fresh skin. Their presented results showed that with the use of glycerol, the peak value of the THz signal transmitted through the fresh tissue and reflected through a metal target get doubled on comparing to that of tissue without glycerol. THz time-of-flight imaging also has its application in the field of conservation of precious artwork. The authors of [81] had used the technique of THz time-of-flight spectroscopy to locate the detachment of the glaze due to the subsurface crack formation or even to detect cavities in the area of the terracotta which in general could not be observed visually. Such system is also useful to guide the restoration of art pieces which are still attached to the old monuments and cannot be transported to tomography setup. Earlier X-ray CT scans were in use for such applications, but THz time-of-flight imaging is more beneficial in use because it does not require ionizing radiation. Moreover, such THz spectrometers can be taken to the field of archeological survey to do the desired measurements and inspections of the on-site/fixed sculptural artwork pieces.

2.5 Computed Tomography

In THz reflection-mode imaging, it is important to have significant depth resolution to determine the precise details of the object under observation. The researchers have found several techniques to improve the depth resolution. One such technique is to use interferometry which improves the depth resolution of THz imaging system and is known as time-domain analog optical coherence tomography (OCT) [82, 83]. In the imaging setup, an interferometer is used to resolve a broadband light pulse temporally through interfering the pulse with a reference pulse. Thus, the time-of-flight is determined from the reflection geometry. However, in THz-TDS, there is no requirement of an interferometer so as to time-resolve the reflected waveform because the detector deployed in the system makes the availability of necessary temporal resolution. Figure 2.5 shows the basic schematic of an interferometer used in normal-incidence reflection imaging system. In this, a beam splitter is used to separate the incident and reflected beams and also directs a portion of the incident beam on to a flat mirror which is used as a reference. At the detector side, the reflected pulse from the sample interferes with the reference pulse.

In general, at the transmitter side the THz pulses are generated and are detected using LT-GaAs photoconductive antennas with the help of femtosecond laser pulse. For reflection-type imaging, a Michelson configuration of high-density polyethylene lenses is used so as to collimate, focus, and bring together the THz beam. To split the THz beam, a high-resistive silicon wafer is used as a beam splitter which divides the THz pulse into a sample and a reference arm. Due to the thickness of the wafer, there are multiple reflections which have arrived within the beam splitter and are delayed relative to the initial THz pulse. Using a lens which is placed in the sample arm of the interferometer, an image of the object under observation is focused. However, the beam in the second arm known as reference arm of the interferometer is allowed to retro-reflected off of a flat mirror using manual translation stage. Moreover, the optical delays of sample arm and reference arm are adjusted equally. In order to get background-free imaging, the lens used in the setup also provides the desired phase shift as when the THz pulse passes through the lens, then the pulse acquires an additional phase known as Gouy phase shift by a focused Gaussian beam of THz field. This additional phase is due to the variation in the wave front curvature when the THz pulse passes through the focus of lens and its value is approximately pi radians [83]. Now, when the two pulses from different arms (sample arm and reference arm) of the interferometer reach the detector then because of the destructive interference, a very small signal is measured at the detector. On the other hand, if the object under observation which is kept in the sample arm has any feature, then it alters the amplitude or phase of the reflected THz pulse. In such situation, the destructive interference gets disrupted, and it results in a large

Fig. 2.5 Basic schematic of an interferometer [83]

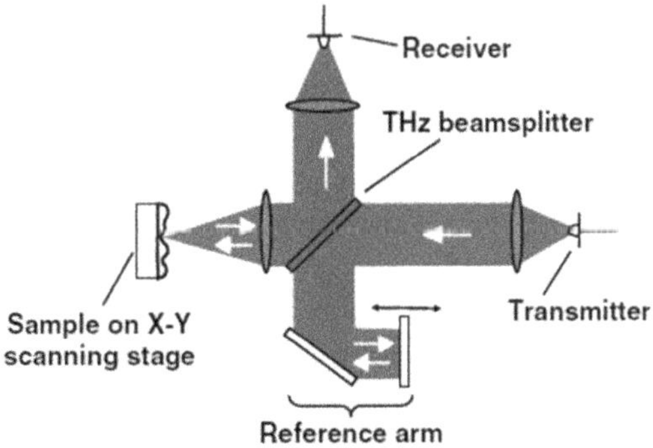

signal which is measured at the detector. Because of these features, topographic imaging has found its applications in synthetic aperture radar, X-ray diagnostics, seismic prospecting, and ultrasonic imaging. One of the main differences between time-of-flight THz imaging and THz tomography is that in former case, the measurements involved a single transmitter and a single receiver, and both are at fixed location. However, in tomographic measurement, there are multiple positions which are set for transmitters and/or detectors so as to illuminate the object under consideration from more than one location. Practically, in case of THz tomographic imaging, a single transmitter and receiver pair is deployed with multiple positions for either transmitter or receiver or both and measurements are done serially, and the measured data subsequently get assembled into an image. With the development of fiber-coupled THz antennas, the emergence of THz tomographic measurement came into existence on large scale since 2001. During initial works, a diffracted field was allowed to be transmitted through 2D patterned aperture at several positions, and after this, back-propagated measured fields were used to reconstruct the aperture pattern [84, 85]. Later the authors [86] extend their research to reconstruct the 3D image of the object using THz tomography. Similarly, several other researchers have also explored the THz computed tomography (THz-CT), wherein the refractive index of the object under consideration along with amplitude and phase information is evaluated. In this technique, the reconstruction algorithms of X-ray computed tomography are adapted to the THz band [87, 88]. Some other THz tomography imaging techniques that are in practice and have been developed while considering the benefits from the available imaging methods from neighboring bands are as follows: (1) THz diffraction tomography and using Fresnel lens THz tomography [89, 90] and (2) THz multistage imaging [91, 92]. All these have helped to open up a new area of capabilities for THz technology in the field of imaging.

2.5.1 Tomography with Pulse Terahertz Radiation

Recently, the researchers have focused to investigate the explosives and biological samples which have unique THz spectral signatures. The pulsed THz sensing involves applying the technique to the study of materials by monitoring transmitted or reflected radiation. As the pulsed THz wave can see through envelopes, this technology has been used by the mail sorters to check the biological and chemical hazardous substances [93]. Another potential application of pulsed THz imaging is the nondestructive characterization of soot from internal combustion engine. It is necessary to check the pattern of the soot in the soot-removal filter of an internal combustion engine to design the highly efficient filtering structure and optimize the timing of filter renewal. Shibuya et al. [94] have described a nondestructive inspection method using the millimeter-to-THz wave imaging and computed tomography. With the use of this technology, the soot removal filter's efficiency is increased. The monitoring of filters helps in deciding the renewal timing needed to optimize fuel consumption and to lower particulate emissions.

2.6 Special Case Imaging Applications

Over the past several years, an imaging in the THz domain has attracted considerable attention in several application areas, such as nondestructive evaluation, security screening, biomedical sensing, and THz frequency-modulated continuous wave (FMCW) radar for automotive application [95–97]. The salient features of THz waves that have been explored by researchers in several imaging applications are as follows: (1) easy penetration of wave into materials like plastics, clothing, and drywall, (2) being nonionized in nature, it is useful in diagnosis human health, and (3) enhanced image resolution due to smaller wavelength in comparison to microwaves. Moreover, the advanced features of THz waves combined with special techniques have been investigated by several researchers to deploy the utilization and benefit of THz waves for many more applications. Some such special cases of imaging applications are hereby discussed for the interested readers and researcher.

2.6.1 Imaging with Compressed Sensing

With the advent of high-power THz sources, it becomes possible to achieve considerable signal-to-noise ratio values in the imaging system. However, such sources depend on use of vacuum electronics which are bulky and expensive. On the other hand, THz images can also be obtained with the help of mechanically operated raster scanning of a single pixel and the use of highly sensitive detector. However, such arrangement results in large acquisition time taken by the system [30]. Similar to the raster scan imaging, a single-pixel compressive imaging has been recommended by the researchers which is used an optically spatial light modulator [98]. In this technique of compressive sensing, the mechanical movement of the raster scan is replaced by spatial encoding of light by random masks. The reconstruction of an image through encoded measurements is done using the method of computational optimization [99]. In [100], the authors have spatially encoded the object wave with the help of random binary masks and by means of quasi-optical system transformed the spatial data into Fourier domain. Moreover, with the help of low-noise detector placed in the center of the Fourier plane records the fundamental component after the transformation. Similarly, the compressed sensing-based THz imaging systems based on photo-excited semiconductor spatial encoding mask are also reported in [101–103]. All these methods of compressive imaging are fast and cost effective, but still there are future prospects related to speed of hardware involved and to reduce the overall cost of real-time THz imaging system [104, 105]. One such technique in this regard is developed based on single-bit sensor deployed with THz compressed sensing [106]. It offers several salient features: (1) robustness to nonlinear noise and so helpful in handling the effects of unwanted vibrations in quasi-optical imaging setup, (2) high performance in low SNR environment, and (3) facilitate the use of low-cost comparator used as an analog to digital converter which increases the speed of acquisition of data and lowers the cost of overall THz imaging system [107, 108]. Further, the optical properties, refractive index and birefringence, and the dielectric permittivity of the materials are very sensitive to phase transitions and hence depend on temperature. Reuter et al. [109] investigated this property at THz frequencies. The semiconductors with moderate carrier concentration have a plasma frequency and damping rate between 0.1 and 2 THz, which translates into the ability to sense the presence or character of different materials. This unique form of compressed sensing has now been used to probe the properties and dynamics of semiconductors [110], superconductors, other correlated electron materials [111], and many other materials [112].

2.6.2 Spectroscopic Imaging

The spectroscopy with high spectral resolution at frequencies in THz regime of the electromagnetic spectrum is a powerful analytical tool for investigating the structure and the energy levels of molecules and atoms. The spectroscopic imaging is capable of revealing and analyzing various biological and chemical conditions. Lee et al. [113] have shown a highly sensitive and selective detection method for residual pesticide molecules using nanoscale metamaterial-based THz time-domain spectroscopy (THz-TDS) system. They have designed the nanoscale metamaterial-based slot-antenna for the strong THz resonance at a certain frequency where the specific molecule has intra-molecular or inter-molecular collective vibration mode. Enhanced THz near-field via a nanoscale metamaterial-based antenna strongly increases the absorption cross-section, and this leads to the detection sensitivity up to parts-per-billion level even in a solution state of pesticide sample. THz spectroscopy and imaging techniques can also be used to analyze works of art [114]. The electromagnetic wave at this frequency can penetrate opaque materials and preparation layers, and thus, THz spectroscopy shows fingerprint-like spectra similar to infrared bands [115]. Time-domain reflection imaging uses THz pulses that act as a probe and propagate through an artwork to get its internal structure without requiring sampling of the specimen. In addition, the energy of the waves in this domain is low enough and considered perfectly noninvasive in practice. The first ever noninvasive cross-sectional image of a tempera masterpiece by Giotto was successfully observed [116].

The THz spectroscopy provides information on the basic structure of molecules which is useful in radio astronomy. The rotational frequencies of light molecules fall in the THz spectral region similar to the vibrational modes of large molecules with many functional groupings. Moreover, many biologic molecules have broad resonances at the THz frequencies. THz spectral region accessible via ultrafast optical pulse generation and detection is used in the investigation on THz vibrational modes of crystalline benzoic acid from 0.1 to 4THz in [117]. Moreover, considering the progress of THz spectroscopy detection technology, several manufacturers of spectrometers have shown the development of handheld spectroscopy devices, such as Terakit-DODS Spectrometer produced by Rainbow Photonics Corporation in Switzerland has the spectral range from 0.1 to 20 THz which has the most wide spectral range at present and its average power is up to 180 mW and Mini-Z Terahertz Time Domain Spectrometer produced by Zomega Corporation of American, whose size is $10.5 \times 6.25 \times 2.75$ cm^3, is having the spectral range 0.1–4 THz, dynamic range is more than 70 dB, and frequency resolution is less than 500 Hz. This spectrometer is very portable and suits for the on-site component analysis of chemicals and drug inspection [118].

The low-frequency end of this spectral region extends to frequencies that are easily achievable using high-precision electronic instruments and long temporal scans are required. To the high-frequency limit above about 10 THz, Fourier transform spectroscopy is used. Pulsed THz spectrometers are inherently broadband systems, a result of using ultrafast optical pulses to generate the THz radiation. THz spectroscopy has determined the far-infrared optical properties of the material as a function of frequency, which yield insight into material characteristics for a wide range of applications. Moreover, the Fourier transform spectroscopy (FTS) is the most commonly used technique for studying molecular resonances which provide an extremely wide bandwidth, enabling material characterization from the THz frequencies to the infrared band. The spectral measurements, which have much higher resolution, have been made using a narrowband system with a tunable THz source or detector. However, both FTS and narrowband spectroscopy are also widely used in passive systems for monitoring the thermal-emission lines of molecules, particularly in astronomy applications. The THz spectroscopy has been classified into three categories: THz time-domain spectroscopy (THz-TDS), time-resolved THz spectroscopy (TRTS), and THz emission spectroscopy (TES) [119].

THz time-domain spectroscopy (THz-TDS): THz time-domain spectroscopy measures the change in the time-resolved electric fields of the THz pulses propagating through a sample and an equal length of free space. In a THz-TDS, both generation and detection occur in the same system. The system uses an ultrafast laser pulse, which gets divided into two optical beams in the system. One generates the THz pump beam, and the other detects the probe beam. The probe beam explores amplitude of the THz pulse over the time. Thus, the THz waveform is time dependent and it is used as a reference. The THz radiation probes the sample to generate sample waveform, and by comparing the two waveforms under Fourier transform, it gives the spectroscopic information. Depending on the sample geometry, the change in the THz pulse shape allows the analytic or numeric calculations of the complex sample index, permittivity, or the conductivity. However, such values are determined when there is a linear interaction of the electric field with the sample under observation for imaging. Although the spectral resolution of THz-TDS is much coarser than narrowband techniques, and its spectral range is much lesser than that of FTS, it has several advantages that have given rise to some important recent applications. The transmitted THz electric field is measured coherently, which provides both high-sensitivity and time-resolved phase information. Moreover, the delay line is a crucial part in every time-domain spectrometer, which needs to carry out the sampling of the electric field over the time. There are a number of uncertainties in TDS measurements, but the delay line uncertainty is significant in the sensing and imaging, and thus, it has been rigorously analyzed by Oberto and Koch [120]. They have modeled the impact of a delay line uncertainty on the acquired THz-TDS data and have emphasized their work on the effect of a small random deviation in delay-line position on the time axis and the measured electric field. The authors have also shown that in a single measurement, a high SNR is achieved by using a high precision delay line. Any small deviation in the positioning of the delay line introduces a small uncertainty in the measured THz transient, which may directly lead to the error in the material property characterization [121]. In principle, also delay lines with a lower precision can yield to an overall acceptable SNR if either averaging or the oversampling of pulse signals can be performed.

Time-resolved THz spectroscopy (TRTS): The TRTS is a photo excitation time-dependent THz spectroscopy in which laser is used to change the carriers and permittivity of the sample. The TRTS measures the dynamic properties of the material which differentiates it from THz-TDS. Baxte et al. [122] have determined the THz absorption coefficient, index of refraction, and conductivity of nanostructured ZnO using TRTS. Likewise, this spectroscopic technique is used to determine the time scales of generation and recombination of charge carriers as well as their transport properties in solution processed perovskite-based solar cells [123].

THz emission spectroscopy (TES): The TES is another spectroscopic technique which uses the sample under investigation as a THz emitter to decide the THz pulse shape. By this technique, information related to the photo-excited carrier is obtained. Pump-probe systems take into account the change in time-resolved electric fields propagating through sample and free space; however, TES investigates the speed and momentum changes. The use of TES technique is made in determination of the carrier-envelope (CE) phase of few-cycle laser pulses in [124]. In this, the authors introduced an approach to determine the CE phase by down-conversion of the laser light to THz frequency range by means of plasma generation in ambient air, an isotropic medium where optical rectification (down-conversion) in the forward direction is only possible if the inversion symmetry is broken by electrical or optical means.

2.6.3 FMCW Radar Imaging

The realization of THz imaging with three-dimensional capabilities depends on its high resolution in range. In general, the high resolution is achieved by using a femtosecond THz pulse of the imaging system. However, a resolution in other dimensions such as azimuth resolution is obtained with the help of scanning the object under consideration using lens-focused beam through electronic or mechanical means. In such traditional methods, the highest achievable resolution is determined from the relation of wavelength to the dimension of the imaging optics which can be either a lens or the diameter of the reflecting mirror. In real-time setup of THz imaging system to achieve large numerical aperture, there is a requirement of large dimensioned optics which are unpractical to procure. Therefore, to overcome this limitation, the researchers have started exploring some other techniques along with THz imaging system in terms of signal processing. The synthetic aperture radar (SAR) is one of such technique based on signal processing, which helps to obtain three-dimensional image of an object from THz imaging system without using large aperture optics. In [125], the authors have used SAR-based simulator and have used an approach of time domain for image reconstruction. Moreover, active THz imaging system having features of high-power coherent illumination and ultra-low-noise heterodyne detection has certain limitations in its operation. One major difficulty with coherent active imaging is that with the help of single frequency, the target recognition depends on object's contrast and brightness which is highly sensitive to (1) the incident angle of the THz beam, (2) clutter signal from the foreground or background, and (3) interference or speckle effect. Therefore, to overcome these limitations of coherent active THz imaging, an all-solid-state active submillimeter imager has been designed in [126] using frequency-modulated continuous wave (FMCW) radar technique to map a three-dimensional image of the object under consideration. Further, FMCW radar also increases the bandwidth of a pulse waveform by rapidly sweeping the carrier frequency within the pulse interval. Using low-cost CMOS chips, FMCW implementation is easy because in FMCW the carrier frequency is linearly swept over a specific bandwidth (B) and the time period (T) being large enough provides a variable slope $\left(B/T\right)$. From the relation, it is clear that by controlling the time period (T), a THz imaging system designer can get different frequency slopes. Moreover, FMCW can also resolve range as well as Doppler by mixing the received signal along with the original transmit signal. However, high multiplication factors should be avoided in FMCW radar system so as to reduce the effect of phase noise in the transmitted signal. The function of radar system is, in general, subject to two fundamental limitations, namely: (a) law of diffraction in which the cross-range resolution of an optical element is determined from the relation $\Delta x \approx \lambda R/d$, where Δx is the cross-range resolution, λ is the wavelength, R is the distance to the target, and d is the diameter of the transmitting antenna or a focusing lens, and (b) relation between bandwidth of the signal (BW) and the signal's shortest possible duration in time which is related to Fourier transform. It is an important relation for an operating radar as it sets the range resolution of radar signal irrespective of the format of the signal modulation, $\Delta r \approx c/2\mathrm{BW}$ where Δr is the range resolution, c is the speed of light, and BW is the signal bandwidth. Now, from these two limiting equations, if the distance (range) is kept fixed and the antenna aperture is constant, then the resolution depends on the wavelength of the radiation and the bandwidth of the signal [127]. In such case, shorter the wavelength kept results in better cross-range resolution and larger bandwidth also provides better range resolution, which is advantageous in the radar-based imaging system [128]. Further, to achieve three-dimensional radar images with high frame rate, there are, in general, two techniques which are in practice. Firstly, using a single transceiver and doing rapid scanning of the beam or transceiver to image the target. Secondly, deploying an array of transceivers to capture an image of the object as is done by digital camera. Combination of both the techniques is an exploring area to design an effective submillimeter wave imaging system. Moreover, as the available output power of a solid-state source falls off sharply with increase in frequency and many devices are not capable to generate pulsed THz wave for imaging purpose, therefore, the use of pulse compression technique such as FMCW is advantageous [129, 130]. Further, the method of FMCW radar is more promising in comparison to other common pulse radars in case when the value of maximum power from the radiation source is too low so as to achieve high SNR for short duration of the pulse. Moreover, the resolution is an important factor in the functioning of FMCW radar which depends on bandwidth (BW) and is independent of SNR. The high resolution with respect to imaging enables to distinguish an individual surface of the object under consideration from clothing, concealed objects, and the body from each other [131]. FMCW radar has been effectively deployed in NASA Jet Propulsion Laboratory's 675 GHz imaging radar, which was assembled to provide fast, reliable source of imaging of person-borne concealed objects. The radar provides 30 GHz BW, and so subcentimeter range resolution was achieved. Further, in order to optimize the radar's range resolution, reliable software calibration procedures were used to compensate for signal distortion from radar waveform nonlinearities. The FMCW radar's design allowed low-distortion fast beam scanning for single-pixel imaging with a frame rate of 1 THz [132].

A heterodyne FMCW radar imaging system is implemented in [126], wherein the use of two frequency sources resolved the phase imbalance and has reduced the demand for the sampling rate. The schematic block diagram representation of the FMCW imaging radar is shown in Fig. 2.6. In this, the two frequency sources 1 and 2 drive the transmit chain and the receive chain individually, and the resultant mixing of these signals from two sources drives the local oscillator chain.

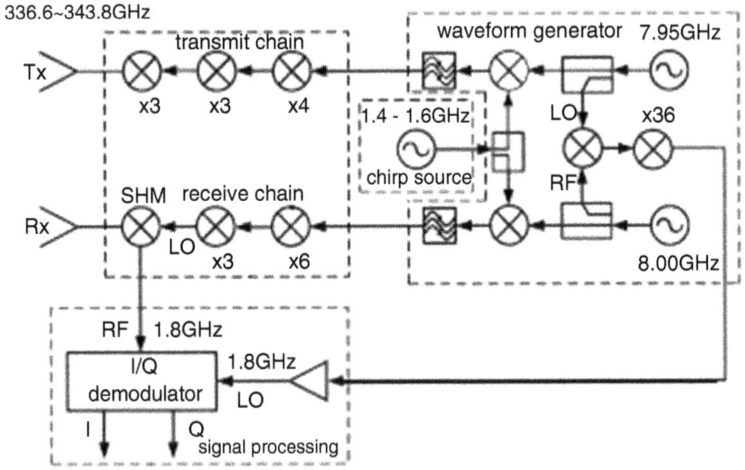

Fig. 2.6 Schematic representation of FMCW imaging radar [126]

Moreover, to achieve high resolution, there is a requirement to increase the bandwidth of the transmitted signal. Therefore, a sweep source is included in the setup. In general, the main building blocks of FMCW radar are wave generator unit, chirp source unit, transceiver unit, and the signal processing unit. The transmitter makes use of a chirped signal to transmit, and the reflected signal is then received by the antenna. After this, the signal is mixed with the help of subharmonic mixer which is pumped by the local oscillator signal. The output of subharmonic mixer is an intermediate frequency, which is then demodulated in I/Q demodulator. This signal is then sampled by the analog-to-digital converter. The authors [126] have claimed to achieve the receiver BW of 200 KHz and its sensitivity of −92 dBm. Similarly, in [133], FMCW emission with BW of 16 GHz has been achieved with the help of MIMO horizontal geometry of antenna array for personal screening application. Therefore, the advantage of FMCW radar in comparison with other pulsed radars including TDS-THz is to facilitate low-cost imaging system with improved resolution.

2.6.4 Near-Field Imaging

In general, for the study of near-field effects, the THz radiation having large wavelength from 30 μm to 2 mm corresponds to a frequency range from 150 GHz to 10 THz. The properties of these THz bands contain several advantages related to near-field imaging. It includes (1) study of structures using optical lithography, (2) identification of perfect conductor/metals such as gold, and (3) artificially engineering the dielectric properties with the help of doped semiconductors such as silicon. The use of THz-TDS in pulsed THz imaging to measure the time-dependent electric field with an ultra-broad bandwidth came into existence almost two decades before. This technique helped to provide not only the intensity but also the phase and the polarization information from the received THz beam within a system. Earlier, the highest resolution of $\lambda/3000$ at 2.54 THz is reported with the help of single-frequency source and a scattering tip [134]. Moreover, in case of broadband system based on single-cycle pulsed source, the resolution is obtained in the region of $\lambda/600$ using separate mapping of different components of the electric near field with its amplitude and phase [135, 136]. Stantchev et al. [137] have performed such single-pixel near-field imaging, which is based on the use of optical modulation. In the setup as shown in Fig. 2.7, an 800 nm, 100 fs optical pump is used onto a highly resistive thick silicon wafer of thickness 115 μm. As the distance traveled by the THz beam in the silicon wafer was kept smaller than the wavelength, so the image was recorded before far field Fraunhofer diffraction limit. A binary pattern generated through reflecting the pump beam onto a digital micrometer chip has the resolution of $\lambda/4$, which made the clear visibility of the pattern of conducting wires localized on the chip.

Similarly, in [138], the authors have used THz near-field imaging system to map the photoconductivity of a graphene sheet and experimentally observed the variation

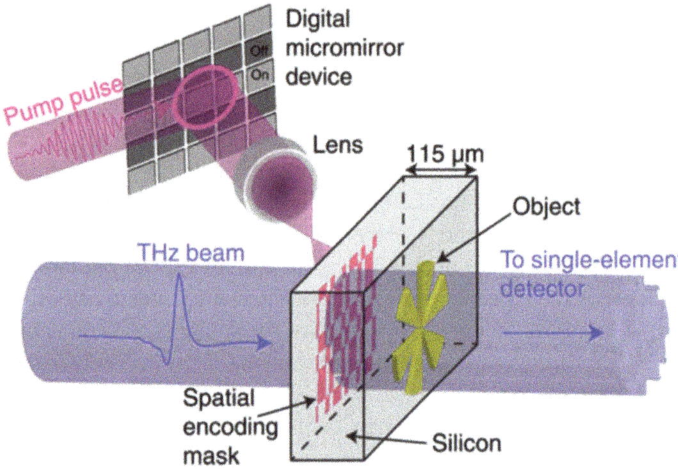

Fig. 2.7 Experimental setup of near-field imaging using a silicon wafer and a digital micromirror device to generate the binary patterns [137]

in resolution of the reconstructed image with different thicknesses of the silicon plate. Moreover, they have achieved to decrease the number of required measurements with the help of adaptive sampling. Further, a coarse edge identification is performed using Haar wavelet decomposition method to decide the location where to sample with higher resolution. The simulation result shows that the use of adaptive sampling outperforms compressive sensing by increasing the SNR value with the same number of measurements.

However, the near-field methods which involve the identification of the evanescent waves (also known as nonradiating fields in field theory) have been explored in [139–141] for optical microscopy so as to break the diffraction barrier. Specifically, near-field probes with tapered waveguide apertures and metallic tips are mostly used in THz-TDS so as to tightly confine the electromagnetic energy. The evanescent waves are limited to sample surfaces; therefore, in such near-field imaging system, such energy confinement is desirable at near-field regions of the surface of the object under consideration. Further, the imaging system which captures the information carried by evanescent waves can overcome the diffraction limit which results in an improvement in resolution. The evanescent waves do not occur in a homogeneous medium; however, they are certainly connected to the interaction of light having inhomogeneities [142]. Moreover, in order to achieve super resolution, the authors in [143] have used THz near-field compressive imaging. As the THz near-field community is flourishing together with the possible applications in biology, nanospectroscopy, or plasmonic, there is still a room for improvement in hardware speed, as well as reducing the cost to realize a ubiquitous real-time THz imaging system. Therefore, the near-field imaging and techniques used to achieve high spatial resolution are further discussed in Chap. 9 of this book in detail.

2.6.5 Far-Field Imaging

A conventional imaging system transforms propagating waves, but does not operate on the decaying evanescent waves; therefore, these waves can only be detected in the near-field imaging. However, the use of super lens amplifies the evanescent waves but does not change their decaying character. Therefore, with a suitable design of a narrow aperture in a plane screen having dimensions much smaller than the wavelength, evanescent waves are converted to propagating waves for ease of detection and processing; however, these waves should not mix with the propagating waves emanating from the object. Moreover, outside the near field, the loss of high spatial frequency information carried by the decaying evanescent waves prohibits the reconstruction of the image of an object with resolution better than $\lambda/2$. As discussed before, the optical imaging in the near field is generally performed using near-field scanning optical microscopy, and the exponentially decaying evanescent waves are detected through scanning probe [144]. Near-field imaging suffers from several limitations such as low throughput, need of substantial postprocessing of the data from the scanning probe, and inability to simultaneously observe the different parts of the object under consideration. However, in many practical applications such as biological microscopy, there is a need of an imaging system which could produce a direct far-field optical image which also includes the subwavelength features. Because of this reason, the use of super lens in THz far-field imaging system is gaining much attention in the research field [145, 146]. With the help of super lens, the propagating waves get focused and the evanescent waves also get amplified in such a way that both the propagating and evanescent fields help to reconstruct an image in the far field with a resolution far below the diffraction limit. However, with this, the subwavelength resolving power of most super lens get affected due to the material losses. It happens because of the enhancement of evanescent waves by super lens [147, 148]. Further, these enhanced evanescent waves cannot be processed with conventional optics. Therefore, for ideal imaging system, there is a need to implement a technique to transfer the information carried by evanescent fields into the fraction part of the propagating spectrum and so would be detected and processed in the far field using conventional imaging methods. In [149], the authors have developed a device using anisotropic metamaterials to form a magnified optical image of a subwavelength object in the far field. The imaging device hyper lens of the system utilizes the cylindrical geometry to enhance the features of the imaged objects above the diffraction limit. Moreover, the output of such hyper lens can be processed easily, and their simulation results show that the material losses do not degrade the performance of the proposed far-field imaging system because of its nonresonant nature.

Over the past few decades, several proposed works were presented on slightly subdiffraction devices for far-field imaging. Super oscillation-based imaging system is one of the techniques which provides significant resolution improvements to the diffraction limit with trade-offs in between the parameters such as the side-lobe level or the power confinement ratio [150]. The phenomenon of super oscillation oscillates the waveform faster than its highest constituent frequency component over a finite spatial duration which further brings subdiffraction imaging capabilities into the far field of an imaging system. The analytical approach for super oscillatory waves had been well designed in [151], and further its use in quantum physics and optical systems have been explored by [152]. Related to the imaging system, in [153], the authors have achieved to illuminate a subwavelength super oscillatory spot upon an object and therefore scanned the spot in subwavelength steps in order to retrieve a super resolution image. Figure 2.8 shows a general setup of optical super microscope which works on the principle of super oscillations so as to construct a high-resolution far-field image [150]. Such set up of super oscillation-based imaging system achieves linear, far-field, subdiffraction imaging in a direct single-capture mode because of simple implementation.

In this setup, a nanometer wavelength laser illuminates the object under consideration and L1 lens transformed the light field projecting on the object into its spectrum at the Fourier plane. After passing through the polarization beam splitter, the spatial light modulator (SLM) performs the reflective filtering. The second lens L2 performs the inverse Fourier transform and so provides a final image in the spatial domain. The working and observation distances can be easily setup as both the

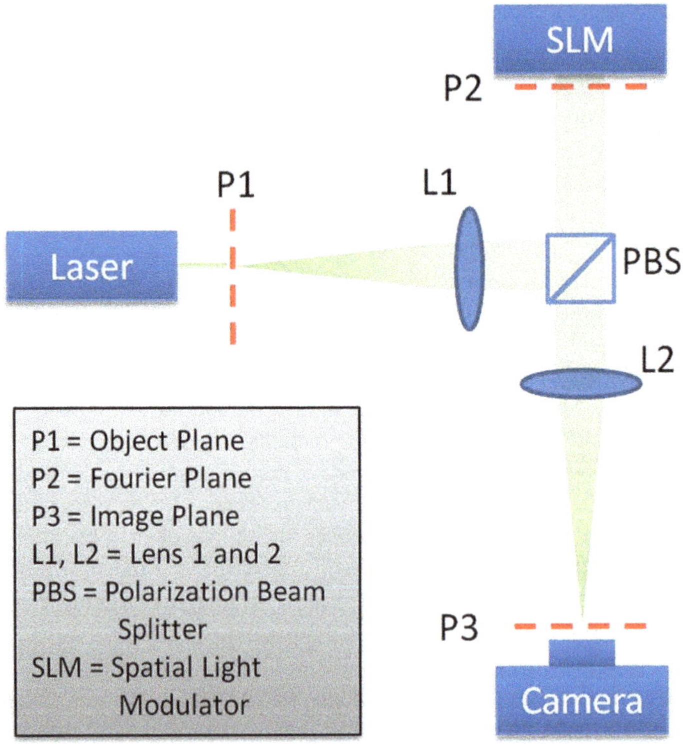

Fig. 2.8 General schematic of super-oscillation-based optical super-microscope for far-field imaging [150]

lenses are kept at 40-cm focal length distance. The width of SLM is related with the numerical aperture (NA) as well as the corresponding diffraction limited spot width. Moreover, high value of NA is desirable for high-resolution imaging. The authors [150] claimed that such system can perform real-time imaging on a wide variety of objects even if they are in motion.

2.7 Summary

Before the mid-1980s, the primary obstacle for THz development was the generation and detection of the radiation. With the remarkable progress of THz science and technology over the last few decades, the detection and imaging for targets of interest is possible by using THz waves. The broad bandwidth of the THz radiation is an additional advantage in a reflection mode, since it permits one to obtain depth information and therefore construct a full three-dimensional representation of an object. According to the detection schemes of THz radiation, the THz imaging system has two main solutions: passive system and active system. Passive system only detects the target scattered signal and heavily conditioned upon the characteristics of a particular scene. Active system has a transceiver which transmits a signal and then receives the reflected or scattered signals from targets. Compared to the passive system, active system is not restricted by the characteristics of the background and can overcome the shortcomings of the passive system. Active system usually has two approaches to generate the THz radiation. One is the down-conversion technology which uses the photomixer to convert from the optics or IR to the THz band and the other method is to use the frequency multipliers and mixers for up-conversion from microwaves or millimeter waves. Based on these two THz radiation generation methods, the target image can be constructed by using raster-scanning technique, focal-plane array technique, or synthetic aperture technique. The raster-scanning technique uses the aperture (or lens at THz range) focusing the beam to a point and scans over the target which is in this focus. Compared to the raster scanning, the focal-plane array technique can acquire an entire image at once due to its two-dimensional array of THz receivers which is similar to an optical camera. This mode does not need to scan in vertical and horizon directions and can realize real-time imaging. Another imaging method is the synthetic aperture imaging technique which uses the signal processing technique rather than large aperture antennas (or lens at THz range) to achieve a target image. The resolution of a target image reconstructed by this technique depends on the transmitted signal bandwidth and the synthetic aperture length. The advantages of this technique are as follows: (1) high resolution, (2) large aperture, (3) relative simple system structure and low cost, and (4) there does not exist a focal plane as that in conventional systems and the target can be imaged over a considerable range of depth. Because of these characteristics, this technique in THz fast imaging field has broad prospects.

Further, one distinction between ultrasound measurements and THz imaging is the possibility for spectroscopic analysis which is offered by the specific interactions of THz radiation with certain materials. Opaque samples having either a high-water content or metal backing can be measured with reflection-type THz-TDS, which can be used not only for spectroscopy but also for tomographic imaging. Since the electromagnetic field of the subpicosecond terahertz pulses is measured directly with THz-TDS, the multilayered structure of a sample can be imaged by detecting the echo pulses reflected from each layer. This technique is valuable in industry because the unique transmission characteristics of terahertz waves enables the inspection of multilayered paints in industrial products or tablet coatings, which are not measurable with optical coherence tomography (OCT) based on an optical light source. Moreover, the pulsed nature and broad bandwidth of THz radiation have made it ideal for imaging applications, such as impulse ranging, biomedical, or chemical specimen identifications. The THz band is advantageous since it includes very short wavelengths and, therefore, provides high resolution compared to traditional electromagnetic tomography at radio frequencies. It is also subject to significantly less scattering than infrared light, which allows for improved reconstruction fidelity compared to infrared tomography techniques.

References

1. D.A. Zimdars, Fiber-pigtailed terahertz time domain spectroscopy instrumentation for package inspection and security imaging, in *Terahertz for Military and Security Applications*, vol. 5070, (International Society for optics and Photonics, 2003), pp. 108–116, Orlando, Florida, United States
2. N.C.J. van der Valk, W.A.M. van der Marel, P.C.M. Planken, Terahertz polarization imaging. Opt. Lett. **30**(20), 2802–2804 (2005)
3. S. Katletz, M. Pflegzater, H. Puhringer, M. Mikulics, N. Vieweg, O. Peters, B. Scherger, M. Scheller, M. Koch, K. Wiesauer, Polarization sensitive terahertz imaging: Detection of birefringence and optical axis. Opt. Express **20**(21), 23025–23035 (2012)
4. M. Koch, S. Hunsche, P. Schumacher, M.C. Nuss, J. Feldmann, J. Fromm, THz-imaging: A new method for density mapping of wood. Wood Sci. Technol. **32**(6), 421–427 (1998)
5. S. Hadjiloucas, L.S. Karatzas, J.W. Bowen, Measurements of leaf water content using terahertz radiation. IEEE Trans. Microwave Theory Technol. **47**(2), 142–149 (1999)
6. E. Pickwell, V.P. Wallace, Biomedical applications of terahertz technology. J. Phys. D. Appl. Phys. **39**(17), R301–R310 (2006)
7. C. Yu, S. Fan, Y. Sun, E. Pickwell-Macpherson, The potential of terahertz imaging for cancer diagnosis: A review of investigations to date. Quant. Imaging Med. Surg. **2**(1), 33–45 (2012)
8. Z.D. Taylor, J. Garritano, S. Sung, N. Bajwa, D.B. Bennett, B. Nowroozi, P. Tewari, J.W. Sayre, J.-P. Hubschman, S.X. Deng, E.R. Brown, W.S. Grundfest, THz and mm-wave sensing of corneal tissue water content: In vivo sensing and imaging results. IEEE Trans. Terahertz Sci. Technol. **5**(2), 184–196 (2015)
9. K. Kawase, Y. Ogawa, Y. Watanabe, H. Inoue, Non-destructive terahertz imaging of illicit drugs using spectral fingerprints. Opt. Express **11**(20), 2549–2554 (2003)
10. Y.C. Shen, T. Lo, P.F. Taday, B.E. Cole, W.R. Tribe, M.C. Kemp, Detection and identification of explosives using terahertz pulsed spectroscopic imaging. Appl. Phys. Lett. **86**(24), 241116/1–241116/3 (2005)

11. H. Zhong, A. Redo-Sanchez, X.C. Zhang, Identification and classification of chemicals using terahertz reflective spectroscopic focal-plane imaging system. Opt. Express **14**(20), 9130–9141 (2006)
12. D.M. Charron, K. Ajito, J.Y. Kim, Y. Ueno, Chemical mapping of pharmaceutical cocrystals using terahertz spectroscopic imaging. Anal. Chem. **85**(4), 1980–1984 (2013)
13. D. L. Woolard, J. O. Jensen, R. J. Hwu (eds.), *Terahertz Science and Technology for Military and Security Applications*, vol 46 (World Scientific Publishing Company Incorporated, Singapore, 2007)
14. J.A. Zeitler, L.F. Gladden, In-vitro tomography and non-destructive imaging at depth of pharmaceutical solid dosage forms. Eur. J. Pharm. Biopharm. **71**(1), 2–22 (2009)
15. M. Hishida, K. Tanaka, Long-range hydration effect of lipid membrane studied by THz time-domain spectroscopy. Phys. Rev. Lett. **106**(15), 158102/1–158102/3 (2011)
16. B.M. Fischer, M. Walther, P.U. Jepsen, Far-infrared vibrational modes of DNA components studied by terahertz time-domain spectroscopy. Phys. Med. Biol. **47**(21), 3807–3814 (2002)
17. B.M. Fischer, M. Hoffmann, H. Helm, R. Wilk, F. Rutz, T.K. Ostmann, M. Koch, P.U. Jepsen, Terahertz time-domain spectroscopy and imaging of artificial RNA. Opt. Express **13**(14), 5205–5215 (2005)
18. D.M. Sheen, T.E. Hall, R.H. Severtsen, D.L. McMakin, B.K. Hatchell, P.L.J. Valdez, Standoff concealed weapon detection using a 350-GHz radar imaging system, in *Proceedings SPIE 7670, Passive Millimeter-Wave Imaging Technology XIII*, vol. 7670, (2010), p. 767008
19. M. Kowalski, M. Kastek, Comparative studies of passive imaging in terahertz and mid-wavelength infrared ranges for object detection. IEEE Trans. Inf. Forensics Secur. **11**(9), 2028–2035 (2016)
20. E.N. Grossman, J. Gordon, D. Novotny, R. Chamberlin, Terahertz active and passive imaging, in *Proceedings of the 8th IEEE European Conference on Antenna and Propagation (EuCAP)*, (The Netherlands, April 2014), pp. 6–11
21. M. Zhang, R. Pan, J. Shen, A fast terahertz imaging: method, simulation, and experiment, in *Proceedings of the 5th Global Symposium on Millimeter Waves (GSMM 2012)*, (Harbin, China, 2012), pp. 431–434
22. D.M. Mittleman, Twenty years of terahertz imaging. Opt. Express **26**(8), 9417–9431 (2018)
23. E. Hack, L. Valzania, G. Gaumann, M. Shalaby, C.P. Hauri, P. Zolliker, Comparison of thermal detector arrays for off-axis THz holography and real-time THz imaging. Sensors **16**(12), 221 (2016)
24. M.A. Patrick, J.A. Holt, C.D. Joye, F.C. DeLucia, Elimination of speckle and target orientation requirements in millimeter-wave active imaging by modulated multimode mixing illumination. J. Opt. Soc. Am. A **29**, 2643–2656 (2012)
25. W. Deal, X.B. Mei, M.K.H. Leong, V. Radisic, S. Sarkozy, R. Lai, THz monolithic integrated circuits using InP high electron mobility transistors. IEEE Trans. Terahertz Sci. Technol. **1**, 25–32 (2011)
26. V. Radisic, M.K.H. Leong, X. Mei, S. Sarkozy, W. Yoshida, W.R. Deal, Power amplification at 0.65 THz using InP HEMTs. IEEE Trans. Microwave Theory Technol. **60**, 724–729 (2012)
27. F. Friederich, W.V. Spiegel, M. Bauer, F. Meng, M.D. Thomson, S. Boppel, A. Lisauskas, B. Hils, V. Krozer, A. Keil, T. Loffler, R. Henneberger, A.K. Huhn, G. Spickermann, P.H. Bolivar, H.G. Roskos, THz active imaging systems with real-time capabilities. IEEE Trans. Terahertz Sci. Technol. **1**(1), 183–200 (2011)
28. R.W. McGowan, R.A. Cheville, D. Grischkowsky, Direct observation of the Gouy phase shift in THz impulse ranging. Appl. Phys. Lett. **76**, 670–672 (2000)
29. P. Kužel, H. Němec, F. Kadlec, C. Kadlec, P. Kuzel, H. Nemec, F. Kadlec, C. Kadlec, Gouy shift correction for highly accurate refractive index retrieval in time-domain terahertz spectroscopy. Opt. Express **18**, 15338–15348 (2010)
30. B.B. Hu, M.C. Nuss, Imaging with terahertz waves. Opt. Lett. **20**(16), 1716–1718 (1995)
31. I. Malhotra, K.R. Jha, G. Singh, Terahertz antenna technology for imaging applications- a technical review. Int. J. Microw. Wirel. Technol. **10**(3), 271–290 (2018)

32. Z.Y. Tan, J.C.C. Tao Zhou, H.C. Liu, Terahertz imaging with quantum-cascade laser and quantum-well photodetector. IEEE Photonics Technol. Lett. **25**(14), 1344–1346 (2013)
33. H. Guerboukha, K. Nallappan, M. Skorobogatiy, Towards real-time terahertz imaging. Adv. Opt. Photon. **10**(4), 843–938 (2018)
34. J.F. O'Hara, W. Withayachumnankul, I. Al-Naib, A review on thin-film sensing with terahertz waves. J. Infrared Millimeter Terahertz Waves **33**, 245–291 (2012)
35. L. Duvillaret, F. Garet, J.-L. Coutaz, A reliable method for extraction of material parameters in terahertz time-domain spectroscopy. IEEE J. Sel. Top. Quantum Electron. **2**, 739–746 (1996)
36. M. Hangyo, T. Nagashima, S. Nashima, Spectroscopy by pulsed terahertz radiation. Meas. Sci. Technol. **13**, 1727–1738 (2002)
37. M. Naftaly, R.E. Miles, Terahertz time-domain spectroscopy for material characterization. Proc. IEEE **95**, 1658–1665 (2007)
38. W. Withayachumnankul, B. Ferguson, T. Rainsford, S.P. Mickan, D. Abbott, Simple material parameter estimation via terahertz time-domain spectroscopy. Electron. Lett. **41**, 800–801 (2005)
39. P. Doradla, K. Alavi, C. Joseph, R. Giles, Detection of colon cancer by continuous-wave terahertz polarization imaging technique. J. Biomed. Opt. **18**(9), 090504 (2013)
40. P.U. Jepsen, B.M. Fischer, Dynamic range in terahertz time-domain transmission and reflection spectroscopy. Opt. Lett. **30**, 29–31 (2005)
41. M. Naftaly, R. Dudley, Methodologies for determining the dynamic ranges and signal-to-noise ratios of terahertz time-domain spectrometer. Opt. Lett. **34**, 1213–1215 (2009)
42. S. Fan, Y. He, B.S. Ung, E. Pickwell-MacPherson, The growth of biomedical terahertz research. J. Phys. D. Appl. Phys. **47**, 374009/1–11 (2014)
43. R.M. Woodward, B.E. Cole, V.P. Wallace, R.J. Pye, D.D. Arnone, E.H. Linfield, M. Pepper, Terahertz pulse imaging in reflection geometry of human skin cancer and skin tissue. Phys. Med. Biol. **47**, 3853–3863 (2002)
44. I. Malhotra, K.R. Jha, G. Singh, Analysis of highly directive photoconductive dipole antenna at terahertz frequency for sensing and imaging applications. Optics Communication **397**, 129–139 (2017)
45. R. Toh, K. Suizu, Y. Tojima, Terahertz pulse reflection imaging using the time-domain correlating synthesis method. IEEE Trans. Terahertz Sci. Technol. **7**(4), 385–392 (2017)
46. F.A. Hegmann, O. Ostroverkhova, D.G. Cooke, Probing organic semiconductors with terahertz pulses, in *Photophysics of Molecular Materials*, (Wiley, Weinheim, 2006), pp. 367–428
47. J. Lloyd-Hughes, T.I. Jeon, A review of the terahertz conductivity of bulk and nano-materials. J. Infrared Millimeter Terahertz Waves **33**, 871–925 (2012)
48. M.C. Nuss, D.H. Auston, F. Capasso, Direct subpicosecond measurement of carrier mobility of photoexcited electrons in gallium arsenide. Phys. Rev. Lett. **58**, 2355–2358 (1987)
49. R. Ulbricht, E. Hendry, J. Shan, T.F. Heinz, M. Bonn, Carrier dynamics in semiconductors studied with time-resolved terahertz spectroscopy. Rev. Mod. Phys. **83**, 543–586 (2011)
50. I. Malhotra, K.R. Jha, G. Singh, Design of highly directive terahertz photoconductive dipole antenna using frequency selective surface for sensing and imaging application. Journal of Computational Electronics **17**(4), 1721–1740 (2018)
51. Q. Li, Y. Zhou, Y.-F. Yang, G.-H. Chen, 2.52 terahertz dual-axis reflection confocal scanning microscope. J. Opt. Soc. Am. A **33**, 637–641 (2016)
52. Y.H. Lo, R. Leonhardt, Aspheric lenses for terahertz imaging. Opt. Express **16**, 15991–15998 (2008)
53. W.L. Chan, J. Deibel, D.M. Mittleman, Imaging with terahertz radiation. Rep. Prog. Phys. **70**, 1325–1379 (2007)
54. E. Betzig, J.K. Trautman, T.D. Harris, J.S. Weiner, R.L. Kostelak, Breaking the diffraction barrier-optical microscopy on a nanometric scale. Sciences **251**, 1468–1470 (1991)
55. S. Hunsche, M. Koch, I. Brener, M.C. Nus, THz near-field imaging. Opt. Commun. **150**, 22 (1998)

56. L. Novotny, B. Hecht, Propagation and focusing of optical fields, in *Principles of Nano-Optics*, (Cambridge University Press, Cambridge, UK, 2006), pp. 45–87
57. A. Rahman, A.K. Rahman, Nanoscale metrology of line patterns on semiconductor by continuous wave terahertz multispectral reconstructive 3-D imaging overcoming the Abbe diffraction limit. IEEE Trans. Semicond. Manuf. **32**(1), 7–13 (2019)
58. J. Durnin, J.J. Miceli, J.H. Eberly, Diffraction-free beams. Phys. Rev. Lett. **58**(15), 1499–1501 (1987)
59. J. Durnin, Exact solutions for nondiffracting beams—I: The scalar theory. J. Opt. Soc. Am. A **4**(4), 651–654 (1987)
60. J.C. Gutiérrez-Vega, M.D. Iturbe-Castillo, S. Chávez-Cerda, Alternative formulation for invariant optical fields: Mathieu beams. Opt. Lett. **25**(20), 1493–1495 (2000)
61. M.A. Bandres, J.C. Gutiérrez-Vega, S. Chávez-Cerda, Parabolic nondiffracting optical wave fields. Opt. Lett. **29**(1), 44–46 (2004)
62. P. Zhang, Y. Hu, T. Li, D. Cannan, X. Yin, R. Morandotti, Z. Chen, X. Zhang, Nonparaxial mathieu and weber accelerating beams. Phys. Rev. Lett. **19**, 109 (2012)
63. J. Turunen, A.T. Friberg, Propagation-invariant optical fields. Prog. Opt. **54**, 1–88 (2009)
64. M.V. Berry, N.L. Balazs, Non spreading wave packets. Am. J. Phys. **47**(3), 264–267 (1979)
65. G.A. Siviloglou, D.N. Christodoulides, Accelerating finite energy airy beams. Opt. Lett. **32**(8), 979–981 (2007)
66. D. McGloin, K. Dholakia, Bessel beams: Diffraction in a new light. Contemp. Phys. **46**(1), 15–28 (2005)
67. A. Bitman, I. Moshe, Z. Zalevsky, Improving depth-of field in broadband THz beams using nondiffractive Bessel beams. Opt. Lett. **37**(19), 4164–4166 (2012)
68. L.T. Niu, K. Wang, Y. Yang, Q. Wu, X. Ye, Z. Yang, J. Liu, H. Yu, Diffractive elements for zero-order Bessel beam generation with application in the terahertz reflection imaging. IEEE Photonics J. **11**(1), 5900212/1–13 (2019)
69. A. Bitman, S. Goldring, I. Moshe, Z. Zalevsky, Computed tomography using broadband Bessel THz beams and phase contrast. Opt. Lett. **39**(7), 1925–1928 (2014)
70. M.S. Kulya, V.A. Semenova, V.G. Bespalov, N.V. Petrov, On terahertz pulsed broadband Gauss–Bessel beam free-space propagation. Sci. Rep. **8**(1), 1390/1–11 (2018)
71. J. Suszek, A. Siemion, M.S. Bieda, N. Blocki, D. Coquillat, G. Cywinski, E. Czerwinska, M. Doch, A. Kowalczyk, N. Palka, A. Sobczyk, 3-D-printed flat optics for THz linear scanners. IEEE Trans. Terahertz Sci. Technol. **5**(2), 314–316 (2015)
72. Z. Zhang, H. Zhang, K. Wang, Diffraction-free THz sheet and its application on THz imaging system. IEEE Trans. Terahertz Sci. Technol. **9**(5), 471–475 (2019)
73. H.E. Kondakci, A.F. Abouraddy, Diffraction-free space–time light sheets. Nat. Photonics **11**(5), 730–740 (2017)
74. M. Baillergeau, K. Maussang, T. Nirrengarten, J. Palomo, L.H. Li, E.H. Linfield, A.G. Davies, S. Dhillon, J. Tignon, J. Mangeney, Diffraction-limited ultrabroadband terahertz spectroscopy. Sci. Rep. **6**, 24811 (2016)
75. D.M. Mittleman, S. Hunsche, L. Boivin, M.C. Nuss, T-ray tomography. Opt. Lett. **22**, 904–906 (1997)
76. V.P. Wallace, A.J. Fitzgerald, E. Pickwell, R. Pye, P.F. Taday, N. Flanagan, T. Ha, Terahertz pulsed spectroscopy of human basal cell carcinoma. Appl. Spectrosc. **60**, 1127–1133 (2006)
77. H.B. Liu, Y. Chen, G.J. Bastiaans, X.C. Zhang, Detection and identification of explosive RDX by THz diffuse reflection spectroscopy. Opt. Express **14**, 415–423 (2006)
78. L. Guo, Y. Hu, Y. Zhang, C. Zhang, Y. Chen, X.C. Zhang, Vibrational spectrum of υ-HNIW investigated using terahertz time-domain spectroscopy. Opt. Express **14**, 3654–3659 (2006)
79. Y. Ino, J.B. Heroux, T. Mukaiyama, M. Kuwata-Gonokami, Reflection-type pulsed terahertz imaging with a phase-retrieval algorithm. Appl. Phys. Lett. **88**, 041114/1–041114/3 (2006)
80. H. Zhong, J. Xu, X.C. Zhang, Nondestructive defect identification with terahertz time-of-flight tomography. IEEE Sensors J. **5**(2), 203–208 (2005)

81. K. Krugener, S.F. Busch, A. Soltani, E.C. Camus, M. Koch, W. Viol, Non-destructive analysis of material detachments from polychromatically glazed terracotta artwork by THz time-of-flight spectroscopy. J. Infrared Millimeter Terahertz Waves **38**(4), 495–502 (2017)
82. J.L. Johnson, T.D. Domey, D.M. Mittleman, Interferometric imaging with terahertz pulses. IEEE J. Sel. Top. Quantum Electron. **7**, 592–599 (2001)
83. J.L. Johnson, T.D. Domey, D.M. Mittleman, Enhanced depth resolution in terahertz imaging using phase- shift interferometry. Appl. Phys. Lett. **78**, 835–837 (2001)
84. A.B. Ruffin, J. Decker, L. Sanchez-Palencia, L. Le Hors, J.F. Whitaker, T.B. Norris, J.V. Rudd, Time reversal and object reconstruction with single-cycle pulses. Opt. Lett. **26**, 681–683 (2001)
85. A.B. Ruffin, J.V. Rudd, J. Decker, L. Sanchez-Palencia, L.L.E. Hors, J.F. Whitaker, T.B. Norris, Time reversal terahertz imaging. IEEE J. Quantum Electron. **38**, 1110–1119 (2002)
86. T. Buma, T.B. Norris, Time reversal three-dimensional imaging using single-cycle terahertz pulses. Appl. Phys. Lett. **84**, 2196–2198 (2004)
87. B. Ferguson, S. Wang, D. Gray, D. Abbott, X.C. Zhang, T-ray computed tomography. Opt. Lett. **27**, 1312–1314 (2002)
88. A. Brahm, M. Kunz, S. Riehemann, G. Notni, A. Tunnermann, Volumetric spectral analysis of materials using terahertz-tomography techniques. Appl. Phys. B **100**(1), 151–158 (2010)
89. J.P. Guillet, B. Recur, L. Frederique, B. Bousquet, L. Canioni, I.M. Honninger, P. Desbarats, P. Mounaix, Review of terahertz tomography techniques. J. Infrared Millimeter Terahertz Waves **35**(4), 382–411 (2014)
90. S.H. Wang, X.-C. Zhang, Tomographic imaging with a terahertz binary lens. Appl. Phys. Lett. **82**, 1821–1821 (2003)
91. T.D. Dorney, J.L. Johnson, J.V. Rudd, R.G. Baraniuk, W.W. Symes, D.M. Mittleman, Terahertz reflection imaging using Kirchhoff migration. Opt. Lett. **26**, 1513–1513 (2001)
92. T.D. Dorney, W.W. Symes, R.G. Baraniuk, D.M. Mittleman, Terahertz multistatic reflection imaging. J. Opt. Soc. Am. A **19**, 1432–1432 (2002)
93. M.K. Choi, A. Bettermann, D.W.V. Wiede, Potential for detection of explosive and biological hazards with electronic THz systems. Phil. Trans. R. Soc. A **362**, 337–349 (2004)
94. T. Shibuya, T. Suzuki, K. Suizu, K. Kawase, Non-destructive characterization of soot in exhaust filters using Millimeter-wave imaging. J. Infrared Millimeter Terahertz Waves **32**(5), 716–721 (2011)
95. M.A. Brun, F. Formanek, A. Yasuda, M. Sekine, N. Ando, Y. Eishii, Terahertz imaging applied to cancer diagnosis. Phys. Med. Biol. **55**(16), 4615 (2010)
96. J.F. Federici, B. Schulkin, F. Huang, D. Gary, R. Barat, F. Oliveira, D. Zimdars, THz imaging and sensing for security applications explosives, weapons and drugs. Semicond. Sci. Technol. **20**(7), S266–S280 (2005)
97. S. Zhong, Y.C. Shen, L. Ho, R.K. May, J.A. Zeitler, M. Evans, P.F. Taday, M. Pepper, T. Rades, K.C. Gordon, R. Muller, Non-destructive quantification of pharmaceutical tablet coatings using terahertz pulsed imaging and optical coherence tomography. Opt. Lasers Eng. **49**(3), 361–365 (2011)
98. W.L. Chan, M.L. Moravec, R.G. Baraniuk, D.M. Mittleman, Terahertz imaging with compressed sensing and phase retrieval. Opt. Lett. **33**(9), 974–976 (2008)
99. C.M. Watts, D. Shrekenhamer, J. Montoya, G. Lipworth, J. Hunt, T. Sleasman, S. Krishna, D.R. Smith, W.J. Padilla, Terahertz compressive imaging with metamaterial spatial light modulators. Nat. Photonics **8**(8), 605–609 (2014)
100. S.N. Saqueb, K. Sertel, Compressive terahertz imaging using a single-bit sensor. IEEE Trans. Terahertz Sci. Technol. **8**(6), 757–764 (2018)
101. M.I.B. Shams, Z. Jiang, S. Rahman, J. Qayyum, L.-J. Cheng, H.G. Xing, P. Fay, L. Liu, Approaching real-time terahertz imaging with photoinduced coded apertures and compressed sensing. Electron. Lett. **50**(11), 801–803 (2014)

102. D. Shrekenhamer, C.M. Watts, W.J. Padilla, Terahertz single pixel imaging with an optically controlled dynamic spatial light modulator. Opt. Express **21**(10), 12 507–512 (2013)
103. A. Kannegulla, J. Zhenguo, S.M. Rahman, M.I.B. Shams, P. Fay, H.G. Xing, L.J. Cheng, L. Liu, Coded-aperture imaging using photo-induced reconfigurable aperture arrays for mapping terahertz beams. IEEE Trans. Terahertz Sci. Technol. **4**(3), 321–327 (2014)
104. S.A.N. Saqueb, K. Sertel, Phase-sensitive single-pixel THz imaging using intensity-only measurements. IEEE Trans. Terahertz Sci. Technol. **6**(6), 810–816 (2016)
105. S. Qaisar, R.M. Bilal, W. Iqbal, M. Naureen, S. Lee, Compressive sensing: From theory to applications: A survey. J. Commun. Netw. **15**(5), 443–456 (2013)
106. S.A.N. Saqueb, K. Sertel, Single-bit compressive imaging system for the mmW and THz bands, in *Proceedigs of the IEEE International Symposium on Antennas and Propagation, USNC/URSI National Radio Science Meeting*, (2017), pp. 843–844
107. P.T. Boufounos, Reconstruction of sparse signals from distorted randomized measurements, in *Proceedings of the IEEE International Conference on Acoustics and Speech Signal Processing*, (Dallas, USA, 2010), pp. 3998–4001
108. J.N. Laska, R.G. Baraniuk, Regime change: Bit-depth versus measurement-rate in compressive sensing. IEEE Trans. Signal Process. **60**(7), 3496–3505 (2012)
109. M. Reuter, N. Vieweg, M. Koch, I. Dierking, THz spectroscopy across liquid crystalline phase transitions. Appl. Phys. Lett. **108**(5), 051908/1–4 (2016)
110. M.C. Beard, G.M. Turner, C.A. Schmuttenmaer, Subpicosecond carrier dynamics in low-temperature grown GaAs as measured by time-resolved THz spectroscopy. J. Appl. Phys. **90**(12), 5915 (2001)
111. R.D. Averitt, G. Rodriguez, J.L.W. Siders, S.A. Trugman, A.J. Taylor, Conductivity artifacts in optical-pump THz-probe measurements of YBa2Cu3O7. J. Opt. Soc. Am. B **17**(2), 327–331 (2000)
112. M. Hangyo, M. Tani, T. Nagashima, THz time-domain spectroscopy of solids: A review. Int. J. Infrared Millimeter Waves **26**(12), 1661–1690 (2005)
113. D.-K. Lee, G. Kim, C. Kim, Y.M. Jhon, J.H. Kim, T. Lee, J.-H. Son, M. Seo, Ultrasensitive detection of residual pesticides using THz near-field enhancement. IEEE Trans. Terahertz Sci. Technol. **6**(3), 389–394 (2016)
114. I. Hosako, K. Fukunaga, THz technology research at NICT from the source to industrial applications. J. Infrared Millimeter Terahertz Waves **32**, 722–731 (2011)
115. K. Fukunaga, Y. Ogawa, S. Hayashi, I. Hosako, THz spectroscopy for art conservation. IEICE Electron. Exp. **4**(8), 258–263 (2007)
116. K. Fukunaga, I. Hosako, I.N. Duling III, M. Picollo, THz imaging systems: A non- invasive technique for the analysis of paintings. Proc SPIE **7391**, 73910D (2009)
117. H. Yan, W.H. Fan, Z.P. Zheng, Investigation on terahertz vibrational modes of crystalline benzoic acid. Opt. Commun. **285**(6), 1593–1598 (2012)
118. S. Galoda, G. Singh, Fighting terrorism with terahertz. IEEE Potential Magazine **26**(6), 24–29 (2007)
119. S.L. Dexheimer, *THz Spectroscopy: Principles and Applications* (CRC Press, Boca Raton, 2008)
120. L. Oberto, M. Koch, On the influence of delay line uncertainty in THz – Time domain spectroscopy. J. Infrared Millimeter Terahertz Waves **37**(6), 605–613 (2016)
121. P. Jepsen, B. Fischer, Dynamic range in THz time-domain transmission and reflection spectroscopy. Opt. Lett. **30**(1), 29–31 (2005)
122. J.B. Baxter, C.A. Schmuttenmaer, Conductivity of ZnO nanowires, nanoparticles and thin films using time-resolved THz spectroscopy. J. Phys. Chem. **110**(50), 25229–25239 (2006)
123. C.S. Ponseca, T.J. Savenije, M. Abdellah, K. Zheng, A. Yartsev, T. Pascher, T. Harlang, P. Chabera, T. Pullerits, A. Stepanov, J.P. Wolf, Organometal halide perovskite solar cell materials rationalized: Ultrafast charge generation, high and microsecond long balanced mobilities and slow recombination. J. Am. Chem. Soc. **136**(14), 5189–5192 (2014)

124. M. Kreβ, T. Loffler, M.D. Thomson, R. Dorner, H. Gimpel, K. Zrost, T. Ergler, R. Moshammer, U. Morgner, J. Ullrich, H.G. Roskos, Determination of the carrier-envelope phase of few-cycle laser pulses with terahertz-emission spectroscopy. Nat. Phys. **2**(5), 327–331 (2006)
125. R. Heremans, M. Vandewal, M. Acheroy, Synthetic aperture imaging extended towards novel THz sensors. IEEE Sens., 438–441 (2008)
126. K.B. Cooper, R.J. Dengler, G. Chattopadhyay, E. Schlecht, A. Skalare, I. Mehdi, P.H. Siegel, A high-resolution imaging radar at 580GHz. IEEE Microwave Wireless Compon. Lett. **18**(1), 64–66 (2008)
127. I. Malhotra, P. Thakur, S. Pandit, K. R. Jha, G. Singh, Analytical framework of small-gap photoconductive dipole antenna using equivalent circuit model. Optical and Quantum Electronics **49**(10), 334/1–23 (2017)
128. R. Appleby, H.B. Wallace, Standoff detection of weapons and contraband in the 100 GHz to 1 THz region. IEEE Trans. Antennas Propag. **55**(11), 2944–2956 (2007)
129. K.B. Cooper, R.J. Dengler, N. Llombart, T. Bryllert, G. Chattopadhyay, E. Schlecht, J. Gill, C. Lee, A. Skalare, I. Mehdi, P.H. Siegel, Penetrating 3-D imaging at 4 and 25 meter range using a submillimeter- wave radar. IEEE Trans. Microwave Theory Technol. **56**(12), 2771–2778 (2008)
130. D.A. Robertson, P.N. Marsh, D.R. Bolton, R.J.C. Middleton, R.I. Hunter, P.J. Speirs, D.G. Macfarlane, S.L. Cassidy, G.M. Smith, 340 GHz 3-D radar imaging test bed with 10 Hz frame rate. Proc. SPIE **8362**, 10 (2012), Art. ID 836206
131. K.B. Cooper, R.J. Dengler, N. Llombart, B. Thomas, G. Chattopadhyay, P.H. Siegel, THz imaging radar for standoff personnel screening. IEEE Trans. Terahertz Sci. Technol. **1**(1), 169–182 (2011)
132. J. Ding, M. Kahl, O. Loffeld, P.H. Bolivar, THz 3-D image formation using SAR techniques: Simulation, processing and experimental results. IEEE Trans. Terahertz Sci. Technol. **3**(5), 606–616 (2013)
133. B. Cheng, Z. Cui, B. Lu, Y. Qin, Q. Liu, P. Chen, Y. He, J. Jiang, X. He, X. Deng, J. Zhang, L. Zhu, 340-GHz 3-D imaging radar with 4Tx-16Rx MIMO array. IEEE Trans. Terahertz Sci. Technol. **8**(5), 509–519 (2018)
134. A.J. Huber, F. Keilmann, J. Wittborn, J. Aizpurua, R. Hillenbrand, Terahertz near-field nanoscopy of mobile carriers in single semiconductor nanodevices. Nano Lett. **8**(11), 3766–3770 (2008)
135. A.J.L. Adam, J.M. Brok, M.A. Seo, K.J. Ahn, D.S. Kim, J.H. Kang, Q.H. Park, M. Nagel, P.C.M. Planken, Advanced terahertz electric near-field measurements at subwavelength diameter metallic apertures. Opt. Express **16**(10), 7407–7417 (2008)
136. M. Wächter, M. Nagel, H. Kurz, Tapered photoconductive terahertz field probe tip with subwavelength spatial resolution. Appl. Phys. Lett. **95**, 041112/1–041112/3 (2009)
137. R.I. Stantchev, B. Sun, S.M. Hornett, P.A. Hobson, G.M. Gibson, M.J. Padgett, E. Hendry, Noninvasive, near-field terahertz imaging of hidden objects using a single-pixel detector. Sci. Adv. **2**(6), e1600190/1–e1600190/7 (2016)
138. S.M. Hornett, R.I. Stantchev, M.Z. Vardaki, C. Beckerleg, E. Hendry, Subwavelength terahertz imaging of graphene photoconductivity. Nano Lett. **16**, 7019–7024 (2016)
139. D. Courjon, *Near-Field Microscopy and Near-Field Optics* (World Scientific Publishing Company, Singapore, 2003)
140. M.D. Serio, R. Zenobi, V. Deckert, Looking at the nanoscale: Scanning near-field optical microscopy. Trends Anal. Chem. **22**, 70–77 (2003)
141. C. Girard, C. Joachim, S. Gauthier, The physics of the near-field. Rep. Prog. Phys. **63**(6), 893–938 (2000)
142. E. Wolf, M. Nieto, Analyticity of the angular spectrum amplitude of scattered fields and some of its consequences. J. Opt. Soc. Am. A **2**(6), 886–889 (1985)
143. R.I. Stantchev, B.P. David, P. Hobson, M.H. Samuel, J.P. Miles, H. Euan, Compressed sensing with near-field THz radiation. Optica **4**(8), 989–992 (2017)

144. B. Hecht, B. Sick, U.P. Wild, V. Deckert, R. Zenobi, O.J.F. Martin, D.W. Pohl, Scanning near-field optical microscopy with aperture probes: Fundamentals and applications. J. Chem. Phys. **112**, 7761–7774 (2000)
145. V.M. Shalaev, W. Cai, U.K. Chettiar, H.-K. Yuan, A.K. Sarychev, V.P. Drachev, A.V. Kildishev, Negative index of refraction in optical metamaterials. Opt. Lett. **30**, 3356–3358 (2005)
146. J.B. Pendry, Negative refraction makes a perfect lens. Phys. Rev. Lett. **85**, 3966–3969 (2000)
147. V.A. Podolskiy, E.E. Narimanov, Near-sighted superlens. Opt. Lett. **30**, 75–77 (2005)
148. R. Merlin, Analytical solution of the almost-perfect-lens problem. Appl. Phys. Lett. **84**, 1290–1292 (2004)
149. Z. Jacob, L.V. Alekseyey, E. Narimanov, Optical hyperlens: Far-field imaging beyond the diffraction limit. Opt. Express **14**(18), 8247–8256 (2006)
150. A.M.H. Wong, G.V. Eleftheriades, An optical super-microscope for far-field, real-time imaging beyond the diffraction limit. Sci. Rep. **3**, 01715/1–6 (2013)
151. D. Slepian, H.O. Pollak, Prolate spherical wave functions, Fourier analysis and uncertainty. Bell Syst. Technol. **40**, 43–63 (1961)
152. Y. Aharonov, J. Anandan, S. Popescu, L. Vaidman, Superpositions of time evolutions of a quantum system and a quantum time-translation machine. Phys. Rev. Lett. **64**, 2965–2968 (1990)
153. T. Brunet, J.L. Thomas, Transverse shift of helical beams and subdiffraction imaging. Phys. Rev. Lett. **105**, 034301/1–034301/3 (2010)

Chapter 3
Terahertz Antenna Technology for Imaging and Sensing Applications

3.1 Introduction

From almost one last decade the researchers have determined several ways to generate THz waves using several techniques wherein each method has some limitation with respect to cost, complexity, the requirement of cryogenic cooling, etc. One such method of generating THz waves is a photosensitive semiconductor having a pair of antennas etched onto its surface [1]. When a voltage on these micrometer-sized antennas gets developed from the biased potential, then it results in a strong electric field across the semiconductor between them. Further, when a laser pulse strikes the semiconductor surface then it results into creation of several pairs of charge carriers. These charge carriers accelerate across the semiconductor through the antennas and such rush of current due to accelerated charges last for about a picosecond, therefore the resulting terahertz wave pulse is weak having an average power in microwatts; however, it is bright enough to produce a still image. However, this terahertz wave pulse has significant features such as: (1) similar to radar signal if the pulse's echo is timed as it bounce-off an object will provide the range to the object which is further useful in processing multilevel T-ray images, (2) it is useful in THz spectroscopy because single pulse comprises of broad swath of T-ray frequencies and the shape of the pulse's echo become useful to determine which frequency get absorbed by the reflecting substance from the respective range of frequencies. Then looking up about the substance that produces such absorption pattern and the identification and classification can easily be performed. In Fig. 3.1, a T-ray scanner is shown wherein the generated pulses drive the photoconductor antenna. From the scanner system, it is evident that in THz imaging system, a THz source plays a significant role and for pulsed THz spectroscopy, the use of photoconductive THz antenna is generally required. Therefore, it is necessary to analyze the developments occurred so far in the field of THz antennas related to THz sensing and imaging applications. An optimum design of THz array antenna helps to enhance the imaging capabilities to address the considerations such as limited depth-of-field (DoF) that is the distance

I. Malhotra, G. Singh, *Terahertz Antenna Technology for Imaging and Sensing Applications*, https://doi.org/10.1007/978-3-030-68960-5_3

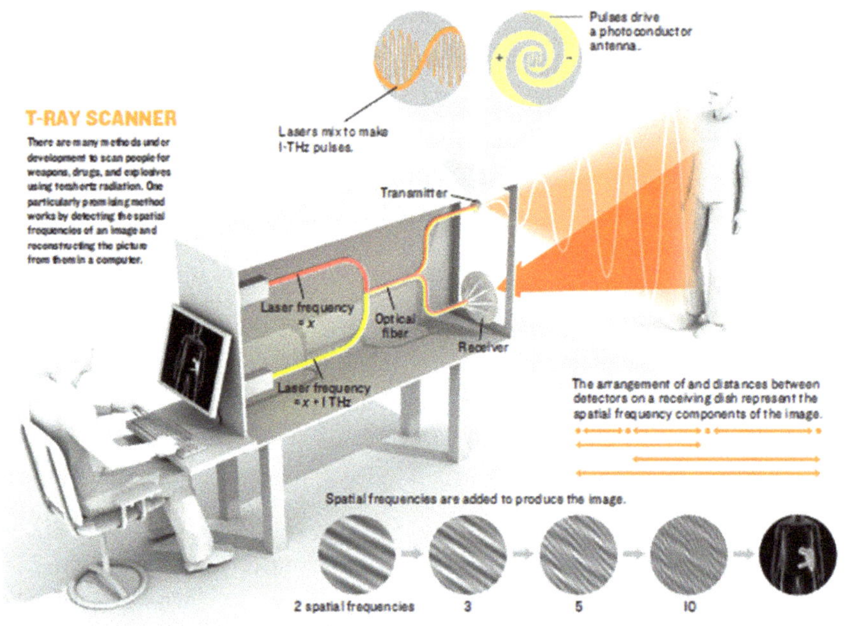

Fig. 3.1 T-ray scanner to scan people for weapons, drugs, or explosives using THz radiations [1]

over which an object is considered in focus and size-weight-and-power (SWaP) of THz source for imaging applications [2]. These are important considerations for applications like stand-off imaging and surveillance of moving targets where the high angular resolution as well as extended depth-of-field is the key for successful detection of hidden explosives and illicit drugs.

3.2 State-of-the-Art Terahertz Antennas Based on Integrated Circuits

Since the THz gap being closed from both electronics and photonics side of the spectrum, therefore with the advent of electronic devices capable of operation at frequencies above 1 THz and similarly with the dawn of long-wavelength photonic devices such as quantum cascade lasers (QCLs) from high-frequency side this gap is becoming narrow. Further, the convergence of such technologies leads to acquire a potential for many new and novel hybrid electronic-photonic devices which are becoming useful for several important applications such as atmospheric science, bio-detection and imaging, and broadband communications. To produce THz radiation using optoelectronic devices [3] through down-conversion of an optical frequency laser pulse, there is a requirement of small semiconductor bandgap of the order of 1000th of eV and such suitable material lacks with the desired small bandgap. However, among the optoelectronics devices, the QCL being the most successful approach is based on radiative inter sub-band transitions between quantum confined states called sub-bands. It has generally reduced dimensional structures such as semiconductor quantum wells and superlattices [3–6]. However, the performance of QCL gets limited to cryogenic temperatures due to the thermal broadening of small inter sub-band energies. Since the research continues from the electronic side also, there are several solid-state technologies which are capable of submillimeter-wave performance and includes up-conversion approaches, heterojunction bipolar transistors (HBTs) [7] and high-electron mobility transistors (HEMTs) [8]. A typical HBT structure comprises of a wide bandgap semiconductor emitter and collector with a narrower gap heavily doped base material named single HBT (SHBT) or a wider bandgap collector named double HBT (DHBT). The use of wide bandgap emitter suppresses minority carrier injection in the base–emitter junction which in turn allow the use of narrow and heavily doped base which further results in reduction in base transit time and base resistance. Moreover, for high-frequency and low-noise applications HEMT has emerged as a preferred device [9, 10]. The HEMTs are made up from a great variety of materials and the efficacy of molecular-beam epitaxy allows strained layers to be incorporated in pseudomorphic HEMTs [11]. Further, the use of In-rich InGaAs in InP-based strained HEMTs provides good performance in submillimeter-wave amplifiers [12]. Moreover, to look into the technological potential of THz radiations, there is a requirement of advanced modeling and simulation tools. Therefore, the diversified research is moving toward to yield new materials and devices to support and be the part of THz systems in the form of sources, detectors and some active devices. However, the parameter space for the assessment of such technologies is quite large. The effectiveness of modeling and simulation is to provide lucrative means of performance and testing new designs prior to manufacturing in addition to the provision of an accurate circuit design for high-frequency components. Further, to treat wave-device-environment there is a need to evolve new global modeling methods. In the following sub-section, more details are shared on the existing technology developed so far and which are being employed to construct high-frequency device structures as THz sources and receivers.

3.2.1 Existing Technology

The size of THz device gets reduced to the nano-scale size because of the development in electronic technologies using nanomaterials. The THz band inspired the use of nano-scale device communication because of the antenna size constraints which are drastically changing in these frequency range. Moreover, the emission frequency range of such THz antennas becomes wider because of the use of graphene nanoribbon (GNR) and carbon nano-tube (CNT) [13]. Further, such nano-devices and sensors are employed in the application areas ranging from health monitoring [14, 15] to defense [16]. Some examples related to the implementation of wireless nano-devices at THz frequencies are mentioned below.

Health Applications

The nano-devices have found their utilization in visual imaging for diagnosis using capsule endoscopes so as to achieve painless imaging of bowel of patients having the problem of recurrent gastrointestinal bleeding [17–21]. Moreover, several researchers are working in the use of biosensors for health monitoring applications to track the prognostics in cancer and DNA mutation [22–29]. Similarly, the nano-devices can effectively be used to deliver the drug to effected metastases in the treatment of metastatic cancer [30–40]. The nanorobots can be designed and accordingly can be deployed inside the tissues to detect and fight against malicious micro-organisms as well as cancer cells with nonoperative issues [41]. However, the use of nano-devices in biomedical applications need to be carefully analyzed in terms of channel and noise modeling because every class of biological tissue carries different amount of water and so they reflect different absorption versus frequency relation [42–46].

Military Applications

The detection of chemical, biological, and nuclear agents used in battlefield can effectively be performed using nano-devices which can communicate in THz bands. Moreover, for the safety of troops the nano-devices can easily be deployed for applications such as brain–machine interface and the research progress reported in BioFluids program of DARPA envisions also shows constructive work in the field of military defense [47–49]. Further, the performance of nano-devices shows high-scale potential application in defense-related industry in terms of control over product quality and making intelligent production lines [50]. However, the lower latency requirement for the production line can be provided using THz networks.

Environmental Pollution Monitoring Application

The THz communication-equipped nano-sensors can be used to track the toxic element density in the environment as well as in potable water reservoirs. This is possible due to the ability of THz nano-sensors to sense chemical compounds present in the contaminated means with one part per billion density [51]. Moreover, using wireless THz nano-sensor networks the air pollution can also be monitored whereby the life quality of people can be raised for smart city applications. Further, the global positioning system (GPS) used in smart vehicles effectively joined with THz nano-sensors can support to collect data in cities while driving and such collected data from smart moving vehicles can then be utilized by THz receiver systems deployed on the roadsides or in-built smart traffic lights for control of traffic or fuel emission of vehicle to control air pollution.

Technology in the Field of Entertainment

Recent developments in the field of visual technologies such as ultra-high-definition movies, 3-D and Blu-ray techniques in visual effects demand huge data rates. Similarly, several state-of-the-art gaming platforms also require high data rate. Moreover, the new emerging technologies such as augmented reality (AR), virtual reality (VR), and high definition holographic video conferencing also require high data rate. Therefore, to fulfill the demand of high data rate, there is a need to deploy ultra-wide THz band communication systems. In [52], the authors have shown the feasibility of downloading high data sized files using kiosks which were practically deployed in shopping malls, and airports. This enabled the users to get high data rates up to 100 Gbps from their handheld devices. In such an establishment, a cavity-like channel has been observed because of the high directional beam under line-of-sight (LOS) condition. It has been proposed that such kiosk-like application can also be accomplished for airplane entertainment systems. Further, instead of transmitting data over wired network, a THz band can also be used which supports in reducing the cost of a system in terms of cable cost, and weight of aircraft will also get reduced. Further, such implementation of THz-based communication also leads to efficiency in fuel consumption by the lightweight aircraft [53]. In high definition holographic video conferencing AR and VR THz communication can provide seamless connection between ultra-high-speed wired networks and wireless devices. In general, when eye and ear are not synchronized then it results in cyber-sickness while using AR and VR systems. Therefore, the THz networks are used to maintain synchronization below 10 ms so as to avoid cyber-sickness [54]. Moreover, the congestion problem in haptic communication systems can be minimized with the help of THz networks [55]. The haptic communication is an audiovisual communication which is supported by haptic modality to provide touching or manipulating the objects [56–59], which helps to enhance the sense of togetherness as well as study performance. However, it needs small delay in audio and video communications because of the use of transmission control protocol (TCP) in distributed haptic-based virtual environments. This delay further enhances due to congestion at the transport layer especially when transmission exceeds available throughput. In such situation, the deployment of THz networks in tactile internet can enable the ultra-low latency with extremely high data rates [60–65].

Satellite Communication Application

In the field of satellite communication, the THz bands is commercially in use such as Atacama Large Millimeter/submillimeter Array (ALMA) in Chile have kept ten bands in the range 0.031–0.95 THz in operation [66]. Moreover, in satellite communication only short distance propagation is experimented because of the limitation of spreading and molecular loses associated with THz waves. The propagation distance can be increased with the help of amalgamation of several approaches such as: (1) to use massive antenna array techniques, (2) to maintain high output power, and (3) to use high gain amplifiers. To counter the losses due to molecular absorption of THz wave by water vapors, the THz bands is more appropriate to be used in dry regions for satellite communication in order to achieve high-speed data links in the sky [67, 68].

In addition, there are some more application scenarios where THz nano-sensors are used as wearable devices. The application areas such as railway technology [69–72], energy harvesting using smart network sensors [73–75], and vehicular networks [76–78] are the upcoming possible adaptive energy-efficient computing areas using THz technology for expansion as well as making inter-device and intra-device communication feasible.

3.2.2 Sources

The THz sources generally categorized into three broad areas such as: (1) vacuum which includes backward-wave oscillators, klystrons, grating-vacuum devices, traveling-wave tubes, and gyrotrons, (2) solid-state devices which include harmonic frequency multipliers, transistors, and monolithic microwave integrated circuits, and (3) laser and photonics which include quantum cascade lasers, optically pumped molecular lasers and optoelectronic RF generators. The salient features of effective THz sources are that it should be powerful enough to overcome the extreme signal attenuation, it must be efficient enough to avoid having to wheel around its own power generator and it must be compact in size or small enough so that it can be deployed in the field without being carried around on a flatbed [79]. However, the low power values of THz sources are acceptable in some application areas wherein the purity of spectrum, tunability, and bandwidth is more important. Figure 3.2 shows the THz sources developed so far in research and commercial purposes and the average transmitting power of each type of source is presented [79]. In Fig. 3.2, the line Pf^2 = constant represents power-frequency slope and $P\lambda$ = constant line represents the expected slope for some commercial lasers for THz systems. Moreover, it is evident from the figure that compact THz sources exhibit low power as well as very small conversion efficiency in most of the cases even less than 1%. Further, in every type of THz source, as the frequency increases toward THz range, the source's output power plunges.

At lower and upper frequencies, the vacuum and lasers devices show the highest average power, respectively. A gyrotron which is a compact vacuum source is quite a stretch over average power values. However, the photonic sources provide high peak power which ranges from hundreds of watts to kilowatts but require high optical-drive power as input. The conventional vacuum traveling wave tube can operate at 1 THz which required an electron beam with a power density of multiple megawatts per square centimeter through an evacuated circuit. Likewise, a THz transistor of nanometer scale also operates at some similar high-power density levels. However, the compact electrical and optical devices need appropriate power dissipation means at high frequencies; therefore, an appropriate material is

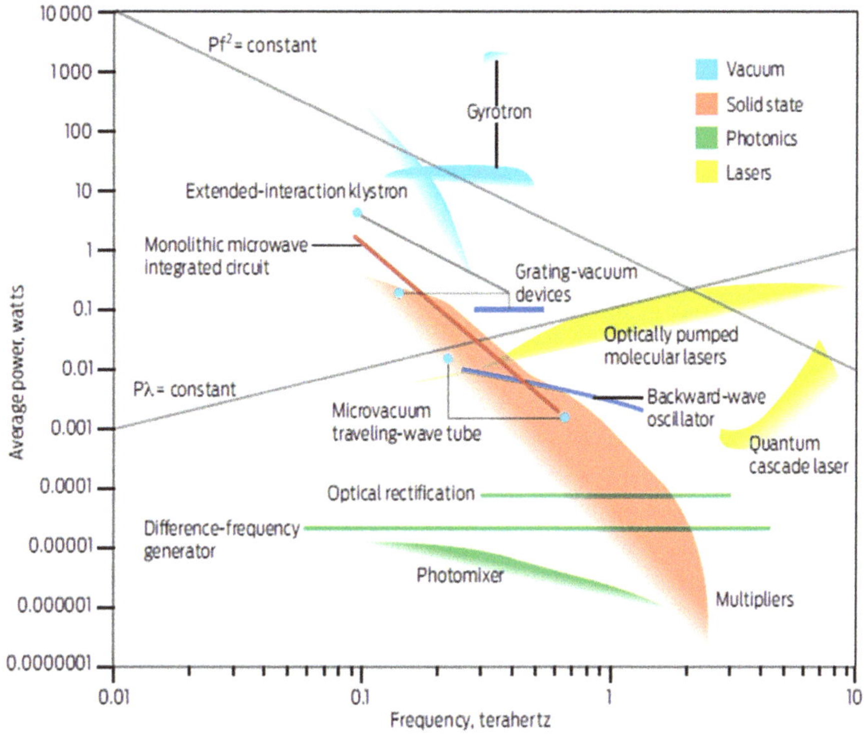

Fig. 3.2 Average power versus THz frequency of operation of source [79]

required to be used in its construction. The power generation at submillimeter waveband can be achieved using oscillatory circuits and such circuits have power extraction at the fundamental frequency which is kept below the maximum frequency; however, the harmonic frequency is above the maximum frequency [80]. A cross-coupled voltage-controlled oscillator has been designed at 91 GHz and reported in [81] which provide an output power of 4.5 dBm with a power consumption of 46 MW. The authors of [81] have used transmission lines instead of inductors in their design of voltage-controlled oscillator which provide an inductive load in the drain of the core transistors. However, for a higher THz frequency range such cross-coupled voltage-controlled oscillators are not so efficient since the degraded quality factors of varactors impact the loaded quality factor of the tank circuit which is responsible for excessive loss and power variation across the tuning range. In [82], a source follower with a resistor and inductor in its base has been proposed to implement a variable inductor so as to minimize the impact of the varactors. On such a way the value of the inductor gets changed by the biased current of the transistor, thereby an achieved output power 7.2 dBm while consuming 30 mW DC power. Moreover, for the design of transistor in THz region, it is desirable to maximize the output power of the oscillator when it has to operate close to the maximum frequency. The large signal transconductance, G_m of a linear two-port device can be optimized using admittance parameters of the transistor and is expressed as:

$$G_{\mathrm{m}} = \frac{P_{\mathrm{R}}}{|V_{\mathrm{in}}||V_{\mathrm{out}}|} = -\left(A^{-1}g_{11} + Ag_{22}\right) - \left|y_{12} + y_{21}^{*}\right| \times \cos\left(\angle\left(y_{12} + y_{21}^{*}\right) + \phi\right) \quad (3.1)$$

where $P_{\mathrm{R}} = \mathrm{Re}\,(P_{\mathrm{out}} - P_{\mathrm{in}})$, $A = V_2/V_1$, g_{11} and g_{22} are the real part of y_{11} and y_{22}, respectively. Using this design equation, the authors in [83] have designed a ring oscillator at 121 GHz with an output power of −3.5 dBm and power consumption of 21 mW. Another method to design high power oscillators is to use the technique of injection locking. In [84], it is proposed that if enough current is injected from one oscillator (I_{inj}) having oscillation frequency (x_1) to the tank circuit of another oscillator with oscillation frequency (x_0) with internal oscillatory current (I_{core}), the oscillation frequencies lie within the locking range such that the second oscillator follows the first oscillator with an equal frequency and a phase shift of:

$$\Delta\Phi = \arcsin\left(2Q\frac{I_{\mathrm{core}}}{I_{\mathrm{inj}}}\frac{\Delta\omega}{\omega}\right), \quad (3.2)$$

where Q is the quality factor of the resonator and $\Delta\omega$ is the frequency difference of the two oscillators. Therefore, from the relation (3.2), it is evident that the frequency of oscillation can change by varying the phase of coupling between the two oscillators.

Moreover, the traditional THz equipment has a major limitation of system cost and compactness which is required for widespread usage in THz applications. Therefore, the field of THz integrated circuits design using CMOS and SiGe HBT technologies becomes functional. Further, an interplay of advances in silicon process technologies, an effective design techniques and availability of microelectronic packaging opened up new opportunities to narrow the gap between the requirements and the reality of system cost and performance of THz components. The silicon technology having the salient features such as scalability, reconfigurability, and signal processing initiated the commencement of research in complex THz integrated circuits. This expands further the functionality of THz systems for new applications, methods, and algorithms. Consequently, in the field to obtain efficient THz sources several new design methodologies for the proper synthesis of passive embedding networks were also get enabled to optimize the harmonic generation [85–87]. This results in the formation of several novel scalable system architectures [88–91].

Since the first demonstration of a CMOS THz source [92], more than a decade ago having the ability to radiate −42 dBm at 410 GHz there is tremendous progress which has been made toward the provision of suitable power levels for imaging applications with THz ICs. The current state-of-the-art CMOS and SiGe HBT radiation sources are grouped into: (1) unlocked oscillator-based sources [93–101], and (2) oscillator or multiplier-chain-based sources that can be locked to an external phase-stable reference signal [102–112]. Further, the researchers continue to explore a large design space as offered by silicon technologies so as to overcome the barriers associated with the limited device speed along with THz wave escape from the chip level [113]. However, above 300 GHz silicon THz sources show nonlinear frequency translation process of strongly driven high-speed devices. Therefore, such circuits are directly implemented either as harmonic N-push oscillator or as a frequency multiplier circuit. Though both the approaches involve conversion loss which increases rapidly with the harmonic order and drive frequency. Specially, the research has moved toward coherent single-chip multielement THz radiators because of the small size of on-chip THz passives and large-scale availability of transistors. Moreover, the baseband processing capabilities of silicon technology are also considered for radiation pattern reconfiguration of multielement sources with the support of power-on control of asynchronous sources to diffuse scene illumination or phased array functionality [114]. Figure 3.3 shows a comparison of THz sources in CMOS and SiGe technologies.

Therefore, the THz integrated circuit designers have the challenges which are interdisciplinary and are related to devices, the electromagnetic design as well as packaging levels which cannot be overcome by simply transferring the classic millimeter-wave design techniques to the THz band. Further, in comparison to other techniques used to generate THz signal such as quantum cascade lasers [115, 116], Impatt diodes [117, 118], Gunn diodes [119, 120], and resonant tunneling diodes [121], the integrated electronic circuits appeared to be more efficient in generating THz frequency signal. The integrated electronic components as shown in Fig. 3.4 generally follow two important trends. The first group of integrated electronic systems are enabled close to 1 THz operation and support applications such as fully integrated spectroscopy and high-resolution imaging systems [122, 123]. The other

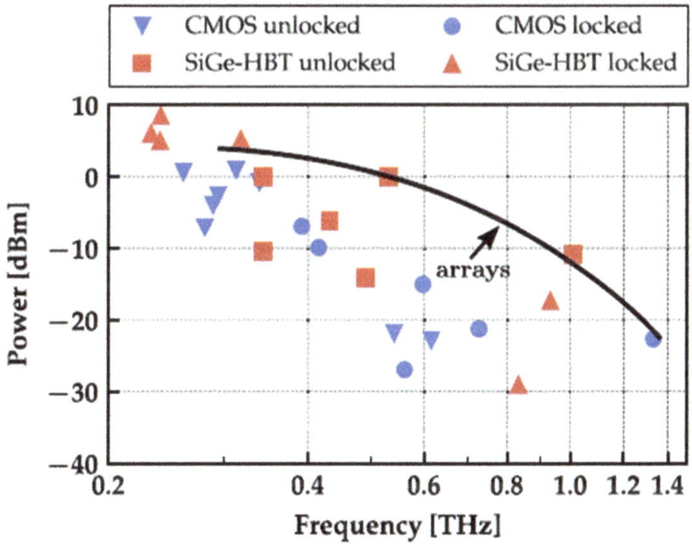

Fig. 3.3 Comparison of state-of-the-art THz sources in CMOS and SiGe technologies [114]

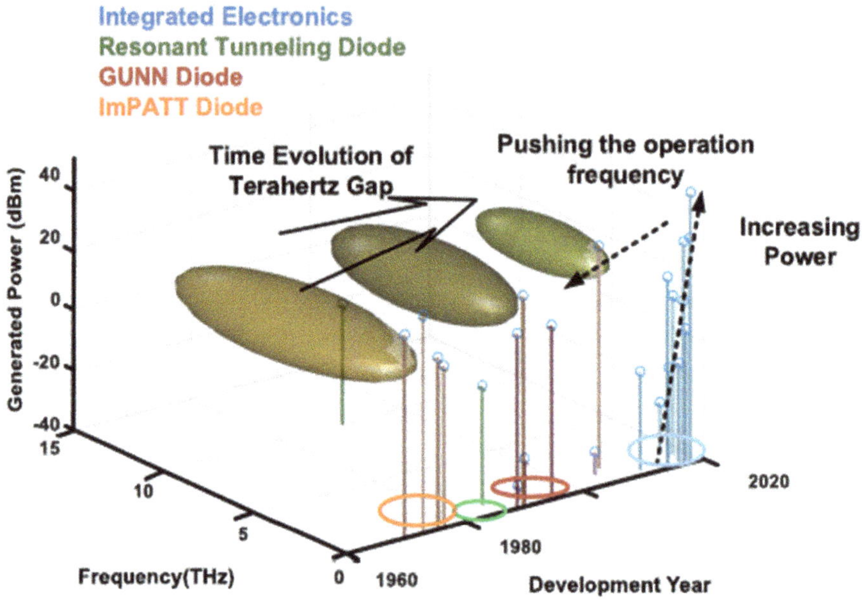

Fig. 3.4 Consecutive advances of millimeter-wave and terahertz power generation [80]

group of integrated electronic systems have explored the millimeter-wave frequency range for the applications such as next generation communication systems, imaging, and sensing radars using 5G, massive MIMO, and automotive radars [124, 125].

Moreover, on the level of fully integrated electronic systems, Si and SiGe platforms exhibit the highest yields and the most cost-efficient solutions when it comes to portable low-power systems.

3.2.3 *Receiver*

The working of THz receivers is broadly classified as an incoherent detection or coherent detection. The incoherent systems make use of direct detection sensors which collect the information about the signal amplitude and are classified as broadband detection devices. However, the coherent systems collect information about amplitude as well as phase of the signal using heterodyne circuit design and are classified as narrow-band detection devices [126]. The direct detection system is simple in implementation and provides the possibility to design large arrays [127, 128]. Similarly, the heterodyne method has an advantage that it converts the signal frequency to a lower intermediate frequency wherein the electronic amplification can be performed easily. Moreover, it provides higher spectral resolution. When the background radiation is weak then direct detection operating in a wider spectral range gives a decent resolution. Further, the direct detection is preferable for imaging application specifically for cases where sensitivity is more important than spectral resolution. However, in imaging application where the scene is illuminated, the heterodyne detection is used to increase the sensitivity.

Similarly, on the basis of mechanism of detection, the THz detectors are also classified as photon detectors or thermal detectors. When the radiated signal strikes the detector material then the surface temperature changes which further results in variation in some physical property to generate electrical output. However, the output signal does not depend upon the photonic nature of the incident wave therefore the thermal effects are frequency independent. The three popular thermal detection methods used in THz detection are bolometer, pyroelectric detector, and Golay cell. In case of bolometer thermal detection, the device is designed with large temperature coefficient in order to get induced with large change in resistance when the THz radiation incident on its surface. It occurs because the radiant power produces heat within the material which results in resistance variation. In the pyroelectric detector, the variation in temperature is responsible to modify the positions of atoms slightly within the structure and it results in the change in polarization of the material and such polarization variation is responsible to generate voltage across the detector. The detection of THz radiation using Golay cells takes place with the help of sealed container which is filled with gas of low thermal conductivity such as xenon gas. The filled gas gets expanded on heating and distorts a flexible membrane on which a mirror is mounted. The mirror's movement deflects a beam of light shining on a photo-resistor and results in output due to the change of photoresistor current. On the other hand, in case of photon detector, the incident radiation is absorbed within the material and the photons interact with electrons which cause the electronic excitation. The electrical output signals result from the changed electronic energy distribution. One of the major advantages of photon detection method is its shorter response time along with much higher sensitivity. However, the limitation is the requirement of cooling in order to reduce thermal noise. The photon detection can be performed using: (1) photoconductive detectors which utilize the variation in electrical conductivity due to change in the number of free carriers generated while photons are absorbed, and (2) photovoltaic detection in which the photon absorption is used to form a voltage difference across a p–n junction to generate photovoltaic current.

The silicon technology carries a significant feature such as high fabrication yield, the availability of on-chip baseband processing circuits. These features enable the integration of detectors into chip-scale implementation. However, due to the lack of low noise preamplification in the THz band, silicon power detectors are designed as antenna-coupled direct detectors. For the THz direct detection, there are two methods which are implemented using silicon technologies which are as follows: (1) non quasistatic self-mixing in cold MOSFET channels and (2) rectification in the base–emitter junction of high-speed HBT [129–133]. The idea of using MOSFET as THz detector was initiated by Dyakonov and Shur in 1993 [134]. Using an appropriate biasing condition, the electron fluid in the channel can be made to act as a resonator for THz wave as plasmon waves [135, 136] and such plasmon waves result in rapid oscillations of the electron density in conducting

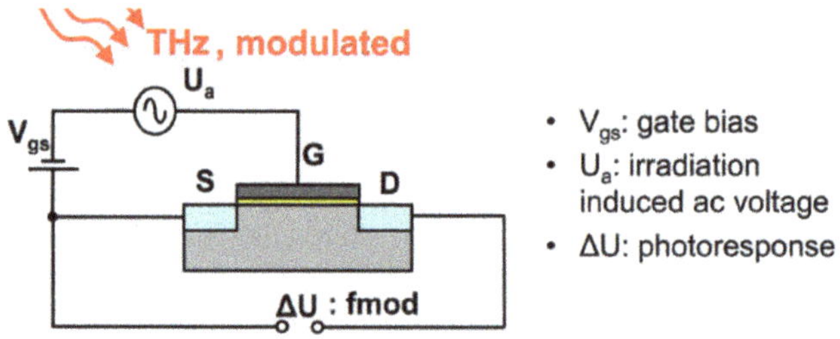

Fig. 3.5 MOSFET THz detection with biased voltage [138]

media such as plasma or metal. Under the proper biasing conditions, the incident THz radiation induces an ac signal between the gate and source terminals. The MOSFET makes the detection because of the nonlinear properties of the transistor which leads to the rectification of the ac current. Moreover, a photoresponse appears as dc voltage between source and drain. The resultant voltage is proportionally related to radiation intensity thereby ensures the validity of THz detection [136]. Further, the THz detection using MOSFET has two modes of operation: (1) resonant in which it is tuned to a specific wavelength, and (2) nonresonant which is the broadband mode of operation. Both are directly tunable by varying the gate voltage [137–142]. Figure 3.5 shows MOSFET THz detection under biased condition. In this, when the channel length is short then a standing plasmon wave gets formed which further provides signal amplification. Under such situation, the device exhibits high responsivity due to internal amplification and so the response is narrowband.

However, when the channel length is long enough then the plasmon wave decays before it reaches the other side of the channel. Even at room temperature, the electron mobility is too low to support the resonant response to happen. A biased voltage between gate and source along with an optimal drain bias current enhances the response. The THz detection using MOSFET counts several advantages such as manufacturing of low cost and high yield THz CMOS devices, low noise equivalent power (NEP) for room temperature operation, compact detector array can be made for frame imaging, sensitivity can be increased with the help of an antenna which help to couple THz radiation to the small absorption element of detector. The antenna is generally connected to the gate and source terminal of the MOSFET. In [143], the authors have explored antenna systems which are based on backside radiating on-chip primary antennas and used external hyper hemispherical silicon lenses to provide 12–14 $\mathrm{pW}/\sqrt{\mathrm{Hz}}$ level NEP across several hundreds of GHz and directivity of about 20 dBi. Contrary to THz detection using MOSFET, the authors in [144, 145] showed the implementation of THz detection using HBTs and reported NEP of 15 $\mathrm{pW}/\sqrt{\mathrm{Hz}}$ and 50 $\mathrm{pW}/\sqrt{\mathrm{Hz}}$ for 130 and 250 nm SiGe HBT technology nodes, respectively.

The commercially available THz systems such as Menlo systems deploy both THz emitters and receivers which uses photoconductive switches based on low temperature grown InGaAs layers to generate and receive THz radiations [146]. With technological advancement in terms of CMOS fabrication and advanced development of HBTs, ultracompact design of THz transmitter and receiver integrates silicon lens, photoconductive switch, bias supply, and low noise current amplifiers into one single compact unit.

3.2.4 Antenna and Its Array Technology

In THz bands, there are mainly three types of antenna configurations which are in use, that is, Graphene, Horn, and planar structured. These are the broadband antennas which are needed to have the advantage of ultra-wideband transmission at the THz band [147]. Graphene-based large antenna arrays mitigate the transmission distance problem which occurs because of high path loss. This has been achieved with small size implementation of graphene-based antenna array with high directivity [148, 149]. Moreover, the realization of nanoscale-sized plasmonic graphene antenna arrays for nano-devices enables the propagation of THz waves for nano-devices [150–155]. In graphene, the speed of surface plasmon polariton (SPP) waves is almost twice as fast as the velocity in the vacuum [156]. Further, the THz frequency can be tuned using material doping as the conductivity of graphene depends on chemical doping, fermi energy as well as electron mobility. In [157], the authors have investigated the performance of graphene-based plasmonic nano antenna arrays in the THz band by considering the effect of reciprocal coupling. The authors have proposed that the near-field coupling can be ignored if the distance between THz antennas is kept smaller than free-space wavelength. Moreover, in [158], the authors have proposed a mathematical framework to analyze the performance of plasmonic nano antennas. Graphene also allows to design reconfigurable directional antenna arrays and provides the features of beamforming as well as beam-scanning for security and defense applications based on MIMO systems [159, 160]. Using graphene-based patch array, a reconfigurable Yagi-Uda MIMO antenna design is presented in [161].

Another THz band antenna which has the capability of providing high directivity is horn antenna [162, 163]. However, the dimension of horn antenna does not allow the utilization for on-chip design. In this regard, planar antennas are the promising option in terms of ease in fabrication and their integration to other planar antennas for array implementation in THz band [164, 165]. Table 3.1 shows the summarized presentation of salient features and limitations in the design and implementation of THz antennas in an array configuration.

The THz antennas are used as scanning antennas in several scanners based on mechanically scanned optical systems such as spherical scanning antenna, the Lewis scanner, the Schwarzchild scanner, and the Rotman lens scanner [166–170]. Most photonic imaging systems utilize a single element of THz antenna with mechanical scanning capability resulting in long acquisition times. Moreover, the system based on THz scanners is not suitable for low-cost real-time imaging systems. Since many commercial imaging systems employ some kind of scanners,

Table 3.1 Features and limitation of THz band antennas

Type of THz antenna	Salient features	Limitation
Graphene	Compact size implementation High directivity Propagation of SPP waves	Technology is in its developing stage but promising Therefore, more studies are required
Horn	High gain Wide bandwidth	Large size Noncompatible with nano-devices
Planar	Ease in fabrication Ease in integration	Low gain and directivity

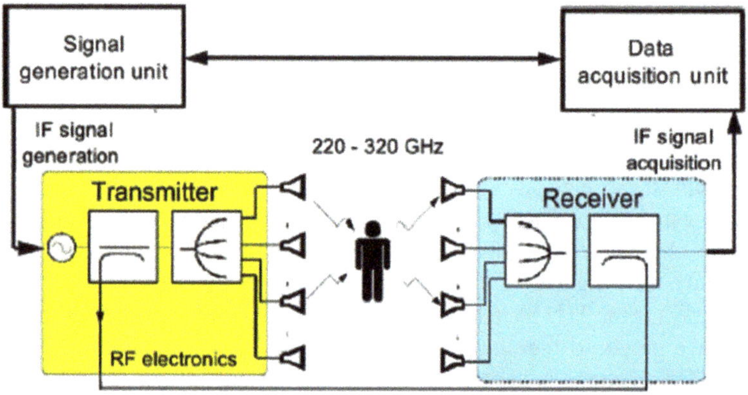

Fig. 3.6 Block diagram of multistatic/multichannel electronic imaging system [172]

however if a number of transmitting and receiving channels can be afforded in an imaging system then optical defocusing and aberrations can be easily kept to a minimum [171].

Realization of multielement electronic imaging system that combines horizontal, linear array of elements with vertically scanning optics comes out to be a promising approach for realizing imaging of moving objects. Figure 3.6 shows an electronic imaging system with transmitting and receiving channels in the multistatic configuration. Such setup is capable of scanning 1–2-m scene at a distance of 7 m as reported in [172]. Using an image refresh rate of 500 ms and 100 lines in the vertical direction leaves 5 ms/line acquisition time. An array of transmitting and receiving channels operates at 220–320 GHz with 100 GHz bandwidth. It also includes data acquisition and signal generation unit.

However, one bottleneck in the development of multichannel THz imaging systems is the data transfer of the acquired data. Also, the real-time computation of such large amounts of data is difficult to achieve using commercial off-the-shelf components. Using superheterodyne architectures alleviates such issue at the expense of large complexity.

3.3 Terahertz Antennas for Imaging Applications

Electronic imaging system is classified as active system or passive system. Both systems have their salient features as well as limitations in their operation of working in imaging application. Active systems operate in several modes such as monostatic, bistatic, or multistatic. Active imaging system provides information on the reflection and transmission properties of the object which are mostly exploited in image processing algorithms. However, passive systems mostly operate in observation mode with a broadband incoherent source and such powerful broadband noise sources are also difficult to realize.

THz antenna technology in passive imaging techniques: Focal plane arrays for direct detection or a heterodyne receiver with scanning system as well as single THz antenna as receiver are mostly employed in passive imaging systems [173–179]. Passive multichip imaging system [180] and direct-detection receivers using monolithic microwave integrated circuit (MMIC) technology with good noise performance [181] have been developed by several researchers. However, for an adequate spatial resolution at large stand-off distances, the antenna aperture needs to be large. Likewise, using the standard CMOS process the focal plane arrays at submillimeter frequencies operated at room temperature is presented in [182, 183]. Moreover, MMIC technology seems to be more promising in passive imaging techniques from cost and performance point of view thereby providing a natural path toward multichannel imaging system.

THz antenna technology in active imaging techniques: In active imaging system a transmitter illuminates the object and a receiver collects the scattered energy. This is generally performed using multiple transmitters and receivers in the imaging system [184–186]. Either focal plane or interferometric arrays having aperture filling techniques are used as transmitters and receiver channels of bolo-metric detection (direct detection) capabilities or heterodyne receiver architecture are in use in such imaging systems. However, heterodyne receiver architecture is more preferable in active imaging system as it provides improved sensitivity in comparison to direct detection method and also provides high resolution of micrometer range with large bandwidth systems and utilizes the phase information. In [187], a fully focused diffraction-limited 3-D image of a person or imaged target is presented wherein an inward-directed vertical array is used to scan around the person or imaged target.

Photonic THz imaging systems use time-of-fight measurements by doing the raster scanning of an imaging object. Most of the systems employ fiber-coupled photoconductive emitters and receivers which acquired data from various positions and subsequently reconstruct an image [188–190]. However, focusing the raw THz data of an imaging system for the applications of security and space is a tough challenge as it requires (1) real-time processing, (2) 3-D mapping, (3) comparable object and antenna aperture size with stand-off distance, (4) relatively large bandwidth, and (5) use of bi-static configuration [191].

Resolution: Theoretically, the range resolution ρ depends on the bandwidth of the transmitted signal B, as $\rho = c/2B$. In a cross-range direction, a passive diffraction-limited system offers an angular resolution of $\alpha = \lambda/D$, where λ is the wavelength of the incident radiation and D is the diameter of the aperture. Moreover, the spatial cross-range resolution is found by multiplying with the range to the object. The angular field of view (FOV) is generally determined by the smaller of two angles, that is, the antenna element beam width and the angular ambiguity spacing which is also known as grating lobe spacing. Therefore, if the beam width exceeds the ambiguity spacing, the imaged object should be sufficiently small to fit in between two ambiguities. The angular ambiguity pacing "θ" depends on the effective element spacing "d," as $\theta = \lambda/d = \lambda N/D$, where N is the effective number of elements. It means large objects need to be scanned by closely spaced elements. The above relation is applicable to a multistatic system for which the ambiguity spacing is twice the monostatic radar system with its effective elements spacing half of the physical element spacing. Therefore, the maximum number of cross-range pixels that can be resolved N_{cr} is the ratio of the ambiguity spacing to the angular resolution $N_{cr} = \theta/\alpha$ which equals to the effective number of array elements.

Aperture filling: In antenna array, to avoid the grating lobes so as to obtain the excellent resolution along with maximum unambiguous scene size with given number of antenna elements, the $N_{TX}N_{RX}$ midpoints must be equidistantly spaced. It can be achieved if the spacing between transmitting antenna elements is $\Delta_{TX} = N_{RX}\Delta_{RX}$, where Δ_{RX} is the receiving antenna element spacing. The overall size of the antenna array, $N_{TX}\Delta_{TX} = N_{TX}N_{RX}\Delta_{RX}$, which offers an angular resolution of $\alpha = \lambda/(N_{TX}N_{RX}\Delta_{RX})$ makes the angular ambiguity separation become $\theta = \lambda/\Delta_{RX} = (N_{TX}N_{RX}\alpha)$. Therefore, a multistatic radar with $(N_{TX} + N_{RX})$ antenna elements can resolve $(N_{TX}N_{RX})$ uncorrelated pixels which is much more than the pixels that can be resolved with a monostatic radar with the same number of elements [192]. Such principle is suitable for imaging of objects that consists of few reflectors on a nonreflecting background.

Range focusing: In order to obtain good SNR along with good resolution, that is, large bandwidth, a large time bandwidth product of the transmitted signal is required. The focused range response can be achieved with the help of fast Fourier transform. Moreover, for a system having multiple transmitters, it is desirable to transmit simultaneously on all transmitters in an array, thereby separating the transmitted signal in frequency. However, the frequency separation Δf_{TX} needs to be small enough compared with the pulse bandwidth B_{IF} such that $B_{IF} \geq N_{TX}\,\Delta f_{TX}$. The ambiguous range is thereafter determined by the transmitter frequency separation as $r_{amb} = c\Delta f_{TX}\tau_p/2B_P$, where τ_p is the uncompressed pulse length and B_P is the modulation bandwidth.

3.4 Terahertz Antennas for Sensing Applications

THz IC design in CMOS and SiGe HBT technologies have surged a progress in THz sensing applications. The advancement in silicon process technology, design techniques, and microelectronic packaging makes reality of system cost and performance of THz components in sensing applications. Earlier the foremost challenge to enable real-world application in THz sensing and spectroscopy was the requirement of efficient and widely reconfigurable chip-scale systems. The THz range is rich in spectroscopic features such as relatively narrow absorption lines of rotational or vibrational excitations in gases and broader lines which result from lattice vibrations, that is, phonons in crystalline solids especially in molecular crystals [193]. To enhance the sensitivity and specificity of gas sensing, the authors in [194] have developed a technique considering the absorption strength of many polar molecules' peaks in Hz range. However, among other parameters, the ability of a sensor to provide highly specific as well as sensitive detection depends on the tuning range of the system. Further, the main challenge for THz spectroscopic sensing is that the spectral lines of interest are generally broader or more widely spaced in comparison to the tuning range of solid-state sources. In [195], the authors have reported a fully integrated spectroscope in silicon with minimum detectable concentration of 11 ppm for carbonyl sulfide. Moreover, to circumvent the need of a tunable THz source, the research is in progress so as to enable the estimation of extremely wide spectrum using spatial sampling of near fields on the capture antenna. In this direction, the authors in [196] have applied the estimation techniques and the regression analysis for source free spectral sensing. Combining the millimeter-wave frequencies, a 3D imaging with both high lateral and depth resolution can be realized at hyperspectral THz range. Therefore, sensor fusion on combining millimeter waves, infrared and optical frequencies is one of the promising approaches developing to have a critical sensing tool for autonomous vehicles and systems. On similar grounds, the advancement can be made at THz using multispectral sources. In [197], the authors have presented chip-scale sources having dynamically programmable spectral content but to create widely reconfigurable sources having the capability of beamforming for fast image acquisition is still a big challenge. Moreover, to enable the features such as quality control and security-based imaging, there can be the use of machine learning and deep neural networks initiated for real-time feature extraction. For real-time applications in THz sensing, the challenges in the form of widely reconfigurable, efficient, scalable, and low-cost technology remain open to handle as new areas of research. Therefore, considering the need of standardizations of the physical and medium access control (MAC) layer for THz range so as to create THz wireless communication in data centers, intrachip/intraboard radio communication, mobile front and backhaul links, a standard by IEEE has been formed. The IEEE 802.15.3d-2017 became the first standard that has addressed frequency bands beyond 0.3 THz range thereby focusing on large span across 252.72–321.84 GHz [198].

Using nearly 256 element-phased array in silicon a scanning beam width of 2–5° is presented in [199]. Similarly, the large-scale THz array is proposed in [200] for satellite communication application as extremely high gain links can be realized at THz frequency range. Because of their scanning capabilities and ability to have automatic alignment, THz antenna arrays are advantageous to be used over free-space optical links. This has come into existence due to integrated photonics platforms with active and passive devices which provide ancillary benefits to the THz sensing systems. Moreover, in order to enable the heterogeneous future, the research and development pathways considering both electronics and photonics will help when converged.

3.5 Summary

One of the potential challenges of approaching THz frequencies is the failure of the lumped parameter representation of the external environment and circuit. The advantage of microscopic modeling of high-frequency technologies such as HEMTs and HBTs is that a complete description of the electron dynamics and device behavior is provided directly from the component materials and device structure which provides a virtual assessment of new material and device technologies prior to actual fabrication. Therefore, once important system issues have been resolved with optimized signal processing, the terahertz pulse imaging-based femtosecond lasers offer wide-ranging analysis in the terahertz spectrum. Moreover, the silicon-based THz ICs show the potential to leverage various THz imaging and sensing applications. Recently, the THz IC design has emerged as a vibrant field of research, achieving breakthroughs in the practical utility, potential system cost, and integration level of THz components—the frontiers that constitute the bottleneck for adoption and commercialization of THz technology. The power generation capability of THz ICs is tightly coupled to the progress in silicon technology. Present foundry-level SiGe HBT technology is just on the verge of enabling fundamental circuit operation in the lower part of THz band, and it continues to show a great development potential regarding device speed. Further, the advances in THz IC technology are driven by the invention of novel circuit and system architectures which exploit the massive scalability and design space of silicon technology.

Therefore, the development of semiconductor technologies opened up the door for integrated electronic circuits to operate at higher frequencies. The efficient combination of integrated electronic circuits and the photonic/optical transceivers has shown promising changes. On the level of fully integrated electronic systems, the Si and SiGe platforms exhibit the highest yields and the most cost-efficient solutions when it comes to portable low-power systems. However, the development of THz integrated systems will be greatly impacted by the applications that develop in the next decade. The capabilities of THz imaging and sensing are promising for nondestructive quality control, 3D imaging, radar, and gesture recognition in autonomous systems, vehicles, robotics, and industrial automation, particularly in the lower part of the spectrum (~0.1–0.3 THz).

References

1. J.F. Federici, D. Gary, R. Barat, Z.H. Michalopoulu, T-rays vs terrorists. IEEE Spectr. **44**(7), 47–52 (2007)
2. I. Malhotra, K.R. Jha, G. Singh, Terahertz antenna technology for imaging applications- a technical review. International Journal of Microwave and Wireless Technologies **10**(3), 271–290 (2018)
3. J. Faist, F. Capasso, D.L. Sivco, S. Sirtori, A.L. Hutchinson, A.Y. Cho, Quantum cascade laser. Science **264**(5158), 553–556 (1994)
4. R. Köhler, A. Tredicucci, F. Beltram, H.E. Beere, E.H. Linfield, A.G. Davies, D.A. Ritchie, R.C. Iotti, F. Rossi, Terahertz semiconductor-heterostructure laser. Nature **417**, 156–159 (2002)
5. M. Rochat, L. Ajili, H. Willenberg, J. Faist, H. Beere, G. Davies, E. Linfield, D. Ritchie, Low-threshold terahertz quantum cascade lasers. Appl. Phys. Lett. **8**(18), 1381–1383 (2002)
6. B.S. Williams, H. Callebaut, S. Kumar, Q. Hu, J. Reno, 3.4-THz quantum cascade laser based on longitudinal-optical-phonon scattering for depopulation. Appl. Phys. Lett. **82**(7), 1014–1017 (2003)
7. W.P. Dumke, J.M. Woodall, V.L. Rideout, GaAs-GaAlAs heterojunction transistor for high-frequency operation. Solid State Electron. **15**(12), 1339–1343 (1972)
8. T. Mimura, S. Hiyamizu, T. Fujii, K. Nanbu, A new field-effect transistor with selectively doped GaAs/n-AlxGa1-xAs heterojunctions. Jpn. J. Appl. Phys. **19**(5), L225–L227 (1980)
9. H. Daembkes, *Modulation-Doped Field Effect Transistors: Principles, Design and Technology* (IEEE Press, New York, 1991)
10. F. Schwierz, J.J. Liou, *Modern Microwave Transistors: Theory, Design and Performance* (Wiley, Hoboken, 2003)
11. R.L. Ross, S.P. Svensson, P. Lugli, *Pseudomorphic HEMT Technology and Applications* (Academic, New York, 1996)
12. D. Dawson, L. Samoska, A.K. Fung, K. Lee, R. Lai, R. Grundbacher, P.-H. Liu, R. Raja, Beyond G-band: A 235 GHz InP MMIC amplifier. IEEE Microw. Wirel. Compon. Lett. **15**(12), 874–876 (2005)
13. J.M. Jornet, I.F. Akyildiz, Graphene-based nano-antennas for electromagnetic nano communications in the terahertz band, in *Proceedings of the IEEE 4th European Conference on Antennas and Propagation (EuCAP)*, 2010, pp. 1–5
14. E. Zarepour, M. Hassan, C.T. Chou, M.E. Warkiani, Characterizing terahertz channels for monitoring human lungs with wireless nanosensor networks. Nano Commun. Netw. **9**, 43–57 (2016)
15. E. Zarepour, N. Hassan, M. Hassan, C.T. Chou, M.E. Warkiani, Design and analysis of a wireless nanosensor network for monitoring human lung cells, in *Proceedings of the 10th EAI International Conference on Body Area Networks, ICST (Institute for Computer Sciences, Social-Informatics and Telecommunications Engineering)*, 2015, pp. 139–145
16. I.F. Akyildiz, J.M. Jornet, Electromagnetic wireless nanosensor networks. Nano Commun. Netw. **1**(1), 3–19 (2010)
17. G. Iddan, G. Meron, A. Glukhovsky, P. Swain, Wireless capsule endoscopy. Nature **405**(6785), 417 (2000)
18. Z. Liao, R. Gao, C. Xu, Z.-S. Li, Indications and detection, completion, and retention rates of small-bowel capsule endoscopy: A systematic review. Gastrointest. Endosc. **71**(2), 280–286 (2010)
19. M. Mimee, P. Nadeau, A. Hayward, S. Carim, S. Flanagan, L. Jerger, J. Collins, S. McDonnell, R. Swartwout, R.J. Citorik, V. Bulovic, An ingestible bacterial-electronic system to monitor gastrointestinal health. Science **360**(6391), 915–918 (2018)
20. M. Turan, Y. Almalioglu, E.P. Ornek, H. Araujo, M.F. Yanik, M. Sitti, Magnetic-visual sensor fusion-based dense 3D reconstruction and localization for endoscopic capsule robots, in

Proceedings of the IEEE International Conference on Intelligent Robots and Systems (IROS), 2018, pp. 1283–1289
21. H. Lay, G. Cummins, B.F. Cox, Y. Qiu, M.V. Turcanu, R. McPhillips, C. Connor, R. Gregson, E. Clutton, M.P. Desmulliez, S. Cochran, In-vivo evaluation of microultrasound and thermometric capsule endoscopes. IEEE Trans. Biomed. Eng. **66**(3), 632–639 (2018)
22. J. Wang, Electrochemical biosensors: Towards point-of-care cancer diagnostics. Biosens. Bioelectron. **21**(10), 1887–1892 (2006)
23. J. Dubach, D.I. Harjes, H.A. Clark, Fluorescent ion-selective nanosensors for intracellular analysis with improved lifetime and size. Nano Lett. **7**(6), 1827–1831 (2007)
24. I.E. Tothill, Biosensors for cancer markers diagnosis. Semin. Cell Dev. Biol. **20**(1), 55–62 (2009)
25. G. Piro, K. Yang, G. Boggia, N. Chopra, L.A. Grieco, A. Alomainy, Terahertz communications in human tissues at the nanoscale for healthcare applications. IEEE Trans. Nanotechnol. **14**(3), 404–406 (2015)
26. K. Gao, Z. Sun, B.G. Pan, X. Qiao, C. Hong, Electrochemical immunosensor for alpha-fetoprotein based on prussian blue-carbon nanotube@polydopamine. Micro Nano Lett. **13**(1), 58–62 (2018)
27. M. Freitas, H.P. Nouws, C. Delerue-Matos, Electrochemical biosensing in cancer diagnostics and follow-up. Electroanalysis **30**(8), 1584–1603 (2018)
28. S. Campuzano, P. Yáñez-Sedeño, J.M. Pingarrón, Electrochemical nucleic acid sensors based on nanomaterials for medical diagnostics, in *Nanotechnology Biosensors*, (Elsevier, Amsterdam, 2018), pp. 319–351
29. T. Garcia-Mendiola, I. Bravo, J.M. López-Moreno, F. Pariente, R. Wannemacher, K. Weber, J. Popp, E. Lorenzo, Carbon nanodots based biosensors for gene mutation detection. *Sensors Actuators B* **256**, 226–233 (2018)
30. M. Arruebo, R. Fernández-Pacheco, M.R. Ibarra, J. Santamaría, Magnetic nanoparticles for drug delivery. Nano Today **2**(3), 22–32 (2007)
31. S. Ganta, H. Devalapally, A. Shahiwala, M. Amiji, A review of stimuli-responsive nanocarriers for drug and gene delivery. J. Control. Release **126**(3), 187–204 (2008)
32. B.P. Timko, T. Dvir, D.S. Kohane, Remotely triggerable drug delivery systems. Adv. Mater. **22**(44), 4925–4943 (2010)
33. O. Veiseh, J.W. Gunn, M. Zhang, Design and fabrication of magnetic nanoparticles for targeted drug delivery and imaging. Adv. Drug Deliv. Rev. **62**(3), 284–304 (2010)
34. C.S. Kumar, F. Mohammad, Magnetic nanomaterials for hyperthermia-based therapy and controlled drug delivery. Adv. Drug Deliv. Rev. **63**(9), 789–808 (2011)
35. S. Yim, M. Sitti, Shape-programmable soft capsule robots for semi-implantable drug delivery. IEEE Trans. Robot. **28**(5), 1198–1202 (2012)
36. F. Munoz, G. Alici, W. Li, A review of drug delivery systems for capsule endoscopy. Adv. Drug Deliv. Rev. **71**, 77–85 (2014)
37. F. Xiong, S. Huang, N. Gu, Magnetic nanoparticles: Recent developments in drug delivery system. Drug Dev. Ind. Pharm. **44**(5), 697–706 (2018)
38. L. Vannozzi, V. Iacovacci, A. Menciassi, L. Ricotti, Nanocomposite thin films for triggerable drug delivery. Expert Opin. Drug Deliv. **15**(5), 509–522 (2018)
39. O.S. Fenton, K.N. Olafson, P.S. Pillai, M.J. Mitchell, R. Langer, Advances in biomaterials for drug delivery. Adv. Mater. **30**(29), 1705328/1–29 (2018)
40. A. Schroeder, D.A. Heller, M.M. Winslow, J.E. Dahlman, G.W. Pratt, R. Langer, T. Jacks, D.G. Anderson, Treating metastatic cancer with nanotechnology. Nat. Rev. Cancer **12**(1), 39–50 (2012)
41. Q.H. Abbasi, K. Yang, N. Chopra, J.M. Jornet, N.A. Abuali, K.A. Qaraqe, A. Alomainy, Nano-communication for biomedical applications: A review on the state-of-the-art from physical layers to novel networking concepts. IEEE Access **4**(1), 3920–3935 (2016)
42. H. Elayan, R.M. Shubair, J.M. Jornet, R. Mittra, Multi-layer intrabody terahertz wave propagation model for nano biosensing applications. Nano Commun. Netw. **14**, 9–15 (2017)

43. P. Johari, J.M. Jornet, Nanoscale optical wireless channel model for intra-body communications: Geometrical, time, and frequency domain analyses. IEEE Trans. Commun. **66**(4), 1579–1593 (2017)
44. H. Elayan, R.M. Shubair, J.M. Jornet, Characterising THz propagation and intrabody thermal absorption in WNSNs. IET Microw. Antennas Propag. **12**(4), 525–532 (2018)
45. H. Elayan, R.M. Shubair, J.M. Jornet, Bio-electromagnetic THz propagation modeling for in-vivo wireless nanosensor networks, in *Proceedings of the IEEE 11th European Conference on Antennas and Propagation (EuCAP)*, 2017, pp. 426–430
46. R. Zhang, K. Yang, Q.H. Abbasi, K.A. Qaraqe, A. Alomainy, Analytical modelling of the effect of noise on the terahertz in-vivo communication channel for body-centric nano-networks. Nano Commun. Netw. **15**, 59–68 (2018)
47. I.F. Akyildiz, F. Brunetti, C. Blázquez, Nanonetworks: A new communication paradigm. Comput. Netw. **52**(12), 2260–2279 (2008)
48. H. Dai, Nanotube growth and characterization, in *Carbon Nanotubes: Synthesis, Structure, Properties, and Applications*, Topics in Applied Physics, ed. by M. S. Dresselhaus, G. Dresselhaus, P. Avouris, vol. 80, (Springer, Berlin, Heidelberg, 2001), pp. 29–54
49. J. Altmann, *Military Nanotechnology: Potential Applications and Preventive Arms Control* (Routledge, London/New York, 2007)
50. I.F. Akyildiz, J.M. Jornet, The internet of nano-things. IEEE Wirel. Commun. **17**(6), 58–63 (2010)
51. J.M. Jornet, I.F. Akyildiz, Channel modeling and capacity analysis for electromagnetic wireless nanonetworks in the terahertz band. IEEE Trans. Wirel. Commun. **10**(10), 3211–3221 (2011)
52. D. He, K. Guan, A. Fricke, B. Ai, R. He, Z. Zhong, A. Kasamatsu, I. Hosako, T. Kürner, Stochastic channel modeling for kiosk applications in the terahertz band. IEEE Trans. Terahertz Sci. Technol. **7**(5), 502–513 (2017)
53. H.-J. Song, H. Hamada, M. Yaita, Prototype of KIOSK data downloading system at 300 GHz: Design, technical feasibility, and results. IEEE Commun. Mag. **56**(6), 130–136 (2018)
54. K. Tekbıyık, A.R. Ekti, A. Görçin, G.K. Kurt, On the advances of terahertz communication for 5G and beyond wireless networks. ed. Hüseyin Arslan, Ertuğrul Başar, Flexible and Cognitive Radio Access Technologies for 5g and Beyond," IET Publication, UK (2020)
55. TERRANOVA Consortium. Wireless terahertz system architectures for networks beyond 5G (2018), https://ict-terranova.eu/wp-content/uploads/2018/07/White-Paper_v15.pdf. Accessed: 2 Dec 2018
56. T. Miyashita, T. Tajika, H. Ishiguro, K. Kogure, N. Hagita, Haptic communication between humans and robots, in *Robotics Research*, ed. by S. Thrun, (Springer, Berlin, 2007), pp. 525–536
57. B.D. Argall, A.G. Billard, A survey of tactile human–robot interactions. Robot. Auton. Syst. **58**(10), 1159–1176 (2010)
58. E. Steinbach, S. Hirche, J. Kammerl, I. Vittorias, R. Chaudhari, Haptic data compression and communication. IEEE Signal Process. Mag. **28**(1), 87–96 (2011)
59. E. Steinbach, S. Hirche, M. Ernst, F. Brandi, R. Chaudhari, J. Kammerl, I. Vittorias, Haptic communications. Proc. IEEE **100**(4), 937–956 (2012)
60. G.P. Fettweis, The tactile internet: Applications and challenges. IEEE Veh. Technol. Mag. **9**(1), 64–70 (2014)
61. M. Simsek, A. Aijaz, M. Dohler, J. Sachs, G. Fettweis, 5G-enabled tactile internet. IEEE J. Sel. Areas Commun. **34**(3), 460–473 (2016)
62. S. Pandit and G. Singh, Spectrum sharing in cognitive radio networks. Berlin, Germani: Springer (2017)
63. A. Aijaz, M. Dohler, A.H. Aghvami, V. Friderikos, M. Frodigh, Realizing the tactile internet: Haptic communications over next generation 5G cellular networks. IEEE Wirel. Commun. **24**(2), 82–89 (2017)

64. K. Antonakoglou, X. Xu, E. Steinbach, T. Mahmoodi, M. Dohler, Towards haptic communications over the 5G tactile internet. IEEE Commun. Surv. Tutor. **20**(4), 3034–3059 (2018)
65. A. Aijaz, Z. Dawy, N. Pappas, M. Simsek, S. Oteafy, O. Holland, Toward a tactile internet reference architecture: Vision and progress of the IEEE P1918.1 standard. *arXiv preprint arXiv*:1807.11915
66. A. Wootten, A. Thompson, The Atacama large millimeter/submillimetre array. Proc. IEEE **97**(8), 1463–1471 (2009)
67. J.Y. Suen, M.T. Fang, S.P. Denny, P.M. Lubin, Modeling of terabit geostationary terahertz satellite links from globally dry locations. IEEE Trans. Terahertz Sci. Technol. **5**(2), 299–313 (2015)
68. R. Zhen, C. Han, Link budget analysis for massive-antenna-array-enabled terahertz satellite communications. J. Shanghai Jiaotong Univ. Sci. **23**(1), 20–27 (2018)
69. K. Guan, G. Li, T. Kürner, A.F. Molisch, B. Peng, R. He, B. Hui, J. Kim, Z. Zhong, On millimeter wave and THz mobile radio channel for smart rail mobility. IEEE Trans. Veh. Technol. **66**(7), 5658–5674 (2017)
70. G. Li, B. Ai, D. He, Z. Zhong, B. Hui, J. Kim, On the feasibility of high-speed railway mm wave channels in tunnel scenario. Wirel. Commun. Mob. Comput. **2017**, 7135896/1–17 (2017)
71. K. Guan, D. He, A. Hrovat, B. Ai, Z. Zhong, T. Kürner, Challenges and chances for smart rail mobility at mmwave and THz bands from the channels viewpoint, in *Proceedings of the 15th IEEE International Conference on ITS Telecommunications (ITST)*, 2017, pp. 1–5
72. K. Guan, X. Lin, D. He, B. Ai, Z. Zhong, Z. Zhao, D. Miao, H. Guan, T. Kürner, Scenario modules and ray-tracing simulations of millimeter wave and terahertz channels for smart rail mobility, in *Proceedings of the 11th IEEE European Conference on Antennas and Propagation (EUCAP)*, 2017, pp. 113–117
73. M. Pierobon, J.M. Jornet, N. Akkari, S. Almasri, I.F. Akyildiz, A routing framework for energy harvesting wireless nanosensor networks in the terahertz band. Wirel. Netw **20**(5), 1169–1183 (2014)
74. S. Mohrehkesh, M.C. Weigle, S.K. Das, Energy harvesting in electromagnetic nanonetworks. Computer **50**(2), 59–67 (2017)
75. X.-W. Yao, C.-C. Wang, W.-L. Wang, J.M. Jornet, On the achievable throughput of energy-harvesting nanonetworks in the terahertz band. IEEE Sensors J. **18**(2), 902–912 (2018)
76. S. Mumtaz, J.M. Jornet, J. Aulin, W.H. Gerstacker, X. Dong, B. Ai, Terahertz communication for vehicular networks. IEEE Trans. Veh. Technol. **66**(7), 5617–5625 (2017)
77. V. Petrov, J. Kokkoniemi, D. Moltchanov, J. Lehtomäki, M. Juntti, Y. Koucheryavy, The impact of interference from the side lanes on mmwave/THz band V2V communication systems with directional antennas. IEEE Trans. Veh. Technol. **67**(6), 5028–5041 (2018)
78. C. Zhang, K. Ota, J. Jia, M. Dong, Breaking the blockage for big data transmission: Gigabit road communication in autonomous vehicles. IEEE Commun. Mag. **56**(6), 152–157 (2018)
79. C.M. Armstrong, The truth about terahertz. IEEE Spectr. **49**(9), 36–41 (2012)
80. H. Aghasi, S.M.H. Naghavi, M.T. Taba, M.A. Aseeri, A. Cathelin, E. Afshari, Terahertz electronics: Application of wave propagation and nonlinear processes. Appl. Phys. Rev. **7**, 021302, 1–35 (2020)
81. A. Tarkeshdouz, A. Mostajeran, S. Mirabbasi, E. Afshari, A 91-GHz fundamental VCO with 6.1% DC-to-RF efficiency and 4.5 dBm output power in 0.13-um CMOS. IEEE Solid State Circuits Lett. **1**(4), 102–105 (2018)
82. P.Y. Chiang, O. Momeni, P. Heydari, A 200-GHz inductively tuned VCO with -7-dBm output power in 130-nm SiGe BiCMOS. IEEE Trans. Microw. Theory Tech. **61**(10), 3666–3673 (2013)
83. O. Momeni, E. Afshari, High power terahertz and sub-millimeter wave oscillator design: A systematic approach. IEEE J. Solid State Circuits **46**(3), 583–597 (2011)
84. R. Adler, A study of locking phenomena in oscillators. Proc. IEEE **61**(10), 1380–1385 (1973)

85. O. Momeni, E. Afshari, High power terahertz and millimeter-wave oscillator design: A systematic approach. IEEE J. Solid State Circuits **46**(3), 583–597 (2011)
86. H. Aghasi, A. Cathelin, E. Afshari, A 0.92-THz SiGe power radiator based on a nonlinear theory for harmonic generation. IEEE J. Solid State Circuits **52**(2), 406–422 (2017)
87. P. Hillger, J. Grzyb, S. Malz, B. Heinemann, U. Pfeiffer, A lens integrated 430 GHz SiGe HBT source with up to −6.3 dBm radiated power, in *Proceedings of IEEE Radio Frequency Integrated Circuits Symposium (RFIC)*, Honolulu, HI, USA, 2017, pp. 160–163
88. Y. Tousi, E. Afshari, A high-power and scalable 2-D phased array for terahertz CMOS integrated systems. IEEE J. Solid State Circuits **50**(2), 597–609 (2015)
89. R. Han, C. Jiang, A. Mostajeran, M. Emadi, H. Aghasi, H. Sherry, A. Cathelin, E. Afshari, A SiGe terahertz heterodyne imaging transmitter with 3.3 mW radiated power and fully-integrated phase-locked loop. IEEE J. Solid State Circuits **50**(12), 2935–2947 (2015)
90. N. Khalid, T. Yilmaz, O.B. Akan, Energy-efficient modulation and physical layer design for low terahertz band communication channel in 5G femtocell internet of things. Ad Hoc Netw. **79**, 63–71 (2018)
91. U.R. Pfeiffer, Y. Zhao, J. Grzyb, R. Al Hadi, N. Sarmah, W. Forster, H. Rucker, B. Heinemann, A 0.53 THz reconfigurable source module with up to 1 mW radiated power for diffuse illumination in terahertz imaging applications. IEEE J. Solid State Circuits **49**(12), 2938–2950 (2014)
92. E. Seok, C. Cao, D. Shim, D.J. Arenas, D.B. Tanner, C.M. Hung, K.O. Kenneth, A 410 GHz CMOS push-push oscillator with an on-chip patch antenna, in *Proceedings of IEEE International Solid-State Circuits Conference-Digest of Technical Papers*, San Francisco, CA, USA, 2008, pp. 472–629
93. K. Guo, Y. Zhang, P. Revnaert, A 0.53-THz subharmonic injection-locked phased array with 63-μW radiated power in 40-nm CMOS. IEEE J. Solid State Circuits **54**(2), 380–391 (2018)
94. R. Han, E. Afshari, A CMOS high-power broadband 260-GHz radiator array for spectroscopy. IEEE J. Solid State Circuits **48**(12), 3090–3104 (2013)
95. J. Grzyb, Y. Zhao, U.R. Pfeiffer, A 288-GHz lens-integrated balanced triple-push source in a 65-nm CMOS technology. IEEE J. Solid State Circuits **48**(7), 1751–1761 (2013)
96. S. Jameson, E. Socher, High efficiency 293 GHz radiating source in 65 nm CMOS. IEEE Microw. Wirel. Compon. Lett. **24**(7), 463–465 (2014)
97. J. Al-Eryani, H. Knapp, J. Kammerer, K. Aufinger, H. Li, L. Maurer, Fully integrated single-chip 305–375-GHz transceiver with on-chip antennas in SiGe BiCMOS. IEEE Trans. Terahertz Sci. Technol. **8**(3), 329–339 (2018)
98. L. Wu, S. Liao, Q. Xue, A 312-GHz CMOS injection locked radiator with chip-and-package distributed antenna. IEEE J. Solid State Circuits **52**(11), 2920–2933 (2017)
99. P. Hillger, J. Grzyb, R. Lachner, U. Pfeiffer, An antenna coupled 0.49 THz SiGe HBT source for active illumination in terahertz imaging applications, in *Proceedings of 10th European Microwave Integrated Circuits Conference*, Paris, France, 2015, pp. 180–183
100. K. Guo, A. Standaert, P. Reynaert, A 525–556-GHz radiating source with a dielectric lens antenna in 28-nm CMOS. IEEE Trans. Terahertz Sci. Technol. **8**(3), 340–349 (2018)
101. B. Khamaisi, S. Jameson, E. Socher, 0.61 THz radiating source with on-chip antenna on 65 nm CMOS, in *Proceedings of 11th European Microwave and Integrated Circuits Conference*, London, UK, 2016, pp. 389–392
102. P. Rodriguez-Vazquez, J. Grzyb, N. Sarmah, B. Heinemann, U.R. Pfeiffer, A 65 Gbps QPSK one meter wireless link operating at a 225–255 GHz tunable carrier in a SiGe HBT technology, in *Proceedings of IEEE Radio Wireless Symposium*, Anaheim, CA, USA, 2018, pp. 146–149
103. K. Sengupta, T. Nagatsuma, D.M. Mittleman, Terahertz integrated electronic and hybrid electronic-photonic systems. Nat. Electron. **1**(12), 622–635 (2018)
104. J. Grzyb, K. Statnikov, N. Sarmah, B. Heinemann, U.R. Pfeiffer, A 210–270-GHz circularly polarized FMCW radar with a single-lens coupled SiGe HBT chip. IEEE Trans. Terahertz Sci. Technol. **6**(6), 771–783 (2016)

105. M. Schroter, U. Pfeiffer, R. Jain, Future of SiGe HBT technology and its applications, in *Silicon-Germanium Heterojunction Bipolar Transistor for mm-Wave Systems: Technology, Modeling and Circuit Applications*, (River Publishers, Gistrup, 2018), p. 309
106. Y. Yang, O.D. Gurbuz, G.M. Rebeiz, An eight-element 370–410-GHz phased-array transmitter in 45-nm CMOS SOI with peak EIRP of 8–8.5 dBm. IEEE Trans. Terahertz Sci. Technol. **64**(12), 4241–4249 (2016)
107. F. Golcuk, O.D. Gurbuz, G.M. Rebeiz, A 0.39–0.44 THz 2 × 4 amplifier-quadrupler array with peak EIRP of 3–4 dBm. IEEE Trans. Microw. Theory Tech. **61**(12), 4483–4491 (2013)
108. Y. Zhao, C. Zuow-Zun, Y. Du, Y. Li, R.A. Hadi, G. Virbila, Y. Xu, Y. Kim, A. Tang, T.J. Reck, M.C.F. Chang, A 0.56 THz phase-locked frequency synthesizer in 65 nm CMOS technology. IEEE J. Solid State Circuits **51**(12), 3005–3019 (2016)
109. B. Khamaisi, S. Jameson, E. Socher, A 0.58–0.61 THz single on chip antenna transceiver based on active X30 LOchain on 65 nm CMOS, in *Proceedings of 11th European Microwave and Integrated Circuits Conference*, London, UK, 2016, pp. 97–100
110. Z. Ahmad, K.O. Kenneth, 0.65–0.73 THz quintupler with an onchip antenna in 65-nm CMOS, in *Proceedings of Symposium on VLSI Circuits*, Kyoto, Japan, 2015, pp. C310–C311
111. E. Ojefors, J. Grzyb, Y. Zhao, B. Heinemann, B. Tillack, U.R. Pfeiffer, A 820 GHz SiGe chipset for terahertz active imaging applications, in *Proceedings of IEEE International Solid-State Circuits Conference*, San Francisco, CA, USA, 2011, pp. 224–226
112. Z. Ahmad, M. Lee, K.O. Kenneth, 20.5 1.4 THz, −13 dBm-EIRP frequency multiplier chain using symmetric- and asymmetric-CV varactors in 65 nm CMOS, in *Proceedings of IEEE International Solid-State Circuits Conference*, San Francisco, CA, USA, 2016, pp. 350–351
113. J. Grzyb, U. Pfeiffer, THz direct detector and heterodyne receiver arrays in silicon nanoscale technologies. J. Infrared Millim. Terahertz Waves **36**(10), 998–1032 (2015)
114. P. Hillger, R. Jain, U.R. Pfeiffer, Terahertz imaging and sensing applications with silicon-based technologies. IEEE Trans. Terahertz Sci. Technol. **9**(1), 1–19 (2019)
115. T. Nagatsuma, G. Ducournau, C.C. Renaud, Advances in terahertz communications accelerated by photonics. Nat. Photonics **10**(6), 371–379 (2016)
116. M.C. Wanke, E.W. Young, C.D. Nordquist, M.J. Cich, A.D. Grine, C.T. Fuller, J.L. Reno, M. Lee, Monolithically integrated solid-state terahertz transceivers. Nat. Photonics **4**(9), 565–569 (2010)
117. M. Mukherjee, N. Mazumder, S.K. Roy, K. Goswami, GaN IMPATT diode: A photo-sensitive high-power terahertz source. Semicond. Sci. Technol. **22**(12), 1258–1267 (2007)
118. C. Dalle, P.A. Rolland, G. Lleti, Flat doping profile double-drift silicon IMPATT for reliable CW high-power high-efficiency generation in the 94-GHz window. IEEE Trans. Electron Devices **37**(1), 227–236 (1990)
119. H. Eisele, G.I. Haddad, High-performance InP Gunn devices for fundamental-mode operation in D-band (110–170 GHz). IEEE Microw. Guid. Wave Lett. **5**(11), 385–387 (1995)
120. H. Eisele, A. Rydberg, G.I. Haddad, Recent advances in the performance of InP Gunn devices and GaAs TUNNETT diodes for the 100–300 GHz frequency range and above. IEEE Trans. Microw. Theory Tech. **48**(4), 626–631 (2000)
121. N. Orihashi, S. Suzuki, M. Asada, One THz harmonic oscillation of resonant tunneling diodes. Appl. Phys. Lett. **87**(23), 233501 (2005)
122. E. Kaymaksut, D. Zhao, P. Reynaert, Transformer-based Doherty power amplifiers for millimeter-wave applications in 40-nm CMOS. IEEE Trans. Microw. Theory Tech. **63**(4), 1186–1192 (2015)
123. Z. Hu, M. Kaynak, R. Han, High-power radiation at 1 THz in silicon: A fully scalable array using a multi-functional radiating mesh structure. IEEE J. Solid State Circuits **53**(5), 1313–1327 (2018)
124. K. Kibaroglu, M. Sayginer, G.M. Rebeiz, A scalable 64-element 280 Hz phased-array transceiver with 50 dBm EIRP and 8–12 Gbps 5G link at 300 meters without any calibration, in *Proceedings of the IEEE MTT-S International Microwave Symposium Digest*, 2018, pp. 496–498

125. A. Babakhani, X. Guan, A. Komijani, A. Natarajan, A. Hajimiri, A 77- GHz phased-array transceiver with on-chip antennas in silicon: Receiver and antennas. IEEE J. Solid State Circuits **41**(12), 2795–2805 (2006)
126. A. Rogalski, F. Sizov, Terahertz detectors and focal plane arrays. Opto-Electron. Rev. **19**(3), 346–404 (2011)
127. A.D. Turner, J.J. Bock, J.W. Beeman, J. Glenn, P.C. Hargrave, V.V. Hristov, H.T. Nguyen, F. Rahman, S. Sethuraman, A.L. Woodcraft, Silicon nitride micromesh bolometer array for submillimeter astrophysics. Appl. Opt. **40**, 4921–4932 (2001)
128. J. Glenn, P.A.R. Ade, M. Amarie, J.J. Bock, S.F. Edgington, A. Goldin, S. Golwala, D. Haig, A.E. Lange, G. Laurent, P.D. Maudkopf, M. Yun, H. Nguyen, Current status of Bolocam: A large-format millimeter-wave bolometer camera. Proc. SPIE **4855**, 30–40 (2003)
129. G. Karolyi, D. Gergelyi, P. Foldesy, Sub-THz sensor array with embedded signal processing in 90 nm CMOS technology. IEEE Sensors J. **14**(8), 2432–2441 (2014)
130. A. Boukhayma, A. Dupret, J.-P. Rostaing, C. Enz, A low-noise CMOS THz imager based on source modulation and an in-pixel high-Q passive switched-capacitor N-path filter. Sensors **16**(3), 325 (2016)
131. Z.-y. Liu, L.-y. Liu, J. Yang, N.-j. Wu, A CMOS fully integrated 860-GHz terahertz sensor. IEEE Trans. Terahertz Sci. Technol. **7**(4), 455–465 (2017)
132. M. Uzunkol, O.D. Gurbuz, F. Golcuk, G.M. Rebeiz, A 0.32 THz SiGe 4 × 4 imaging array using high-efficiency on-chip antennas. IEEE J. Solid State Circuits **48**(9), 2056–2066 (2013)
133. M. Bozanic, S. Sinha, Emerging transistor technologies capable of terahertz amplification: A way to re-engineer terahertz radar sensors. Sensors **19**(11), 2454 (2019)
134. M. Dyakonov, M. Shur, Shallow water analogy for a ballistic field effect transistor: New mechanism of plasma wave generation by dc current. Phys. Rev. Lett. **71**(15), 2465 (1993)
135. A. Rostami, H. Rasooli, H. Baghban, *Terahertz Technology: Fundamentals and Applications*, vol 77 (Springer Science & Business Media, Berlin, Heidelberg, 2011)
136. G. Fertig, Evaluation of MOSFETs for terahertz detector arrays. Master dissertation, Rochester Institute of Technology, 2014
137. W. Knap, V. Kachorovskii, Y. Deng, S. Rumyantsev, J.Q. Lu, R. Gaska, M.S. Shur, G. Simin, X. Hu, M.A. Khan, C.A. Saylor, Nonresonant detection of terahertz radiation in field effect transistors. J. Appl. Phys. **91**(11), 9346–9353 (2011)
138. A. El Fatimy, Abdelouahad, F. Teppe, N. Dyakonova, W. Knap, D. Seliuta, G. Valusis, A. Shchepetov, Y. Roelens, S. Bollaert, A. Cappy, S. Rumyantsev, Resonant and voltage-tunable terahertz detection in InGaAs/InP nanometer transistors. Appl. Phys. Lett. **89**(13), 131926 (2016)
139. Y.M. Meziani, N. Dyakonova, W. Knap, D. Seliuta, E. Simulis, J. Devenson, G. Valusis, F. Boeuf, T. Skotnicki, Non resonant response to terahertz radiation by submicron CMOS transistors. IEICE Trans. Electron. **89**(7), 993–998 (2006)
140. G.C. Dyer, J.D. Crossno, G.R. Aizin, J. Mikalopas, E.A. Shaner, M.C. Wanke, J.L. Reno, S.J. Allen, A narrowband plasmonic terahertz detector with a monolithic hot electron bolometer, in *Terahertz Technology and Applications II. International Society for Optics and Photonics*, vol. 7215, 2009, p. 721503
141. W. Knap, M. Dyakonov, D. Coquillat, F. Teppe, N. Dyakonov, J. Lusakowsk, K. Karpierz, M. Sakowicz, G. Valusis, D. Seliuta, I. Kasalynas, Field effect transistors for terahertz detection: Physics and first imaging applications. J. Infrared Millim. Terahertz Waves **30**(12), 1319–1337 (2009)
142. W. Knap, D. Coquillat, N. Dyakonova, F. Teppe, O. Klimenko, H. Videlier, S. Nadar, J. Lusakowski, G. Valusis, F. Schuster, B. Giffard, Plasma excitations in field effect transistors for terahertz detection and emission. C. R. Phys. **11**(7), 433–443 (2013)
143. R. Al Hadi, H. Sherry, J. Grzyb, Y. Zhao, W. Forster, H.M. Keller, A. Cathelin, A. Kaiser, U.R. Pfeiffer, A1 k-pixel video camera for 0.7–1.1 terahertz imaging applications in 65-nm CMOS. IEEE J. Solid State Circuits **47**(12), 2999–3012 (2012)

144. R. Al Hadi, J. Grzyb, B. Heinemann, U.R. Pfeiffer, A terahertz detector array in a SiGe HBT technology. IEEE J. Solid State Circuits **48**(9), 2002–2010 (2013)
145. J. Grzyb, B. Heinemann, U.R. Pfeiffer, Solid-state terahertz super resolution imaging device in 130-nm SiGe BiCMOS technology. IEEE Trans. Microw. Theory Tech. **65**(11), 4357–4372 (2017)
146. B. Sartorius, H. Roehle, H. Künzel, J. Böttcher, M. Schlak, D. Stanze, H. Venghaus, M. Schell, All-fiber terahertz time-domain spectrometer operating at 1.5 mm telecom wavelengths. Opt. Express **16**, 9565–9570 (2008)
147. I. Malhotra, P. Thakur, S. Pandit, K. R. Jha, G. Singh, Analytical framework of small-gap photoconductive dipole antenna using equivalent circuit model. Optical and Quantum Electronics **49**(10), 334/1–23 (2017)
148. S. Dash, A. Patnaik, Material selection for THz antennas. Microw. Opt. Technol. Lett. **60**(5), 1183–1187 (2018)
149. M. Roshanaei, H. Karami, P. Dehkhoda, H. Esfahani, F. Dabir, A highly directive graphene antenna embedded inside a Fabry–Perot cavity in terahertz regime. Superlattices Microstruct. **117**, 356–363 (2018)
150. M. Tamagnone, J. Gomez-Diaz, J. Mosig, J. Perruisseau-Carrier, Analysis and design of terahertz antennas based on plasmonic resonant graphene sheets. J. Appl. Phys. **112**(11), 114915/1–4 (2012)
151. E. Lallas, Key roles of plasmonics in wireless THz nanocommunications-a survey. Appl. Sci. **9**(24), 5488 (2019)
152. I.F. Akyildiz, J.M. Jornet, Graphene based plasmonic nanoantenna for terahertz band communication. US Patent 9,643,841, 2017
153. S.E. Hosseininejad, E. Alarcón, N. Komjani, S. Abadal, M.C. Lemme, P.H. Bolívar, A. Cabellos-Aparicio, Study of hybrid and pure plasmonic terahertz antennas based on graphene guided-wave structures. Nano Commun. Netw. **12**, 34–42 (2017)
154. S.A. Naghdehforushha, G. Moradi, Plasmonic patch antenna based on graphene with tunable terahertz band communications. Optik **158**, 617–622 (2018)
155. S.A. Naghdehforushha, G. Moradi, High directivity plasmonic graphene-based patch array antennas with tunable THz band communications. Optik **168**, 440–445 (2018)
156. J.M. Jornet, I.F. Akyildiz, Graphene-based plasmonic nano-antenna for terahertz band communication in nanonetworks. IEEE J. Sel. Areas Commun. **31**(12), 685–694 (2013)
157. L. Zakrajsek, E. Einarsson, N. Thawdar, M. Medley, J.M. Jornet, Design ofgraphene-based plasmonic nano-antenna arrays in the presence of mutualcoupling, in *Proceedings of the 11th IEEE European Conference on Antennas and Propagation (EuCAP)*, 2017, pp. 1381–1385
158. M. Nafari, J.M. Jornet, Modeling and performance analysis of metallic plasmonic nano-antennas for wireless optical communication in nanonetworks. IEEE Access **5**, 6389–6398 (2017)
159. M. Esquius-Morote, J.S. Gómez-D, J. Perruisseau-Carrier, Sinusoidally modulated graphene leaky-wave antenna for electronic beam scanning at THz. IEEE Trans. Terahertz Sci. Technol. **4**(1), 116–122 (2014)
160. F. Liang, Z.-Z. Yang, Y.-X. Xie, H. Li, D. Zhao, B.-Z. Wang, Beam-scanning microstrip Quasi-Yagi–Uda antenna based on hybrid metal-graphene materials. IEEE Photonic Technol. Lett. **30**(12), 1127–1130 (2018)
161. Z. Xu, X. Dong, J. Bornemann, Design of a reconfigurable MIMO system for THz communications based on graphene antennas. IEEE Trans. Terahertz Sci. Technol. **4**(5), 609–617 (2014)
162. A. Standaert, P. Reynaert, A 400 GHz transmitter integrated with flip-chiped 3D printed horn antenna with an EIRP of 1.26 dBm, in *Proceedings of the IEEE/MTT-S International Microwave Symposium (IMS 2018)*, 2018, pp. 141–144
163. K.R. Jha, G. Singh, Terahertz Planar Antennas for Next Generation Communication. Springer: Switzerland (2014)

164. I. Malhotra, K.R. Jha, G. Singh, Design of highly directive terahertz photoconductive dipole antenna using frequency selective surface for sensing and imaging application. J. Comput. Electron. **17**, 1–20 (2018)
165. I. Malhotra, K.R. Jha, G. Singh, Design of highly directive lens-less photoconductive dipole antenna array with frequency selective surface for terahertz imaging applications. Optik **173**, 206–219 (2018)
166. B. Shen, W.L. Stutzman, A scanning spherical tri-reflector antenna with a moving flat mirror. IEEE Trans. Antennas Propag. **43**(3), 270–276 (1995)
167. N.A. Agapov, Two-mirror scanning system. Sov. J. Opt. Technol. **53**(6), 324–325 (1986)
168. V.G. Borkar, B.K. Mukhopadhyay, J.V. Prasad, V.M. Pandharipande, R. Ethiraj, A millimeter-wave monopulse twist reflector scanning antenna system: Theory and experiment. IETE J. Res. **43**(6), 415–421 (1997)
169. S. Campana, Passive electro-optical systems, in *The Infrared and Electro-Optical Systems Handbook*, vol. 5, (SPIE, Bellingham, 1993)
170. X. Feng, J.R. Schott, T.W. Gallagher, Modelling the performance of a high-speed scan mirror for an airborne line scanner. Opt. Eng. **33**(4), 1214–1222 (1994)
171. I. Malhotra, K.R. Jha, G. Singh, Beam steering characteristics of highly directive photoconductive dipole phased array antenna for terahertz imaging application. Optical and Quantum Electronics, **51**(27), 1–19 (2019)
172. V. Krozer, T. Loffler, J. Dall, A. Kusk, F. Eichhorn, R.K. Olsson, J.D. Buron, P.U. Jepsen, V. Zhurbenko, T. Jensen, Terahertz imaging systems with aperture synthesis techniques. IEEE Trans. Microw. Theory Tech. **58**(7), 2027–2039 (2010)
173. A. Luukanen, A.J. Miller, E.N. Grossman, Active millimeter-wave video rate imaging with a staring 120-element microbolometer array, in *Proceedings of SPIE—International Society of Optical Engineering*, vol. 5410, Radar Sensor Technol. VIII and Passive Millim.-Wave Imaging Technology VII, 2004, pp. 195–201
174. R. Doyle, B. Lyons, J.Walshe, P. Curtin, A.H. Lettington, T. McEnroe, J. McNaboe, Low cost millimeter wave camera imaging up to 140 GHz, in *Proceedings of 34th European Microwave Conference*, 2004, pp. 1285–1288
175. R. Appleby, Passive millimeter-wave imaging and how it differs from terahertz imaging. Philos. Trans. R. Soc. Lond. A Math. Phys. Eng. Sci. **362**, 379–394 (2004)
176. A.H. Lettington, D. Dunn, M. Attia, I.M. Blankson, Passive millimeter-wave imaging architectures. J. Opt. A Pure Appl. Opt. **5**(4), S103 (2003)
177. J.W. Archer, M.G. Shen, 176–200 GHz receiver module using indium phosphide and gallium arsenide MMICs. Microw. Opt. Technol. Lett. **43**, 458–462 (2004)
178. J.W. Archer, M.G. Shen, W-band transmitter module using gallium arsenide MMICs. Microw. Opt. Technol. Lett. **42**, 210–213 (2004)
179. R. Doyle, B. Lyons, A. Lettington, T. McEnroe, J.Walshe, J. McNaboe, P. Curtin, S. Bleszynski, Stand-off detection of hidden threat objects on personnel at checkpoints and in public areas using active millimeter-wave imaging, in *Proceedings of SPIE—International Society of Optical Engineering*, vol. 5619, Passive Millimeter Wave Terahertz Imaging Technology, 2004, pp. 90–97
180. Y.A. Pirogov, V.V. Gladun, S.P. Chzhen, D.A. Tischenko, A.L. Timanovskiy, Radio thermal images of natural objects in 8-mm and 3-mm ranges. Proc. SPIE Int. Soc. Opt. Eng. **4719**, 318–326 (2002)
181. J.W. May, G.M. Rebeiz, Design and characterization of W-band SiGe RFICs for passive millimeter-wave imaging. IEEE Trans. Microw. Theory Tech. **58**(2), 1420–1430 (2010)
182. E. Öjefors, U.R. Pfeiffer, A. Lisauskas, H.G. Roskos, A 0.65 THz focal-plane array in a quarter-micron CMOS process technology. IEEE J. Solid State Circuits **44**(7), 1968–1976 (2009)
183. A. Lisauskas, U. Pfeiffer, E. Ojefors, P.H. Bolivar, D. Glaab, H.G. Roskos, Rational design of high-responsivity detectors of terahertz radiation based on distributed self-mixing in silicon field-effect transistors. J. Appl. Phys. **105**(11), 114511–114518 (2009)

184. K.B. Cooper, R.J. Dengler, N. Llombart, T. Bryllert, G. Chattopadhyay, I. Mehdi, P.H. Siegel, An approach for sub-second imaging of concealed objects using terahertz (THz) radar. J. Infrared Millim. Terahertz Waves **30**(12), 1297–1307 (2009)
185. K. Watabe, K. Shimizu, M. Yoneyama, K. Mizuno, Millimeter-wave active imaging using neural networks for signal processing. IEEE Trans. Microw. Theory Tech. **51**(5), 1512–1516 (2003)
186. V.I. Fedorov, S.S. Popova, A.N. Pisarchik, Dynamic effects of submillimeter wave radiation on biological objects of various levelsof organization. Int. J. Infrared Millim. Waves **24**(8), 1235–1254 (2003)
187. D.M. Sheen, D.L. McMakin, T.E. Hall, Cylindrical millimeter-wave imaging technique and applications, in *Proceedings of SPIE—International Society of Optical Engineering*, vol. 6211, Passive Millimeter Wave Imaging Technology IX, 2006, pp. 6357–6365
188. I. Malhotra, K.R. Jha, G. Singh, Analysis of highly directive photoconductive dipole antenna at terahertz frequency for sensing and imaging applications. Optics Communication, **397**, 129–139 (2017)
189. X. Yin, B. Ng, D. Abbott, *Terahertz Imaging for Biomedical Applications: Pattern Recognition and Tomographic Reconstruction* (Springer Science & Business Media, New York, 2012)
190. N. Vieweg, N. Krumbholz, T. Hasek, R. Wilk, V. Bartels, C. Keseberg, V. Pethukhov, M. Mikulics, L. Wetenkamp, M. Koch, Fiber-coupled THz spectroscopy for monitoring polymeric compounding processes. Polym. Test. **28**(1), 30–35 (2009)
191. X. Shen, C.R. Dietlein, E. Grossman, Z. Popovic, F.G. Meyer, Detection and segmentation of concealed objects in terahertz images. IEEE Trans. Image Process. **17**(12), 2465–2475 (2008)
192. S.S. Ahmed, A. Schiessl, L.-P. Schmidt, Near field mm-wave imaging with multistatic sparse 2D-arrays, in *Proceeding of 6th European Radar Conference*, 2009, pp. 180–183
193. B. Fischer, M. Hoffmann, H. Helm, G. Modjesch, P.U. Jepsen, Chemical recognition in terahertz time-domain spectroscopy and imaging. Semicond. Sci. Technol. **20**, S246–S253 (2005)
194. A. Sampaolo, P. Patimisco, M. Giglio, M.S. Vitiello, H.E. Beere, D.A. Ritchie, G. Scamarcio, F.K. Tittel, V. Spagnolo, Improved tuning fork for terahertz quartz-enhanced photoacoustic spectroscopy. Sensors **16**(4), 439 (2016)
195. C. Wang, B. Perkins, Z. Wang, R. Han, Molecular detection for unconcentrated gas with ppm sensitivity using 220-to-320-GHz dual frequency-comb spectrometer in CMOS. IEEE Trans. Biomed. Circuits Syst. **12**, 709–721 (2018)
196. X. Wu, K. Sengupta, Single-chip source-free terahertz spectroscope across0.04–0.99 THz: Combining sub-wavelength near-field sensing and regression analysis. Opt. Express **26**, 7163–7175 (2018)
197. X. Wu, K. Sengupta, Dynamic waveform shaping with picosecond time widths. IEEE J. Solid State Circuits **52**, 389–405 (2017)
198. IEEE standard for high data rate wireless multi-media networks – amendment 2: 100 Gb/s wireless switched point-to-point physical layer, in *IEEE Std 802.15.3d-2017 (amendment to IEEE Std 802.15.3–2016 asamended by IEEE Std 802.15.3e-2017)*, (IEEE, 2017), pp. 1–55
199. S. Shahramian, M. Holyoak, A. Singh, B.J. Farahani, Y. Baeyens, A fully integrated scalable W-band phased-array module with integrated antennas, self-alignment and self-test, in *Proceedings of the IEEE International Solid State Circuits Conference*, San Francisco, CA, USA, February 11–15, 2018, pp. 74–76
200. T. Nagatsuma, K. Oogimoto, Y. Inubushi, J. Hirokawa, Practical considerations of terahertz communications for short distance applications. Nano Commun. Netw. **10**, 1–12 (2016)

Chapter 4
Small-Gap Photoconductive Dipole Antenna for Imaging and Sensing

4.1 Introduction

Various proposed applications exploit the unique capabilities of THz radiation to penetrate the packaging materials and therefore provide their spectroscopic information. However, there are certain emerging issues related to THz for imaging and sensing applications such as the water content present inside the human body eludes transmission-type imaging [1]. Moreover, in comparison with optical light, the THz radiations offer lesser spatial resolution of images. In the real-world scenario, there are nine low-attenuation windows in the range of 0.1–3 THz of the spectrum: (1) 0.1–0.55 THz, (2) 0.56–0.75 THz, (3) 0.76–0.98 THz, (4) 0.99–1.09 THz, (5) 1.21–1.41 THz, (6) 1.42–1.59 THz, (7) 1.92–2.04 THz, (8) 2.05–2.15 THz, and (9) 2.47–2.62 THz, respectively [2], which can be considered for effective transmission-type THz imaging. These frequency ranges are determined by considering the measurements at 23 °C temperature and 26% of relative humidity. These transmission bands are important to consider because several commonly used solid-state explosives and related compounds (ERCs) have spectral fingerprints in 0.1–2.8 THz range and such fingerprints occur from the intramolecular and intermolecular vibrational modes or phonon modes of these explosive materials [3]. Therefore, the THz sensing and imaging in transmission /low attenuation windows is necessary for the detection of the hidden explosives. In THz imaging system, the electromagnetic waves are used to spectroscopically detect the presence of concealed explosives such as research department explosive (RDX) and high melting explosive (HMX) from the characteristic transmission or reflectivity spectra shown by these explosives in the THz range [4]. A detection technique to RDX pellet using THz frequency is as shown in Fig. 4.1 [5]. In this setup, a femtosecond pulsed laser (Ti:Sapphire laser) is used for the generation of signal. The optical beam from the laser is split into two beams with the help of a beam splitter (BS). One beam is utilized for exciting photoconductive antenna, which works as emitter, and the other beam is used for measuring the THz signal at ZnTe crystal detector for modulation.

I. Malhotra, G. Singh, *Terahertz Antenna Technology for Imaging and Sensing Applications*, https://doi.org/10.1007/978-3-030-68960-5_4

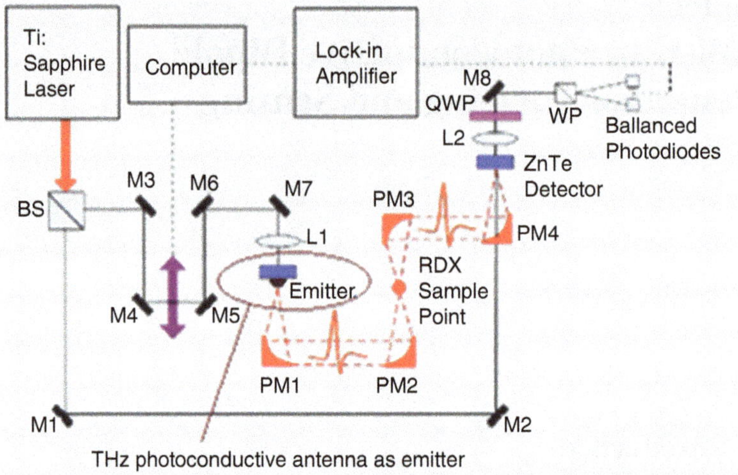

Fig. 4.1 The THz frequency system based on the principle of photoconductivity, wherein LT-GaAs emitter is used as a THz photoconductive antenna [5]

Various beam splitters are used for the beam deflection. The parabolic optical mirror combination is used for the optical beam to strike the photoconductive material of the emitter to generate the THz beam, which is focused on sample point. The sample (RDX pellet) is positioned at the focus point of THz wave and is perpendicular to the incident beam. The transmitted THz beam after passing through the sample is then collected and focused with the help of other pair of off-axis parabolic mirrors (PM) onto the ZnTe crystal where the probe beam from the laser detect the THz field using electro-optic sampling. When such radiations are allowed to pass through the sample, they get absorbed at their respective features and the absorption coefficients provide the information related to detection of hidden explosives at detector side.

From Fig. 4.1, it is clear that the photoconductive antenna is one of the most important components in a THz sensing and imaging system as it plays significant role for both the impedance matching and power source. In [5], a terahertz time-domain spectroscopy system has been deployed in 0.2–3.4 THz range of frequency. The bandwidth of a photoconductive antenna is generally considered as the range of frequencies over which the signal strength of the measured frequency domain exceeds the noise level of the system. However, an enlarged bandwidth is significant for the applications that make use of distinct spectral characteristics in materials such as in THz sensing and imaging applications for the detection of hidden powdered explosives, and this necessitates the capability to detect narrow absorption peaks in the THz band. Accordingly, increasing the bandwidth of a THz photoconductive antenna-based THz imaging system will set aside additional vibrational modes to be quantified [6]. Several companies such as TeraView Ltd., Picometrix LLC, Advantest, and Menlo Systems provide absolute THz imaging and spectroscopy systems, which employ photoconductive antennas as emitters and detectors. Some commercially available photoconductive antenna-based THz spectroscopy systems by TeraView company with model numbers TPS Spectra 3000 and TeraPulse 4000 have a bandwidth (maximum frequency) of 4 THz and 6 THz, respectively. Similarly, Picometrix and Advantest have developed the spectroscopy systems T-Ray 5000 and TAS each having a bandwidth of 4 THz. The company Mento Systems have developed TERA K8 and TERA K15 spectroscopy systems that are photoconductive antenna-based system with a bandwidth of 3.5 THz and 4 THz, respectively.

For the imaging purpose, it is required to have a highly directive low-profile photoconductive antenna that generates the desired THz radiation at operating frequency with high directivity and optimum radiation efficiency with broad bandwidth. However, one of the key issues of the various reported photoconductive dipole antennas is that the antenna efficiency is very small, which makes the difficulty for the THz imaging system to achieve high-power THz waves. The photoconductive dipole antenna is unable to transfer the laser source power to the THz power efficiently because the highest power conversion efficiency as reported in literature is much less than 0.1% for THz pulsed systems [7]. Therefore, the researchers generally increase the illumination power, as well as the applied bias to yield higher output power. However, in such situations, the phenomena such as saturation of charges, velocity overshoot, field, and thermal breakdown occur in the THz pulsed systems. The thermal and field breakdowns are required to be avoided under every condition. Furthermore, three main causes for low efficiency of the photoconductive dipole antennas are as follows [8]:

- Space-charge also known as coulomb screening effect and screening effects due to radiation field.
- Spatial nonuniformities.
- Inadequate field strength due to the insufficient acceleration of the charges.

Due to the space-charge field generated by the photo-excited electron–hole pairs, there is an occurrence of coulomb screening effect. The movement of free charge carrier's under the influence of the applied bias field creates a static field, which is responsible for moderately screening of the applied bias field. However, the occurrence of radiation screening effect is because of the THz near-field radiation, which is also responsible for screening of the applied bias field. With the use of THz photoconductive dipole antenna in continuous wave operation mode for high peak optical pulse intensity the effect of radiation screening gets negligible [8]. Further, the spatial nonuniformities occur in photoconductive dipole antenna since the thickness of the substrate is generally larger than the wavelength of the THz waves, which results in the generation of surface/substrate modes. The antenna performance is sensitive to the substrate thickness. If the thickness of substrate is increased, the dipole antenna couples the power to higher order substrate modes. Moreover, in several cases even more than 90% of power gets trapped within the substrate [9]. Moreover, when a photoconductive dipole antenna is fabricated on a thick substrate it results in the excitation of a surface wave mode, which depends on several characteristics such as frequency, thickness of the substrate, and the relative dielectric constant of the substrate. It is clear that the total power radiated from the photoconductive dipole structure does not get directly transferred into the medium; however, a fraction of the radiated power gets trapped within the substrate and therefore create an effect on the radiation pattern.

4.2 Related Work and Problem Formulation

With the development of ultrashort pulse femtosecond lasers like Ti-Sapphire laser and quantum cascade lasers, it is possible to generate THz signals. The electro-optic (EO) rectification, which is based on different frequency mixing technique, is also used to generate THz radiation, but it has certain limitations. This method cannot produce THz frequency signals of broad frequency range and is also very sensitive to optical and thermal noises. Moreover, for the application of sensing and imaging, there is a requirement of a broadband short-pulse THz source for spectroscopic techniques such as time domain spectroscopy (TDS) or THz pulsed imaging (TPI). Therefore, the photoconductive antenna is one of the simple and stable devices for THz photonics used for sensing and imaging applications at THz frequency. The photoconductive antenna is relatively stable against optical and thermal noises in comparison to the electro-optic rectification [10]. However, the total antenna efficiency which includes optical laser to THz conversion efficiency, impedance matching efficiency, and radiation efficiency, is small. Various photoconductive dipole antenna structures are projected and are used for the THz frequency range [11]. On the basis of the architecture of photoconductive antennas for THz pulsed systems, they are classified as an aperture antenna, spiral antennas, bowtie antennas, and dipole antennas. In the large aperture photoconductive antenna, the distance between the electrodes is much larger than the center wavelength of THz wave with a range of few hundred micrometers. However, in the small-gap photoconductive dipole antenna, the antenna gap distance is only few micrometers. Moreover, in the photoconductive dipole antenna, to increase the conversion efficiency, the electrodes with sharp tip ends can be used and the efficiency can further be improved by putting them in a laterally offset format [12]. In this case, the THz emission can be improved with less optical power because of better overlap among the laser spot and high electric field point and also stronger fringing field effects between the electrodes. For small-gap photoconductive antennas, the fabrication of such tiny sharp tips is not easy. Moreover, an appropriate configuration of the electrodes may double the efficiency of the antennas, with a consequence in the bandwidth of the radiated pulse. However, the first photoconductive antenna was accounted by Mourou et al. [13], which operate in the GHz range, then the design is extended into the THz region by Auston et al. [14, 15]. Grischkowsky et al. [16, 17] have reported the application of optical technique for the generation of diffraction limited THz beams with a relatively large size of the source. This technique was further developed at Bell Labs and the IBM Watson Research Centre, which is now commercially available as a product by Picometrix Inc., MI [18]. Moreover, it is important to determine the antenna's spatial response as it is related with the power collected by the antenna. Therefore, the measurements of spatial response of infrared dipole and bow-tie lithographic antennas are discussed by Fumeauxet et al. [19]. A dipole antenna placed in a pyramidal horn cavity, which is impressed in silicon operated at 0.8THz, is reported in [20]. A stripline dipole antenna for a broad frequency range up to 5 THz on semi-infinite and lens substrates is discussed in terms of input impedance, as well as radiation characteristics in [21].

Further, the emission efficiency of photoconductive dipole antenna is discussed by Tani et al. [22], wherein the authors have considered the saturation effect that occurs due to the field screening effect caused by the photo-generated carriers. The key features about the near-field patterns on the photoconductive antenna are discussed by Hughes et al. [23]. They introduced a finite-difference time-domain method for the pulsed laser excited vector THz fields from photoconductive antenna. Berry and Jarrahi [24] have evaluated the criteria to optimize the impedance matching in photoconductive antenna. Moreno et al. [25] presented the mobility model to describe the carrier dynamics for the analysis of radiating semiconductor photoconductive devices in the THz regime. The biased electric field analysis of photoconductive antenna for THz generation is reported by Yang et al. [26]. In their simulation results, it is illustrated that the stripline photoconductive antenna and photoconductive dipole antenna cannot withstand high-biased voltage because of the small value of breakdown electric field of the substrate material. Another exciting new technology for THz antennas is the idea of active surface correction for improving the beam efficiency [27]. However, the high directivity with high front-to-back ratio, optimum radiation efficiency, broad bandwidth, and tuning/phase scanning are the significant challenges related to the design issues of photoconductive dipole antenna. A broadband end-fire photoconductive antenna for photomixing [28] has been explored in [29], which does not require hyper-hemispherical silicon lens to collect and collimate THz radiation.

Moreover, in time domain THz wave detection system having GaAs-based photoconductive antennas, the noise power spectral density needs to be determined quantitatively [30]. Such noise value depends on resistance of the photoconductive antenna, its circuit parameters, and the frequency of operation. In [31], the authors have investigated two types of noise in photoconductive antenna, wherein the thermal noise (also known as Johnson-Nyquist noise) is determined and modelled for THz antenna resistance and another noise density is evaluated from frequency with its intensity dependent on the properties of GaAs and the metallization. Recently, the photoconductive antennas have been designed using graphene ribbons and metals, which supports surface plasmon polaritons [32]. However, since 2D nature of graphene and its electronic modelling, as well as electromagnetic modelling differ from 3D bulk metals; therefore, certain design challenges are remaining to handle with respect to electronic and electromagnetic solvers.

4.3 Parametric Estimation of Photoconductive Dipole Antenna

In this section, using a simple synthesis technique the physical parameters of THz small-gap photoconductive dipole antenna are determined corresponding to 1.5 THz resonance frequency. We are interested to design a THz photoconductive dipole antenna for sensing and imaging to detect the hidden explosives (like RDX and HMX), which shows their spectral fingerprints with absorption peak position in THz regime in the range of 1–2THz. Therefore, a THz photoconductive antenna is designed at 1.5THz to detect such explosives easily. Further, the main obstacle in THz free-space sensing and imaging is the atmospheric attenuation, which is dominated by the absorption of THz wave due to the presence of the water vapor; however, the choice of using 1.5 THz operating frequency is also supported by the transmission windows mentioned in Sect. 4.1. In addition to this, to analyze the performance of designed THz photoconductive dipole antenna, the frequency range 1–2THz is chosen because many home-made ammonium nitrate bombs, as well as some other improvised explosive devices, have featureless THz spectra below 3THz, which are also posing potential challenges to THz security applications. Therefore, this proposed simple antenna design can also be used for the detection of such explosives having THz spectra lower than 3THz. The basic structure of THz photoconductive dipole antenna is shown in Fig. 4.2, which consists of a dipole

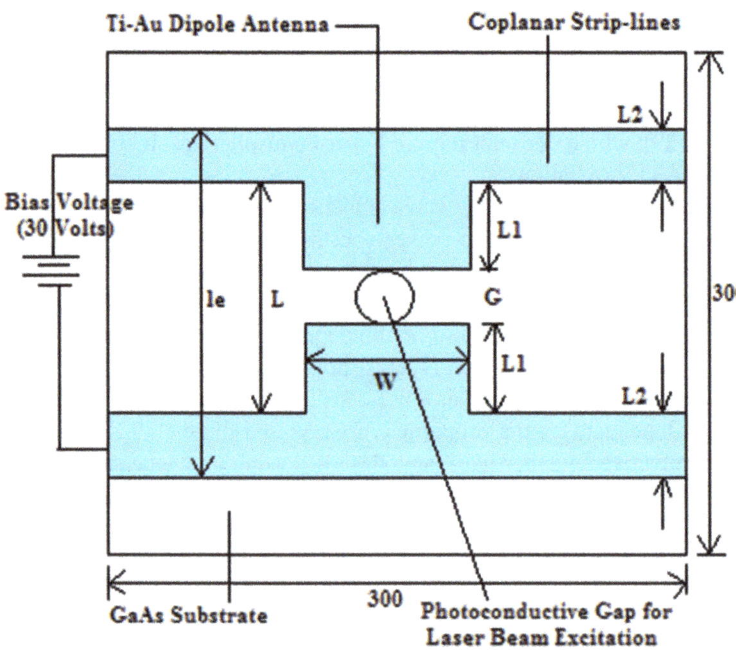

Fig. 4.2 The basic structure of THz photoconductive dipole antenna

antenna structure, a photoconductive substrate, and a ground plane placed under the substrate.

At the center of the dipole antenna of effective length (l_e), there is a photoconductive gap (G) biased by a voltage source (30 volts), which is illuminated at the center with a femtosecond laser pulse.

4.3.1 *Working Phenomenon of Small-Gap Photoconductive Dipole Antenna*

By illuminating the photoconductive layer of the antenna using femtosecond laser pulse, the electron–hole pairs are generated within the photoconductive gap of the antenna. This happens because of the higher photon energy of laser pulse in comparison with the band-gap energy of the photoconductive material. When a biased electric field (E_{bias}) is applied across the antenna electrodes through transmission lines, then the photo-excited carriers get accelerated. A macroscopic electron-hole field (E_{e-h}) gets created in the reverse direction because of the physical separation of charges. As more of electron–hole pairs are generated, there is also an increase in the electron-hole field and after sometime the total electric field at the location of carriers near the dipole electrodes (defined as $E_{\text{field}} = E_{\text{bias}} - E_{e-h}$) is screened. This results in the reduction of the effective electric field across the photoconductive gap of the antenna. Due to the sudden change in the total electric field E_{field}, there is a creation of the transient current, which is responsible for the THz radiations from the photoconductive antenna [33]. The generated transient current decays with time constant, which is determined from the carrier lifetime in the photoconductive substrate used for the antenna. The radiation efficiency of a photoconductive dipole antenna is proportional to the carrier mobility of photoconductive substrate as presented in [1] but does not directly depend on the carrier lifetime. Though, a short carrier lifetime is preferred to reduce the noise generated at the detector side of the THz system due to the thermal motion of the carrier. Likewise, the antenna efficiency is also proportional to the substrate resistivity because it increases linearly with the applied biasing voltage. Moreover, in addition to a low carrier lifetime along with high resistivity of photoconductive material, there is a requirement for the (1) continuance of relatively high carrier mobility, (2) suitable band gap, and (3) high breakdown voltage with suppression of zero bias photocurrent. All these factors play an important role as they influence the antenna's (1) output power, (2) maximum optical pump power, (3) maximum bias voltage, (4) bandwidth, and (5) SNR values [34–36]. Owing to these properties, gallium arsenide (GaAs), low-temperature-grown GaAs, bulk indium gallium arsenide (In GaAs), indium aluminum arsenide (InAlAs), radiation damaged silicon on sapphire, alternating nanoscale-multilayers of InGaAs, and amorphous silicon are the most promising substrate materials for THz photoconductive antennas [37]. The proposed antenna in this work usages GaAs substrate, which has a photocarrier lifetime $\tau_c = 0.25$ ps, a high mobility 200 cm^2/Vs, a high breakdown field 4 × 10^{5}V/cm, and has a room temperature bandgap of 1.424 eV (871 nm) [38] and makes the antenna well-matched to Titanium-doped Sapphire (Ti: Sapphire) femtosecond pulsed laser source generally used to excite photoconductive antennas. If the thickness of substrate is kept 1 μm, then nearly 30% of the light gets absorbed into the substrate; therefore, to reduce the absorption (i.e., below 5%) in substrate at least 3 μm thickness of the substrate is essential. Therefore, the thickness of the substrate is taken as 10 μm in this design, and its dimension is 300 μm× 300 μm × 10 μm. A lossy-metal Ti-Au is used as the material for the dipole antenna structure placed on the GaAs substrate. Due to the good ohmic contact between Ti-Au and GaAs, they are used as the antenna and substrate material, respectively. Graphene is also a promising material for the miniaturized resonant THz antennas design [39, 40]. However, only few initial works considered the use of graphene in THz antennas. Firstly, the graphene has been considered as a parasitic layer below a dipole antenna made of gold (Au) and radiated at millimeter wave frequency of 120 GHz [41]. The scattering of an incident wave impinging on graphene rectangular patches was studied in [42], wherein it has been concluded that the graphene patches support surface plasmonic resonances in the THz range. In [43], the graphene has been used as an actual antenna radiator where radiation is attained by placing a THz continuous wave (CW) photomixer as source in the middle of the graphene patch. The photomixer excites the graphene patches that are DC biased and thus enables its surface to radiate.

4.3.2 Antenna Physical Parameter Estimation Technique

The first physical parameter in the design of photoconductive dipole antenna is the photoconductive gap size (G), where the femtosecond laser pulse strikes to generate the photo carriers.

$$E_{\mathrm{THz}}^{\mathrm{peak}} = e\mu T_{\mathrm{int}} \frac{(1-R)}{h\vartheta} \frac{P_{\mathrm{in}}}{G} \frac{E_{\mathrm{bias}}}{G} \tag{4.1}$$

where e, μ, T_{int} R, $h\vartheta$, P_{in}, G, and E_{bias} are electron charge, carrier mobility, pulse interval, reflection from the photoconductive substrate, photon energy, average pump laser power, photoconductive gap size, and bias voltage, respectively. From (4.1), this is clear that the emission efficiency is inversely proportional to gap size of the photoconductive dipole antenna while the bias field (E_{bias}), as well as the pump-power (P_{in}) are kept constant. Consequently, it is important to keep the photoconductive gap as small as possible and thereby focusing the laser beam (optical excitation) very closely to the small gap. Moreover, when the pump laser power is small, then the efficiency saturates at higher pump intensities. Therefore, in such condition the antenna gap needs to adjust such as to minimize the screening effect (the cancellation of a portion of the bias field by the transient current at the surface when the near field is generated) and as a consequence improve the efficiency in THz photoconductive dipole antenna, which persuades the initial spatial distribution of photo-excited carriers on photoconductive substrate. The photoconductive gap is taken as 5 μm for small-gap photoconductive dipole antenna in all three antenna-designed configurations and its value is optimized using the simulation software (CST Microwave Studio). The length of coplanar stripline is generally lay-down to be long enough to evade the reflection at the line end. Therefore, the length of coplanar stripline taken is 300 μm, width 10 μm, and thickness 0.35 μm. The length of the dipole is determined using the relation of resonant frequency as $f_r = c/2nL$, where c, L, and n are the speed of light in vacuum, separation between two coplanar striplines and refractive index of the material, respectively. For resonance, $L = m \times \lambda n/2$, where m = 1,2,3…., and the wavelength λ_n in the material depends on the refractive index n, which is given by $\lambda_n = \lambda/n$. If we take $m = 1$, then $L = \lambda/2n$. The refractive index of the semiconductor antenna material for GaAs at THz frequencies is 3.4 and for f = 1.5 THz, the value of L (length of dipole) is: $L = C/(2n\, fr)$ = 30 μm. The other equally important physical parameter of the photoconductive dipole antenna is the width (W) of dipole because of two reasons: (1) the directivity and (2) radiation efficiency of the photoconductive dipole antenna, which depends on the relative dimensions of the dipole as both enhances appreciably by increasing the aspect ratio ($\eta \equiv L/W$). The emission efficiency also get increased with increasing the aspect ratio (L/W) of THz photoconductive dipole antenna as reported in [44]. However, an optimum value of aspect ratio for photoconductive dipole antenna can be determined by considering the following factors:

- With the decrease in the dipole length (L), the emission intensity gets decreased significantly.
- The peak frequency of emission spectra shifts to lower frequency values when the dipole length (L) is increased.
- With an increase in the dipole length (L), the bandwidth of antenna happens to be narrower.
- The peak intensity of the antenna decreases considerably with an increase in the dipole width (W), which reveals that the emission efficiency of dipole antenna is proportional to the aspect ratio.

From the aforementioned points, it is concluded that for the better performance of THz photoconductive dipole antenna, the length of dipole must be greater than the width of dipole. In the proposed antenna, we have also checked the performance of the antenna with different values of the aspect ratio by keeping the length of dipole constant to 30 μm and varying the width of the dipole to set the aspect ratios as (L/W) 0.5, 1, 1.5, 2, with the values of width of dipole (W) as 60, 30, 20, 15 (all in μm), respectively. It is observed from the simulated results discussed in sect. V that by keeping aspect ratio 1.5, the performance of THz photoconductive dipole antenna is interesting in terms of directivity in both the E- and H-planes. Further, the other physical parameter of photoconductive dipole antenna is the effective length (l_e), and its value has been computed considering following points.

- If the effective length of dipole is assumed to be same as the distance between coplanar striplines, that is, $l_e = L$, then in this case, $l_e = 30$ µm.
- If the effective length of the dipole antenna is taken into the account: $l_e = 2\,L_2 + L$, and in this case, $l_e = 50$ µm with L2, the width of the stripline, which is 10 µm.

The choice of considering the effective length will affect the resonating frequency as $f_r = c/(2l_e[(1 + \varepsilon_d)/2]^{1/2})$, where f_r and $\in_d$ are the resonating frequency and the dielectric permittivity of GaAs substrate, respectively. If the effective length of dipole is assumed to be same as the distance between the coplanar striplines, that is, $l_e = L$, then by using $c = 3 \times 10^8$ m/s, $l_e = 30$ µm and $\varepsilon_d = 12.9$ for GaAs, the resonance frequency $f_r = 1.89$ THz is obtained. If the effective length of dipole antenna is chosen by considering the width of coplanar striplines, that is, $l_e = 2\,L_2 + L$, then the resonance frequency is $f_r = 1.14$ THz . A THz photoconductive dipole antenna of Ti-Au (lossy-metal) material with conductivity $\sigma = 1.6 \times 10^7$ S/m and thickness 0.35 µm, is designed on a photoconductive substrate GaAs (lossy) having dielectric permittivity $\varepsilon = 12.94$, magnetic permeability $\mu = 1$, and loss tangent $\delta = 0.006$ S/m using the determined physical parameters mentioned before. The antenna designed with these specifications is named as Design-A in the manuscript. However, we are interested to use this THz photoconductive dipole antenna for sensing and imaging application, wherein high directivity is required; therefore, for this purpose, two more designs are simulated. The Design-B is configured by using a superstrate that helps to enhance the radiation performance and efficiency of THz photoconductive dipole antenna [45]. A very thin superstrate of low-temperature-grown GaAs (LT-GaAs) with thickness 1 µm is placed in between the GaAs substrate and the antenna structure. In other antenna configuration, that is Design-C, a silicon lens is used to enhance the directivity of the THz photoconductive dipole antenna. A small hemispherical silicon lens is placed beneath the ground plane, which is the direction of propagation of generated THz wave from the photoconductive dipole antenna. The potential importance of using thin LT-GaAs superstrate and a silicon lens in the THz photoconductive dipole antenna is discussed as follows.

Use of Thin LT-GaAs Superstrate

With the illumination of the photoconductor by using a shot optical pulse from a femtosecond laser pulse, a current surge as shown in Fig. 4.3 is noticed, which results in the generation of the THz radiation. However, in case of long carrier lifetime of the photoconductive substrate, the generated current keeps flowing even after the withdrawn of excitation short optical pulse. This results in broadening of the photocurrent pulse, which further broaden the output pulse and therefore decrease the overall THz frequency bandwidth. Moreover, in case of arrival of next shot optical pulse before the current dies out (generated by previous short optical pulse) and excites the photoconductive dipole antenna, a new THz pulse is created, although at this time the presence of already existing background current due to previous short optical pulse may affect the generation of THz radiation. Therefore,

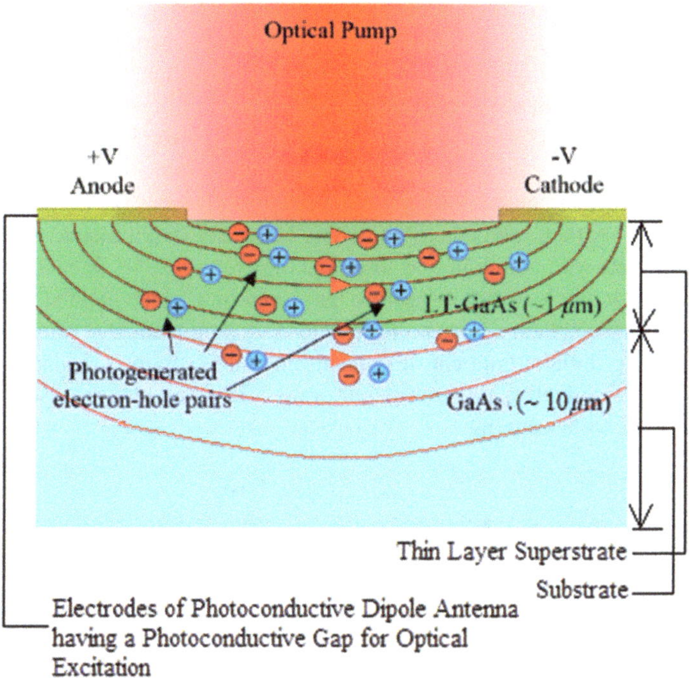

Fig. 4.3 Optical carrier generation at the photoconductive dipole gap of a LT-GaAs superstrate-based THz photoconductive dipole antenna and the red arrows represent the flux lines of the electric field [46]

to prevent this, photoconductors with subpicosecond carrier lifetime such as low-temperature-grown gallium arsenide (LT-GaAs) having very small thickness in comparison to the GaAs substrate may be used.

Moreover, the THz photoconductive dipole antenna with a thin layer (nearly 1 μm) of short carrier lifetime LT-GaAs as superstrate can avoid saturation at high frequency, which helps to achieve peak power spectral density. The LT-GaAs has the relevant features such as very high electric breakdown field (~ 500KV/cm), short photocarrier lifetime (0.1 ps) along with high mobility of charges (> 200 cm^2/Vs); therefore, it is a good material for fabrication as a superstrate in the photoconductive antenna.

Use of Silicon Hemispherical Lens

For directivity enhancement of an antenna and because of the ease of built-up, an extended hemispherical dielectric lens (such as silicon lens) is beneficial in its use with photoconductive dipole antenna. Moreover, the silicon lens reduces loss due to the reflection and refraction of radiation at the substrate–air interface [47]. A silicon lens, which is a hemisphere with radius, R is generally located directly on the ground plane of THz photoconductive dipole antenna. The directivity (*D*) of a lens antenna in the direction of the main lobe of THz radiation can be found using the formula for the directivity of a circular aperture having radius R and uniform distribution of the electric field [48], as $D = 20 \log ((2\pi R)/\lambda 0)$, where λ_0 is the wavelength in free-space and *R* is the lens radius. The width of the beam at −3 dB of the directivity of the integrated lens antenna is estimated as $\Omega = 59^{\circ}(\lambda 0/2R)$. The use of a lens with THz photoconductive dipole antenna also provides the electronic beam-steering capability to the antenna by switching between two photoconductive dipole antenna elements with common bias lines and the ground plane. In such a case, the photoconductive dipole antennas need to be placed on a plane focal surface of the lens. The electronic beam steering allows antenna to automatically adjust the beam direction during initial alignment of transmitting and receiving antennas [49], and thus, the photoconductive dipole antenna array with lens may be used for THz pulsed imaging with scanning.

4.4 Simulation Model

In the small-gap photoconductive dipole antenna, the gap is of only few micrometers (μm), and thus, there is an insufficient acceleration of the charges, which results in the inadequate field strength to generate THz radiations in comparison to that of the large aperture photoconductive antenna, which results in low antenna efficiency. Therefore, to increase the radiation efficiency of photoconductive dipole antenna, it is essential to consider the factors which affecting the total antenna efficiency. In the photoconductive dipole antenna, three types of efficiencies are considered to evaluate the total antenna efficiency (η_t), which are as follows:

- Laser-to-electrical conversion efficiency (η_{LE}).
- Impedance matching efficiency (η_m).
- Radiation efficiency (η_r).

The total efficiency of photoconductive dipole antenna (also known as optical-to-THz power conversion efficiency) represents the multiplication of these three efficiencies:

$$\eta_t = \eta_{LE} * \eta_m * \eta_r \tag{4.2}$$

4.4.1 Computation of Laser-to-Electrical Conversion Efficiency

When a short-duration optical pulse incident onto the photoconductive gap of the proposed antenna, the induced photocurrent is expressed as [15]:

$$I = \frac{e\,E_{\text{bias}}\mu_e\,\tau\eta_L P_L}{hf_L G^2} \tag{4.3}$$

where e, E_{bias}, μ_e, τ, h, G, f_L, P_L,and $\eta_L$are electron charge, the applied bias voltage, free-carrier mobility inside the photoconductor, photocurrent decay time, Planck's constant, gap length, laser frequency, laser power incident onto the photoconductive gap, and the illumination efficiency, respectively. Illumination efficiency considers many issues such as (a) the reflection of laser on the surface of the substrate and (b) the quantum efficiency. The optical-pump laser is focused onto the feeding gap, wherein the photon energy of the laser is equal to or slightly greater than the band gap of the semiconductor substrate so as to ensure that free electrons are efficiently created. Such electro-optical procedure converts the laser power P_L to the electrical power P_E. To determine an expression for the electrical power, initially it is important to obtain the associated resistance, R, such as

$$R \approx \frac{h\,c\,f_R\,G^2}{\eta_L e\,\mu_e P_L \lambda_L} \tag{4.4}$$

where $f_R$is the laser repletion frequency. From (4.3) and (4.4), the induced electric power on the photoconductor is

$$P_E = I^2\,R \approx \left(\frac{e\,E_{\text{bias}}\mu_e\,\tau\eta_L P_L}{hf_L G^2}\right)^2 \frac{h\,c\,f_R\,G^2}{\eta_L e\,\mu_e P_L \lambda_L} \tag{4.5}$$

Therefore, the laser-to-electrical power conversion efficiency is estimated as:

$$\eta_{LE} = \frac{P_E}{P_L} \approx \frac{e\,E_{\text{bias}}{}^2\,\mu_e \tau^2 \eta_L f_R}{hf_L G^2} \tag{4.6}$$

From (4.6), it is clear that besides the laser source, the laser-to-electrical conversion efficiency factor also includes the bias voltage. Therefore, an electric power is not exclusively produced by the laser source.

4.4.2 Calculation of Impedance Matching Efficiency

The impedance matching efficiency is

$$\eta_m = 1 - \left(\frac{Z_a - Z_s}{Z_a + Z_s} \right)^2 \tag{4.7}$$

where $Z_a$and Z_s are the antenna impedance and source impedance, respectively. The value of source impedance is determined from the time-varying source conductance. The time-varying source conductance $G_S(t)$ in the THz photoconductive dipole antenna depends on the values of length (G) and width (W) of the gap. Moreover, the inverse of $G_S(t)$ of the photoconductive material is the time-variant resistance $R(t)$, which is used to determine the time or frequency variant impedance matching efficiency of the photoconductive dipole antenna.

4.4.3 Computation of Radiation Efficiency

The radiation efficiency of photoconductive dipole antenna is the ratio of gain and directivity of the antenna at the chosen frequency of operation, which is determined using an electromagnetic simulation tool CST Microwave Studio.

However, the low radiation efficiency is the major challenge to the present photoconductive dipole antenna, and it occurs due to the excessive ohmic losses at THz frequencies. Further, the impedance matching efficiency of antenna must be considered for proper impedance matching of laser source with photoconductive material of antenna. It is essential to mention here that the THz photoconductive dipole antenna is simulated using the CST Microwave Studio. It offers a simulation platform for all kind of electromagnetic field applications. We have used the transient solver, which is based on the Finite Integration Technique (FIT) in which direct time-domain analysis is applied with broadband computation of S-parameters through one single calculation run on applying DFTs to the time signals. The simulation run for the proposed antenna is also supported by the adaptive-mesh refinement in 3D using S-parameters supported by the transient solver of CST Microwave Studio. In our work, we have determined the length (and effective lengths) of the photoconductive dipole antenna for a specific terahertz frequency, and the corresponding resonant frequencies have been computed numerically. The structure parameters considered for the proposed antenna is presented in Table 4.1. We have used the CST Microwave Studio to obtain the S_{11} parameter from which the resonant peaks can be determined and is compared with the values of resonant frequencies obtained mathematically. If we refer Fig. 4.6, “The S-parameter (dB) for three proposed antenna design configurations,” the three resonant peaks obtained for each antenna design are nearly same as that of the theoretical obtained values.

Table 4.1 The structure parameters for the proposed antenna

Parameter	Value
Dipole antenna (Ti-au) and ground	
Conductivity (S/m)	1.6×10^7
Superstrate (LT-GaAs)	
Carrier lifetime, majority carriers (psec)	0.1
Mobility (cm^2/Vs)	Greater than 200
Dielectric permittivity	13.26 [1]
Magnetic permeability	1
Loss tangent (S/m)	0.006
Electric breakdown field (V/cm)	Greater than 5×10^5
Substrate (GaAs)	
Carrier lifetime, majority carriers (psec)	0.25
Mobility (cm^2/Vs)	200
Dielectric permittivity	12.9
Magnetic permeability	1
Loss tangent(S/m)	0.006
Electric breakdown field (V/cm)	Nearly 4×10^5
Silicon lens	
Permittivity	11.9
Permeability	1
Loss tangent (S/m)	0.00025
Voltage source	
DC voltage (V)	30

However, with the help of multiphysics finite-element solver (COMSOL), the response of the photoconductor of the designed dipole to the incident short-optical pulse can be estimated. In such case, the photo-generated carrier density is derived using the drift-diffusion model from the calculated optical intensity in the photo-absorbing substrate along with the bias electric field data to compute the induced photocurrent [24]. Using Ti:Sapphire laser having characteristics of 800 nm central wavelength, repetition rate of 76 MHz with pulse width of 100 fsec, an optical pump beam from the laser is allowed to get tightly focused onto the photoconductive antenna gap and is also positioned near the anode contact electrode to maximize the radiated power [50]. The generated terahertz power from photoconductive emitter can be measured using a pyroelectric detector.

Moreover, to determine the photoconductive dipole emitter characterization with the help of experimental setup in terms of device alignment, output power measurement, and radiation spectral characterization, the stages that are required are mentioned in brief as follows:

Stage 1: Device Alignment

(a) Initially, place the photoconductive THz emitter in the aluminium washer and using Ti: Sapphire mode-locked laser such as MIRA 900D V10 XW OPT 110 V, focus the optical laser pulse on the photoconductive gap of the antenna.

(b) With the help of the parametric analyzer, provide the bias voltage to the coplanar striplines of photoconductive dipole antenna to measure the induced electric current.
(c) To modulate the optical pump from the laser focused onto the gap of photoconductive dipole antenna, an optical chopper such as Thorlabs MC2000 can be used.

Stage 2: Output Power Measurement

(a) Using pyroelectric detector such as Spectrum Detector, Inc. SPI-A-65 THz, measure the output power of photoconductive dipole antenna.
(b) To pull through terahertz power data at low noise levels, connect the output of pyroelectric detector with lock-in amplifier such as Stanford Research Systems SR830 along with the reference frequency of optical chopper.

Stage 3: Radiation Spectral Characterization

(a) Using a beam splitter, split the output optical beam of mode-locked Ti:Sapphire laser into pump beam and probe beam.
(b) To generate the THz radiations, focus the pump beam of laser on the photoconductive gap of photoconductive dipole antenna.
(c) With the help of polyethylene spherical lens, collimate the THz beam generated from the photoconductive dipole antenna as emitter.
(d) Merge the collimated THz beam coming from the polyethylene spherical lens with the probe beam using ITO-coated glass filter.
(e) At the combined focus of the optical and THz beam, place a ZnTe crystal (of 1 mm thickness).
(f) To vary the time delay occurring between the optical beam and the THz pulse, which is interacting with ZnTe Crystal, place-in a controllable optical delay line in the path of optical probe beam with the help of a motorized linear stage such as Thorlabs NRT100.
(g) With a Wollaston prism, split the optical beam, and using balance detectors linked to lock-in amplifier, measure the optical beam power in each branch.
(h) Join other end of motorized delay line, as well as lock-in amplifier, to the computer in which a Matlab script is encoded in such a way that iteration can be performed to move the motorized delay line, pause and read the magnitude of the signal from lock-in amplifier.
(i) By dividing the total optical delay length with speed of light, convert the stage position to time-domain and further, using Matlab obtain the frequency domain data from discrete Fourier transform.

In the photoconductive antenna, the optical source is a femtosecond laser pulse, which has a Gaussian distribution at its output; therefore, in the CST MSW simulation software, we have applied the Gaussian beam excitation into the substrate from the gap of photoconductive dipole antenna. The antenna configurations, which are designed in CST Microwave studio, are shown in Fig. 4.4.

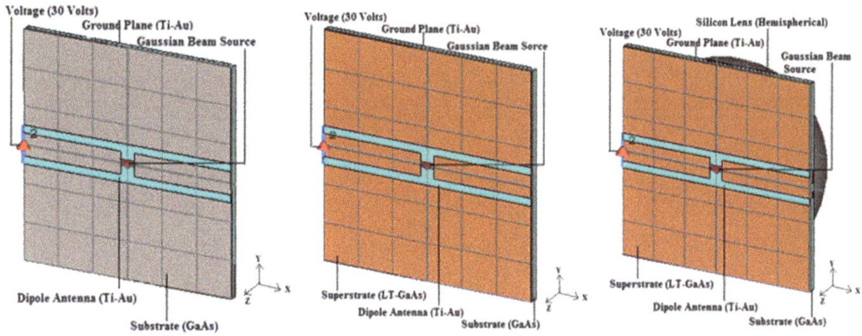

(i) Design-A (ii) Design-B (iii) Design-C

Fig. 4.4 Three configurations (i) Design-A: Basic THz photoconductive dipole antenna, (ii) Design-B: THz photoconductive dipole antenna with LT-GaAs superstrate, and (iii) Design-C: THz photoconductive dipole antenna with LT-GaAs superstrate and silicon lens [51]

4.5 Simulation Results and Discussions

Initially, the comparison of THz photoconductive dipole antenna (Design-A) with different aspect ratio (*L/W*) is performed, choose and optimize the L/W for the proposed reference antenna (Basic THz photoconductive dipole antenna). The length and width of the co-planar striplines are assumed constant 300 μm and 10 μm, respectively. The thickness of substrate and ground plane is 10 μm and 0.35 μm, respectively, with the antenna thickness 0.35 μm. The distance between striplines (*L*) is kept constant 30 μm and the width of the gap W is varied to observe the effect of aspect ratio on the performance of THz photoconductive dipole antenna. The values of *W* are chosen as 60, 30, 20, 15 (all in μm), with fixed *L* as 30 μm. The comparison for several aspect ratios in terms of gain and directivity is presented in Table 4.2. The antenna efficiency is another important parameter, which needs to be enhanced to use the antenna in the sensing and imaging applications. Figure 4.5 shows the effect of variation in aspect ratio on radiation efficiency of Design-A antenna.

From Table 4.2 and Fig. 4.5, it is observed that the selection of aspect ratio is chosen either $L/W = 1$ or $L/W = 1.5$. If we choose the aspect ratio less than 1, then the gain and directivity of Design-A are interesting with respect to simple dipole design, but the radiation efficiency is too small. On the other hand, if we choose the aspect ratio 2, the gain reduces. Furthermore, we have chosen the aspect ratio as 1.5, and for this *L* and *W* are 30 μm and 20 μm, respectively. Further the reason for this selection of aspect ratio is supported by the case, wherein on providing adequate power from the laser beam onto the photoconductive gap the peak intensity gets decreased considerably with an increase in the dipole width, which results in constant total input current. Therefore, the width of dipole must be smaller than the length of the dipole. With the aspect ratio 1.5, the dipole length 30 μm and width of

Table 4.2 Comparison of gains directivity in both E and H plane photoconductive dipole antenna for several aspect ratio

Aspect ratio (L/W) of THz photoconductive dipole antenna	Gain (dB) E-plane	Gain (dB) H-plane	Directivity (dBi) E-plane	Directivity (dBi) H-plane
0.5	3.24	8	5.18	9.94
1.0	3.31	9.03	4.02	9.74
1.5	3.13	8.71	4.11	9.69
2.0	3.06	8.59	4.17	9.71

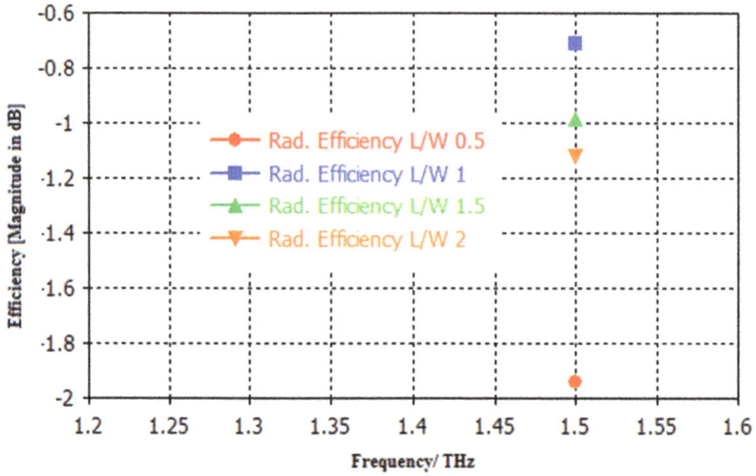

Fig. 4.5 Radiation efficiency at different values of aspect ratio of Design-A (THz photoconductive dipole antenna)

dipole 20 μm are considered for the design of basic photoconductive dipole antenna Design-*A*. The S-parameter (in dB) for: (1) Design-A, (2) Design-B, and (3) Design-C are compared for the S-parameter (in dB) as shown in Fig. 4.6. The resonant peaks observed in the results are in accordance with the expressions of resonance frequency discussed in sect. III with small deviations from their theoretical values. For all the three designed configurations, three bands are observed below −10 dB as shown in Fig. 4.6, and the comparison is performed corresponding to the highest resonant peak at $f = 1.587$ THz.

In Design-A, the S_{11} parameter at 1.3 THz, 1.593 THz, and 1.9 THz are −10.22 dB, −20.32 dB, and −24.568 dB, respectively. The 10 dB impedance bandwidths obtained for each center frequency are 50 GHz, 39 GHz, and 39 GHz, respectively. For Design-B, the S_{11} parameter at frequencies 1.301 THz, 1.587 THz, and 1.868 THz are −11.65 dB, −18.37 dB, and − 16.30 dB, respectively. The 10 dB impedance bandwidths obtained for each center frequency are 14 GHz, 36 GHz, and 20 GHz, respectively. Similarly, for the Design-C, the return loss values at frequencies 1.301 THz, 1.587 THz, and 1.868 THz are 11.55 dB, 26.47 dB, and 14.09 dB,

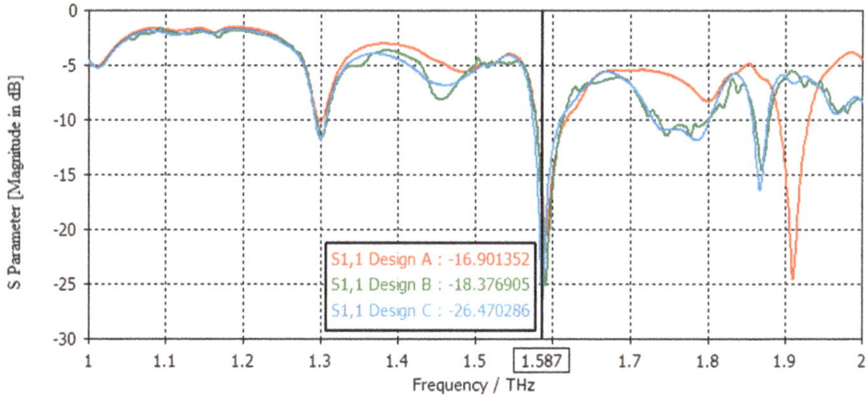

Fig. 4.6 The S-parameter (dB) for three proposed antenna design configurations, Design-A: Basic THz photoconductive dipole antenna, Design-B: THz photoconductive dipole antenna with LT-GaAs superstrate, and Design-C: THz photoconductive dipole antenna with LT-GaAs superstrate and silicon lens

respectively, and the 10 dB impedance bandwidths are 13 GHz, 35 GHz, and 20 GHz, respectively. The potential reasons for small deviation in resonant peaks are as follows:

- The resonance phenomena appear when the quality (Q) factor of the antenna is large enough. Moreover, the THz field does not propagate or reflect for a large distance along the antenna because of the significant losses associated with the antenna such as radiation loss and dielectric loss. Consequently, the Q-factor of the THz antenna is small (no standing wave), and therefore the resonance effect is not as effective as it is being observed in the microwave frequencies.
- Another reason is the slow decay time of the generated photocurrent and is approximately 0.5 ps in case of GaAs photoconductive substrate, which limits the generation of higher-frequency components of radiation.
- Moreover, the absence of resonance peaks at the position of expected frequencies ought to be attributed to some other factors also, such as there is broadening of the resonance peak because of the large antenna width or damping of the switching response because of the dominance of the capacitance effect across the photoconductive gap at high frequency.

The performance of proposed antenna designs (Design-A, Design-B, and Design-C) are also compared for Gain (dB) and Directivity (dBi), in the principal plane patterns. The simulated results are shown in Figs. 4.7 and 4.8, respectively.

For Design-A, the gain is 3.13 dB in E-plane with main lobe direction at 0°. The angular width (3-dB) is 34.2° and the side lobe level is −0.8 dB and in H-plane, the gain is 8.71 dB with its main lobe direction 40.0°. The angular width (3-dB) is 18.2° having a side lobe level −5.6 dB. The directivity at E-plane is 4.11 dBi with main lobe direction at 0°. The angular width (3-dB) is 34.2°, and the side lobe level is

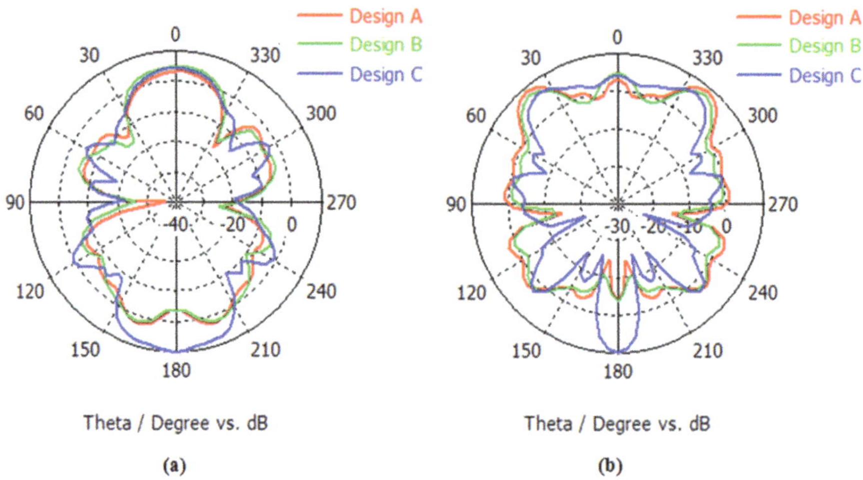

Fig. 4.7 The antenna gain characteristics of all three configurations in (**a**) E-plane, (**b**) H-plane, Design-A: Basic THz photoconductive dipole antenna, Design-B: THz photoconductive dipole antenna with LT-GaAs superstrate, and Design-C: THz photoconductive dipole antenna with LT-GaAs superstrate and silicon lens

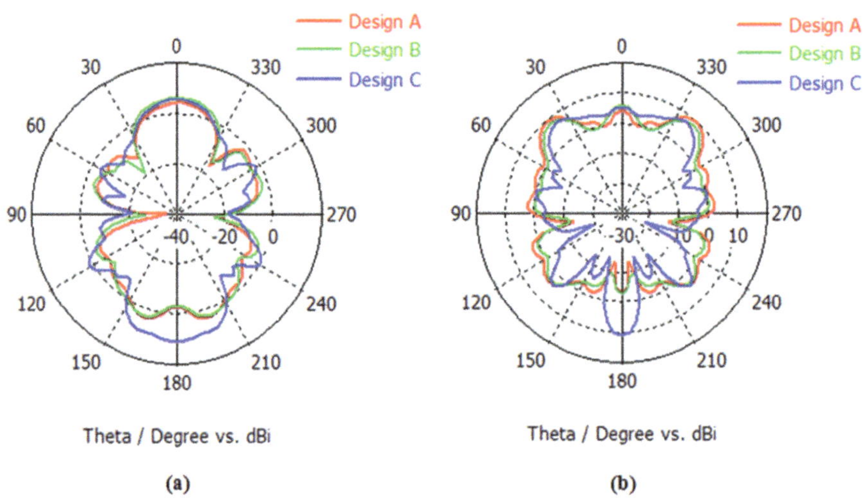

Fig. 4.8 The antenna directivity of all three configurations in (**a**) E-plane, (**b**) H-plane, Design-A: Basic THz photoconductive dipole antenna, Design-B: THz photoconductive dipole antenna with LT-GaAs superstrate, and Design-C: THz photoconductive dipole antenna with LT-GaAs superstrate and silicon lens

−0.8 dB, and for H-plane, the directivity is 9.69 dBi with its main lobe direction 40°. The angular width (3-dB) is 18.2° having a side lobe level −5.6 dB. The radiation efficiency obtained in E-plane and H-plane are 0.76 and 0.89 with power flow 5.12×10^{10} VA/m^2. As the radiation efficiency in E-plane is less than 80% and to increase the radiation efficiency of the THz photoconductive dipole antenna, a thin superstrate is used in this reference design. The simulated results for Design-B show the gain of 4.85 dB in E-plane with main lobe direction at 0°. The angular width (3-dB) is 39.6° and the side lobe level is −3.0 dB. However, for H-plane, the gain is 7.3 dB with its main lobe direction 40°. The angular width (3-dB) is 18.5° having a side lobe level −2.5 dB. The directivity in E-plane is 5.78 dBi with main lobe direction at 0°. The angular width (3-dB) is 39.6°, and the side lobe level is −3.0 dB, and for H-plane, the directivity is 8.24 dBi with its main lobe direction 40°. The angular width (3-dB) is 18.5° having a side lobe level −2.5 dB. The radiation efficiency obtained in E-plane and H-plane are 0.84 and 0.89 with power flow 6.24×10^{10} VA/m^2. It is clearly observed from the simulated results that the radiation efficiency has been increased in the E-plane from 76% to 84% along with the increase in the directivity from the 3.13 dBi to 5.78 dBi for the same direction of main lobe, that is, 0°. Moreover, other interesting point, which has been observed, is that with the use of thin superstrate (LT-GaAs) along with the substrate (GaAs); the power flow has also been increased.

For the purpose of THz sensing applications, there is a need of high-directional scanning photoconductive antennas with inexpensive steerable integrated lens. As the proposed photoconductive dipole antenna is a simple configuration, thus, it allows us to use the focusing property of the dielectric lens to excite the directed radiation. Therefore, to increase the directivity of THz photoconductive dipole antenna, we have used a silicon lens placed beneath the ground plane that is Design-C from where the THz radiations are generated. The simulation results for Design-C show the value of gain 9.8 dB in E-plane with main lobe direction at 180.0 degree. The angular width (3-dB) is 42.8°, and the side lobe level is −5.5 dB. For the H-plane, the gain remains same, which is 9.8 dB with its main lobe direction same as in E-plane, that is, 180°. The angular width (3-dB) is 13.4° having a side lobe level −3.5 dB. The directivity in the E- and H-planes is also having same values, 10.7 dBi, with main lobe direction at 180° for both cases. The angular width (3-dB) is 42.8° in the E-plane, and the side lobe level is −5.5 dB. However, for H-plane, the directivity has the angular width (3-dB) as 13.4 degree having a side lobe level −3.5 dB. The radiation efficiency obtained in E-and H-planes are 91.59% with power flow 5.93×10^{10} VA/m^2. It is clearly observed from the results that the radiation efficiency has been increased and is uniform in both the E-and H-planes. Therefore, the use of silicon lens helps to achieve enhanced gain and directivity values with uniformity in both the principal planes. Moreover, according to the dipole approximation, the radiated electric field from a point source is proportional to the time derivative of the point current in the far-field. Using this relation, the electric field due to the distributed current is computed by taking the volume integration of time-derivative of the current density, $J(r',t')$ [14]. Therefore, the THz electric field emitted from the photoconductive dipole antenna is expressed in terms of the current density, $J(r',t')$ defined at a point r′ and time t′, is expressed as:

$$E_{\mathrm{THz}}(r,t) = -\frac{1}{4\pi\varepsilon c^2}\int_0^{l_e}\left[\frac{\partial J(r',t')}{\partial t'}\right]\frac{\sin\theta}{|r-r'|}\mathrm{d}^3x' \tag{4.8}$$

where $E_{\mathrm{THz}}(r, t)$ is the THz electric field at the observation point r, and the observation time t. ε is the dielectric constant of the medium, and Θ represents the angle between the direction of current and the direction of observation.

In photoconductive dipole antenna, when a biased field E_{bias} is applied across the co-planar striplines, then it acts as an initial driving force for the photo-generated carriers to move toward the biased antenna electrodes. This results in the formation of time-varying photocurrent within the gap of the antenna and is given by

$$J_{\mathrm{PC}}(t) = \frac{\sigma(t) E_{\mathrm{bias}}}{\left[\dfrac{\sigma(t) Z_0}{1+n} + 1\right]} \tag{4.9}$$

where $\sigma(t)$ is the conductivity of the photoconductive material, Z_0 represents characteristic impedance of the free-space, and n is the refractive index of the substrate. The numerator of (4.9) corresponds to the Ohm's law, and the denominator corresponds to the saturation effect, which occurs due to the field screening by the charged carriers. Using simulation software, we have also illustrated the current density distribution of each of proposed antenna configurations (Design-A, Design-B, and Design-C) as shown in Fig. 4.9. In the Design-A, a very high current density 4×10^5 A/m is observed because the use of thick photoconductive substrate (GaAs) with thickness 10 μm. For the Design-B and Design-C, the value of current density is 1.38×10^5 A/m and 1.42×10^5 A/m, respectively. The lower values of current density are due to the use of thin layer of superstrate LT-GaAs placed over the GaAs substrate. It is also observed from Fig. 4.9 that in the Design-C, the current density is more at the center of the substrate near the photoconductive gap in comparison of the sides of substrate. This may be due to the use of silicon lens, which is making the THz field highly directive in both the principal plane patterns in the main lobe direction at 180°.

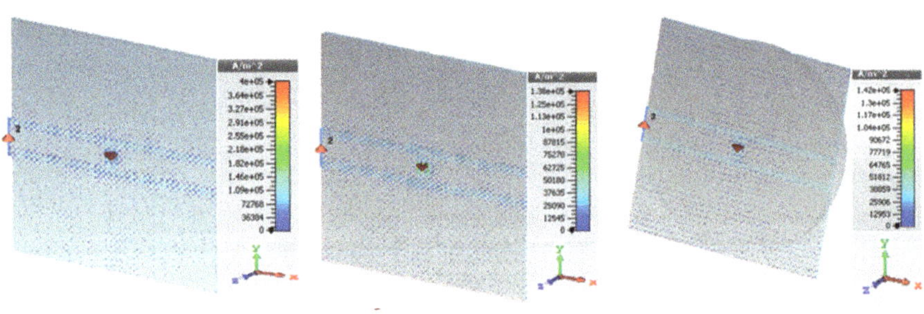

(i) Design A (ii) Design B (iii) Design C

Fig. 4.9 The current density distribution on the planar surface of three configurations presented using CST Microwave Studio, (i) Design-A: Basic THz photoconductive dipole antenna, (ii) Design-B: THz photoconductive dipole antenna with LT-GaAs superstrate, and (iii) Design-C: THz photoconductive dipole antenna with LT-GaAs superstrate and silicon lens [51]

4.6 Summary

In this chapter, we have presented a simple synthesis technique to determine the physical parameters of photoconductive dipole antenna, which is used for THz sensing and imaging applications. For the Design-A, the choice of simple dipole antenna with small-gap geometry has been proposed because of its simplicity in fabrication. However, the basic photoconductive dipole antenna illustrates low values of directivity and radiation efficiency. Therefore, by using thin superstrate (LT-GaAs) in the Design-B, the radiation efficiency increases in the E-plane from 76% to 84% along with the increase in directivity from 3.13 dBi to 5.78 dBi. Further, the proposed basic geometry of photoconductive dipole antenna with silicon lens is presented, which enhances the antenna performance that is potentially useful for THz sensing and imaging application like the detection of hidden explosives (RDX, HMX, PETN, and TNT) along with some commonly used explosive-related compounds. This proposed antenna (Design-C) shows significantly high directivity up to 10.7dBi and radiation efficiency of 91.59% in both E-plane and H-plane at 1.5 THz. A THz imaging system using THz photoconductive dipole antenna provides relatively good signal-to-noise ratio (SNR), large dynamic range with optimum bandwidth; however, the optical-to-THz conversion efficiency of the system is very low. Therefore, it is necessary to analyze the practical constraints on choosing the antenna parameters using equivalent circuit approach, which helps to improve the antenna total efficiency.

References

1. M. Tani, M. Herrmann, K. Sakai, Generation and detection of terahertz pulsed radiation with photoconductive antennas and its application to imaging. Meas. Sci. Technol. **13**(11), 1739–1745 (2002)
2. H.-B. Liu, H. Zhong, N. Karpowicz, Y. Chen, X.-C. Zhang, Terahertz spectroscopy and imaging for defense and security applications. Proc. IEEE **95**(8), 1514–1527 (2007)
3. J. Chen, Y. Chen, H. Zhao, G.J. Bastiaans, X.-C. Zhang, Absorption coefficients of selected explosives and related compounds in the range of 0.1–2.8 THz. Opt. Express **15**(19), 12060–12067 (2007)
4. J.F. Federici, B. Schulkin, F. Huang, D. Gary, R. Barat, F. Oliveira, D. Zimdars, THz imaging and sensing for security applications—explosives, weapons and drugs. Semicond. Sci. Technol. **20**(7), S266–S280 (2005)
5. S. Galoda, G Singh, Fighting terrorism with terahertz. IEEE Potential Magazine, **26**(6), 24–29 (2007)
6. L. Ho, M. Pepper, P. Taday, Terahertz spectroscopy: Signatures and fingerprints. Nat. Photonics **2**(9), 541–543 (2008)
7. Y. Huang, N. Khibani, Y. Shen, D. Li, Terahertz photoconductive antenna efficiency. In *Proceedings of International Workshop on Antenna Technology (iWAT)*, pp. 152–156, 7–9 March, 2011, Hong Kong
8. S.-G. Park, A.M. Weiner, M.R. Melloch, C.W. Siders, J.L.W. Siders, A.J. Taylor, High-power narrow-band terahertz generation using large-aperture photoconductors. IEEE J. Quantum Electron. **35**(8), 1257–1268 (1999)

9. D. Jackson, N. Alexopoulos, Microstrip dipoles on electrically thick substrates. J. Infrared Millim. Terahertz Waves **7**(1), 1–26 (1986)
10. M. Tani, Y. Hirota, C.T. Que, S. Tanaka, R. Hattori, M. Yamaguchi, S. Nishizawa, M. Hangyo, Novel terahertz photoconductive antennas. J. Infrared Millim. Terahertz Waves **27**(4), 531–546 (2006)
11. T.K. Nguyen, W.T. Kim, B.J. Kang, H.S. Bark, K. Kim, J. Lee, I. Park, T.-I. Jeon, F. Rotermund, Photoconductive dipole antennas for efficient terahertz receiver. Opt. Commun. **383**, 50–56 (2017)
12. N. Zhu, R.W. Ziolkowski, Photoconductive THz antenna designs with high radiation efficiency, high directivity, and high aperture efficiency. IEEE Trans. Terahertz Sci. Technol. **3**(6), 721–730 (2013)
13. G. Mourou, C. Stancampiano, A. Antonetti, A. Orszag, Picosecond microwave pulses generated with a subpicosecond laser-driven semiconductor switch. Appl. Phys. Lett. **39**(4), 295–296 (1981)
14. D. Auston, K. Cheung, P. Smith, Picosecond photoconducting Hertzian dipoles. Appl. Phys. Lett. **45**(3), 284–286 (1984)
15. P.R. Smith, D.H. Auston, M.C. Nuss, Subpicosecond photoconducting dipole antennas. IEEE J. Quantum Electron. **24**(2), 255–260 (1988)
16. C. Fattinger, D. Grischkowsky, Terahertz beams. Appl. Phys. Lett. **54**(6), 490–492 (1989)
17. D. Grischkowsky, I. Duling III, J. Chen, C.-C. Chi, Electromagnetic shock waves from transmission lines. Phys. Rev. Lett. **59**(15), 1663–1666 (1987)
18. J.V. Rudd, M.W. Warmuth, S.L. Williamson, D.A. Zimdars, Compact fiber pigtailed terahertz modules. *Washington, DC, U.S. Patent 6,816,647*, issued November 9, 2004
19. C. Fumeaux, G.D. Boreman, W. Herrmann, F.K. Kneubühl, H. Rothuizen, Spatial impulse response of lithographic infrared antennas. Appl. Opt. **38**(1), 37–46 (1999)
20. G.M. Rebeiz, D.B. Rutledge, Integrated horn antennas for millimeter-wave applications. *In Annales des télécommunications*, vol. 47, no. 1–2, pp. 38–48, Springer-Verlag, 1992
21. T.K. Nguyen, F. Rotermund, I. Park, A traveling-wave stripline dipole antenna on a substrate lens at terahertz frequency. Curr. Appl. Phys. **14**(8), 998–1004 (2014)
22. M. Tani, S. Matsuura, K. Sakai, S.-I. Nakashima, Emission characteristics of photoconductive antennas based on low-temperature-grown GaAs and semi-insulating GaAs. Appl. Opt. **36**(30), 7853–7859 (1997)
23. S. Hughes, M. Tani, K. Sakai, Vector analysis of terahertz transients generated by photoconductive antennas in near-and far-field regimes. J. Appl. Phys. **93**(8), 4880–4884 (2003)
24. C.W. Berry, M. Jarrahi, Principles of impedance matching in photoconductive antennas. J. Infrared Millim. Terahertz Waves **33**(12), 1182–1189 (2012)
25. E. Moreno, M. Pantoja, F. Ruiz, J. Roldán, S. García, On the numerical modeling of terahertz photoconductive antennas. J. Infrared Millim. Terahertz Waves **35**(5), 432–444 (2014)
26. J. Yang, W. Fan, B. Xue, Biased electric field analysis of a photoconductive antenna for terahertz generation. Nucl. Instrum. Methods Phys. Res., Sect. A **637**(1), S165–S167 (2011)
27. K.R. Jha, G. Singh, Analysis and design of ring-resonator integrated hemi-elliptical lens antenna at terahertz frequency. Opt. Commun. **285**(16), 3445–3452 (2012)
28. E. Ohmichi, T. Fujimoto, K. Minato, H. Ohta, Terahertz electron paramagnetic resonance spectroscopy using continuous-wave frequency-tunable photomixers based on photoconductive antennae. Appl. Phys. Lett. **116**, 051101 (2020)
29. M.F. Abdullah, A.K. Mukherjee, R. Kumar, S. Preau, Vivaldi end-fire antenna for Thz photomixers. Int. J. Infrared Millim. Terahertz Waves, 1–12 (2020)
30. D. Palaferri, Y. Todorov, D. Gacemi, S. Barbieri, L.H. Li, A.G. Davies, E.H. Linfield, C. Sirtori, Noise characterization of patch antenna THz photodetectors. Appl. Phys. Lett. **113**, 161105 1–161105 5 (2018)
31. M. Nitta, R. Nakamura, Y. Kadoya, Measurement and analysis of noise spectra in terahertz wave detection utilizing low-temperature-grown GaAs photoconductive antenna. J. Infrared Millim. Terahertz Waves **40**(11), 1150–1159 (2019)

32. R. Emadi, R. Safian, A.Z. Nezhad, Transmitting and detecting THz pulses using graphene and metals-based photoconductive antenna. J. Opt. Soc. Am. B **35**(1), 113–121 (2018)
33. N. Khiabani, Y. Huang, Y.-C. Shen, Discussions on the main parameters of THz photoconductive antennas as emitters. *Proceedings of the 5th European Conference on Antennas and Propagation (EUCAP)*, pp. 462–466, 11 April 2011, Rome, Italy
34. K. Moon, J. Choi, J.-H. Shin, S.-P. Han, H. Ko, N. Kim, J.W. Park, Y.J. Yoon, K.Y. Kang, H.C. Ryu, K.H. Park, Generation and detection of terahertz waves using low-temperature-grown GaAs with an annealing process. ETRI J. **36**(1), 159–162 (2014)
35. T.-A. Liu, M. Tani, M. Nakajima, M. Hangyo, K. Sakai, S.-I. Nakashima, C.-L. Pan, Ultrabroadband terahertz field detection by proton-bombarded InP photoconductive antennas. Opt. Express **12**(13), 2954–2959 (2004)
36. L. Hou, W. Shi, S. Chen, Noise analysis and optimization of terahertz photoconductive emitters. IEEE J. Sel. Top. Quantum Electron **19**(1), 8401305/1–8401305/5 (2013)
37. M. Venkatesh, K. Rao, T. AbM. Nitthilash, S. Tewari, A. Chaudhary, Optical characterization of GaAs photoconductive antennas for efficient generation and detection of terahertz radiation. Opt. Mater. **36**(3), 596–601 (2014)
38. S.L. Chuang, *Physics of Photonic Device* (Wiley, 2nd ed, 2008) Hoboken, New Jersey
39. H. Skulason, H. Nguyen, A. Guermoune, V. Sridharan, M. Siaj, C. Caloz, T. Szkopek, 110 GHz measurement of large-area graphene integrated in low-loss microwave structures. Appl. Phys. Lett. **99**(15), 153504/1–153504/4 (2011)
40. P.-Y. Chen, C. Argyropoulos, A. Alu, Terahertz antenna phase shifters using integrally-gated graphene transmission-lines. IEEE Trans. Antennas Propag. **61**(4), 1528–1537 (2013)
41. M. Dragoman, A. Muller, D. Dragoman, F. Coccetti, R. Plana, Terahertz antenna based on graphene. J. Appl. Phys. **107**(10), 104313/1–104313/3 (2010)
42. I. Llatser, C. Kremers, D.N. Chigrin, J.M. Jornet, M.C. Lemme, A. Cabellos-Aparicio, E. Alarcon, Characterization of graphene-based nano-antennas in the terahertz band. *6th European Conference on Antennas and Propagation (EUCAP),* pp. 194–198, 26–30 March, 2012, Prague, Czech Republic
43. M. Tamagnone, J. Gomez-Diaz, J. Mosig, J. Perruisseau-Carrier, Analysis and design of terahertz antennas based on plasmonic resonant graphene sheets. J. Appl. Phys. **112**(11), 114915/1–114915/4 (2012)
44. F. Miyamaru, Y. Saito, K. Yamamoto, T. Furuya, S. Nishizawa, M. Tani, Dependence of emission of terahertz radiation on geometrical parameters of dipole photoconductive antennas. Appl. Phys. Lett. **96**(21), 211104/1–211104/3 (2010)
45. I. Malhotra, P. Thakur, S. Pandit, K.R. Jha, G. Singh, Analytical framework of small-gap photoconductive dipole antenna using equivalent circuit model. Optical and Quantum Electronics, **49**(10), 334/1–23 (2017)
46. N.M. Burford, M.O. El-Shenawee, Review of terahertz photoconductive antenna technology. Opt. Eng. **56**(1), 010901–010901 (2017)
47. W. Dou, Z. Sun, Ray tracing on extended hemispherical and elliptical silicon dielectric lenses. Int. J. Infrared Millim. Terahertz Waves **16**(11), 1993–2002 (1995)
48. A. Artemenko, A. Mal'tsev, R. Maslennikov, A. Sevastyanov, V. Sorin, Studying integrated silicon-lens antennas for radio communication systems operated in the 60 GHz frequency band. Radiophys. Quantum Electron. **55**(8), 511–519 (2013)
49. A.A. Artemenko, V.N. Sorin, R.O. Maslennikov, A.V. Mozharovskiy, Lens antenna with electronic beam steering capabilities, U. S. Patent Application 14/593,552, filed January 9, 2015
50. P.C. Upadhya, W. Fan, A. Burnett, J. Cunningham, A.G. Davies, E.H. Linfield, J. Lloyd Hughes, E. Castro-Camus, M.B. Johnston, H. Beere, Excitation-density-dependent generation of broadband terahertz radiation in an asymmetrically excited photoconductive antenna. Opt. Lett. **32**(16), 2297–2299 (2007)
51. I. Malhotra, K.R. Jha, G. Singh, Analysis of highly directive photoconductive dipole antenna at terahertz frequency for sensing and imaging applications. Opt. Commun. **397**, 129–139 (2017)

Chapter 5
Analytical Framework of Small-Gap Photoconductive Dipole Antenna: An Equivalent Circuit Model

5.1 Introduction

For the prospective demand to achieve large depth-of-field (DoF) to get a better image resolution in the THz imaging application, there is a necessity to design a highly directive, compact, and planar THz antenna source with its capability of on-chip fabrication. The photoconductive dipole antenna has the significant use in THz imaging application because the antenna can be used at the emitter side as well as at the receiver side with same dimensions. However, the only difference is that at the receiver side no biased voltage is applied across the antenna electrodes. Further, the performance of the THz imaging system is mainly affected by photoconductive antenna as an emitter because the detected power level of the THz imaging system at the receiver side is mainly governed by the total radiated power of the emitter. However, there are certain modalities for improving the photoconductive dipole antenna performance that are required to identify to achieve high THz average radiated power and improved total efficiency. Therefore, the unit-cell small-gap photoconductive dipole antenna radiation power enhancement techniques need to be optimized with respect to the antenna design parameters along with the selection of photoconductive material by means of theoretical simulation. Further, to enhance the radiation power of small-gap photoconductive dipole antenna, it is necessary to improve the coupling efficiency of THz wave with air in addition to the femtosecond laser incident efficiency. Accordingly, it is crucial to have a detailed analytical analysis of the THz photoconductive dipole antenna as a radiating source.

When the semiconductor gap in the middle of the photoconductor contact electrodes is uniformly illuminated by the optical pump, then the conjugate matching (for the maximum power transfer) of the photoconductor impedance to the antenna impedance maximizes the radiated power from the photoconductive dipole antenna [1]. However, concerning the extension of the gap size, the photoconductive antennas are classified as follows:

I. Malhotra, G. Singh, *Terahertz Antenna Technology for Imaging and Sensing Applications*, https://doi.org/10.1007/978-3-030-68960-5_5

1. The narrow-gap photoconductive antenna (also known as small-gap) in which the gap width is generally smaller than that of the radiation wavelength (with gap size order of 5–50 μm) [2].
2. The large-gap photoconductive antennas (also known as large-aperture) in which the gap width is kept larger than that of the radiation wavelength (with gap size order of 0.5–5 mm) [3].
3. The semilarge-gap photoconductive antennas in which the gap width lies in between the two abovementioned types (with gap size order of 0.1–2 mm) [4].

The large-gap photoconductive antennas are popular configurations particularly as emitters since higher radiated powers can be attained. Moreover, in large-gap photoconductive antennas the decrease in effective electric field across the antenna gap (phenomenon also known as screening effect) occurs later in comparison to that of the small-gap antennas. Further, in large-aperture antennas with constant laser power, the distance among the free carriers increases due to the large-sized photoconductive area, and as a result, there is a significant decrease in the screening field. However, these large-aperture antennas require a higher bias voltage, which is generally in the kilovolt range [5]. On the other hand, the design of semilarge-gap photoconductive dipole antenna helps to set down antenna electrodes of sufficient thickness, which facilitates to accommodate the skin depth and also reduce the heating effects. Conversely, the achievable bandwidth in case of semilarge-gap photoconductive antenna is narrower than that of small-gap antennas. However, for the pulsed THz imaging system, a broad frequency spectrum is desired for scanning purpose. Therefore, the small-gap photoconductive antenna has the superiority over the large-gap and semilarge-gap photoconductive antennas in terms of radiation efficiency, broad bandwidth, and stability. Additionally, in small-gap photoconductive dipole antenna, the geometry of the antenna electrodes has their significance on the spectral characteristics of the antenna [6]. Further, with the small-gap photoconductive dipole antenna the THz radiation over the broad frequency spectrum can be generated wherein the antenna electrode structure acts as a filter and establishes the radiated frequency spectrum. In photoconductive dipole antenna, the resonating frequency depends on the length of the electrodes, $f_r = c/(2nL)$, where c, n, and L are the speed of light in vacuum, refractive index of the material, and total length of antenna electrodes, respectively. Moreover, in comparison with other types of THz emitting devices, for example, electro-optic (EO) crystals which needs the pump power of higher magnitude of 172.9 mW, the photoconductive dipole antenna gets triggered with the help of mode-locked Ti-sapphire laser providing 10 mW average incident power on the photoconductive gap [7].

Furthermore, the technique of photodetection using photoconductive antenna is comparatively stable against optical as well as thermal noises, while the electro-optic (EO) sampling detection is especially sensitive to such noises. These noises occur generally due to the fluctuation caused by the vibrations in the cavity length, shot noise on photon detection, and thermal-mechanical noise of the passive cavity components through the laser output on the EO crystal. Because of the increasing interest in THz systems in addition to the requirement for efficient imagers, interferometers, and broad bandwidth spectrometers for the imaging systems, various researchers are investigating the small-size antenna concepts. In the small-gap photoconductive dipole antenna, the categorization of the multiphysical phenomenon which is taking place can be listed as follows:

1. Light–matter interaction.
2. Photoexcited carrier dynamics.
3. Full-wave propagation of the THz radiation.

To generate desired THz radiation from photoconductive dipole antenna, there is a requirement to have sufficient incoming laser power, a superior photoconductive material, and enough bias voltage along with a well-designed antenna configuration. A biased voltage source in addition to the laser optical source drives the photoconductive gap acting as current source to produce THz radiations. However, the main issue with the photoconductive dipole antenna is the low THz output power and low antenna efficiency. To counter such issue, the optical illumination power and the applied bias voltage are increased in such a manner to yield higher output power. However, under such situations, the observable phenomenon, such as saturation, field breakdown, velocity overshoot, and thermal breakdown, happens within the system [8].

Moreover, in the material selection for photoconductive antenna, there is a requirement of the following: (1) maintenance of relatively high carrier mobility with suitable band gap, (2) control of zero bias photocurrent, (3) high breakdown voltage, and (4) low carrier lifetime with high resistivity. These parameters influence the performance of photoconductive antennas in terms of the following: (1) output power, (2) bandwidth, (3) maximum optical pump power, and (4) signal-to-noise ratio (SNR), respectively [9]. For the pulse THz imaging application, to increase the SNR of spectroscopic system, the THz radiation with high emission intensity is required, which facilitates to perform fast scan of the object. The value of emission intensity of the THz radiation considerably increases with an enhancement in the dipole length [10]. Moreover, the emission efficiency of the THz radiation depends on the relative dimensions of the dipole. The efficiency increases appreciably by increasing the aspect ratio (the ratio of length and width of the antenna electrode, L/W). Similarly, another parameter which can be adjusted to reduce the screening effect, and as a consequence, to improve the radiation efficiency in small-gap photoconductive dipole antenna is through the excitation spot size which persuade the initial spatial distribution of the photoexcited carriers on the surface of the photoconductive substrate. On increasing bias voltage, higher THz output power can be achieved as under such situation the acceleration of the photoexcited carriers gets increased. However, under such situation the breakdown voltage of device limits to give up maximum radiation power. To increase the level of breakdown voltage so that the antenna can withstand high biased voltages across the antenna electrodes, the properties of several photoconductive materials such as gallium arsenide (GaAs), low-temperature grown GaAs, alternating nanoscale multilayers of InGaAs, bulk indium gallium arsenide (InGaAs), indium aluminum arsenide (InAlAs), radiation-damaged silicon on sapphire, and amorphous silicon are investigated by numerous researchers for THz photoconductive antennas [11].

The small-gap photoconductive dipole antenna is simple in fabrication and furthermore shows its potential use as THz source for pulsed broadband system of THz imaging application. However, there are certain prospective challenges to improve the conventional THz photoconductive dipole antenna as discussed earlier. Therefore, considering both the optical excitation effects and the antenna geometrical effects on the performance of a photoconductive antenna, a systematic analysis technique and a simulation procedure is necessary to design so as to have the better understanding of the phenomenon, such as (1) static electric field creation, (2) photoexited transient current generation in the near field, and (3) the THz radiation in both near and far field. Further, an analytical framework for modeling of pulsed THz small-gap photoconductive dipole antenna has been proposed to find out several factors contributing to the antenna performance parameters like THz average radiated power and the total efficiency. In this chapter, a systematic procedure employing precise mathematical expression leading to the physical behavior of small-gap photoconductive dipole antenna is projected. Moreover, the effect of biased lines on the antenna performance parameters is also considered and is presented in the proposed equivalent circuit model. Further, the effect of gap size of proposed photoconductive dipole antennas on the THz radiated power as well as on total radiation efficiency has been explored.

5.2 Related Work and Problem Formulation

The main disparity in analyzing THz antennas in comparison to the microwave antennas is the optoelectronic characteristics that arise due the optical excitation and accordingly the response of the photoconductive material. Therefore, it demands the development of some novel simulation as well as an analysis procedure, which incorporate the effects generated from both optical excitation and antenna geometry. In this respect, several researchers have explored different physical models to present the better understanding about the working of the photoconductive antenna. However, such physical models can be categorized under three different approaches as briefly mentioned below with each category having its own advantage as well as limitation.

- *Drude–Lorentz Model*: In this approach, the photoexcited carrier dynamics inside the semiconductor is described using Drude–Lorentz model [12]. It is a useful method to evaluate the dependence of THz radiation on the material's properties like doping density, carrier lifetime, carrier mobility, and laser source's intensity with its pulse width [13]. The use of this model can compute the photoexcited current inside the semiconductor in the near field. However, the far-field radiation can only be realized in an approximate way. Therefore, one of the major limitations of this model is the different radiation properties such as radiated power, electric field strength, and efficiency of the antennas with various shapes that cannot be distinguished. In addition, this model can hardly simulate the space-related phenomenon like the effect of asymmetrical illumination of the laser spot within the photoconductive gap.

- *Full-Wave Finite-Difference Time-Domain (FDTD) Model*: The FDTD model couples the carrier's dynamics with the full-wave interaction along with propagation to simulate the phenomenon occurring in both near field and far field [14]. In this, the continuity equation is used to describe the carrier dynamics. Moreover, the drift–diffusion equation is used to depict the corresponding transient current wherein the photoexcited current is considered as the driving source of a photoconductive antenna [15]. However, the full-wave analysis is incapable to analyze the THz antenna with a large silicon lens which is used to collimate the THz radiation in one particular direction. This is because the size of the lens is much larger than the wavelength [16]. Further, the analytical method such as diffraction theory also required to implement with FDTD for analyzing propagation behavior with large elements.
- *Equivalent Circuit Model (ECM)*: In this approach, the photoconductive dipole antenna is considered as a special lumped element which comprises a combination of voltage or current source with time-varying resistance and antenna impedance [17]. The laser-induced resistance of the source is evaluated by means of carrier dynamics. This method is useful as it takes up the existing antenna theory for the analysis of photoconductive antenna as well as all the antenna-related aspects in its simulation. Moreover, with ECM it is simple to determine the impedance matching efficiency amid photoexcited source and radiating antenna [18]. Further, the ECM helps to study the antenna-related properties to design photoconductive antenna with better performance. This approach provides a realistic description of the antenna by means of direct link among the parameters of the equivalent circuit model, the structure of the antenna, and the photoconductive material properties. In this respect, there are the following two types of equivalent circuit models which have been explored and discussed by the researchers in literature:
- 1 ECM obtained from electrical engineering perspective and based on large signal analysis. This method considers the following: (1) the constant capacitance representing the gap capacitance of the photoconductive antenna, (2) a time-dependent resistance of the photoconductive material, and (3) the antenna resistance [19]. In this model, the approach used is simple to implement because of the deployment of lumped element components. However, a major limitation of this model is to precisely envisage the local fields of photoconductive antenna.
- 2 ECM based on real physical performance of antenna using carrier dynamics. It is essential to consider the carrier dynamics in the generation of THz waves as it helps to find out the real physical behavior of the antenna. This method involves the following: (1) the antenna resistance, (2) a time-varying field due to the space-charge screening effect, and (3) the time-varying

source resistance [20]. The main benefit of this model in comparison to simplified lumped element model is improvement of the accuracy since more physical aspects are considered. However, there is a complexity in the derivation of equations. In addition, there are certain assumptions in its analysis which have been made and limit its applications. The photoconductive gap between the antenna electrodes is generally assumed to be fully and uniformly illuminated by the laser beam. Further, while calculating source resistance the exponential decay of the photocurrent caused due to the recombination of photoexcited carriers is ignored due to which the laser pulse information is lost. Moreover, it is also assumed that when there is no laser illumination (under the dark conditions), the photoconductive antenna is considered as a charged capacitor.

- Several researchers have mentioned the use of these models in the literature for the photoconductive antennas [21–23]. In [24], the screening contribution because of the Coulomb and radiation screening effects of electromagnetic field in the photoconductive antenna is presented. The authors have used the Monte Carlo method to analyze the consequences of excitation spot size as well as the excitation level onto the emitted THz radiations. Jepsen et al. [24] have anticipated a model based on the Drude–Lorentz theory of carrier transport to explain the details of the ultrashort carrier dynamics occurring in the small-gap photoconductive antenna. However, the authors have considered only the effect of space charge screening for carrier dynamics. Due to the statistical approach which has been used, the results obtained require high computational effort. Similarly, the use of semiclassical Monte Carlo model is presented in [25] to explain fast carrier dynamics occurring in the photoconductive antenna due to photoexcitation. In [26], the authors have shown the theoretical model based on Bloch–Floquet theorem using Maxwell's equations to analyze the periodic dielectric structures along with the periodic low-temperature-grown GaAs strips within the gap between the antenna electrodes. They have proposed this model to increase the THz radiation power of photoconductive antenna. Khiabani et al. [27] have presented a theoretical model for the time-variant conductance. In equivalent circuit model, they have used the lumped element approach to retain the simplicity of the model. However, the accuracy has been achieved in [27] by considering each lumped element component value based on complex physical mechanism. The only limitation of this model is that the nonuniformity of the applied bias field as well as the illuminated region has been ignored. In [28], saturation behavior of the DC photocurrent in addition to velocity overshoot phenomenon has been observed. The authors discussed the hot carriers effect using energy balance transport model. A characteristic equivalent electrical circuit of the plasma medium for photoconductive dipole antenna is shown in [28]. It includes a combination of circuits under dark as well as illuminated conditions of photoconductive gap with an assumption of no antenna losses (radiation efficiency of antenna equal to 1).

However, the realization of plasma medium for small-gap photoconductive antenna [29, 30] is difficult in real-time application scenario. Therefore, in our proposed antenna, the LT-GaAs photoconductive superstrate of 1 μm thickness is used which has high mobility and high breakdown field. Moreover, at THz range of electromagnetic spectrum, the ohmic losses increase, therefore the antenna losses need to be considered. Further, in the basic geometry of small-gap photoconductive dipole antenna there is fixed bias line connecting the biased voltage source and the antenna electrodes. The length of these biased lines is generally half of the wavelength of operating frequency of the antenna [31]. Therefore, such biased line contributes to the emergence of constant physical parameters in the form of distributed elements on the line and need to be considered in the equivalent circuit model of small-gap photoconductive dipole antenna to achieve high matching efficiency [32, 33]. Moreover, at half of the wavelength of the biased lines, the phase delay as well as the interference of any reflections on the line becomes significant which can bring about unpredictable behavior in the performance of the system. Further, the biased lines are also significant for their use to further extend beyond the antenna electrodes to facilitate the provision of antenna array implementation [34].

5.3 Circuit Modeling Using Numerical Equations

The working principle of a THz small-gap photoconductive antenna is generally based on (1) the laser excitation, (2) the antenna structure over photoconductive material, and (3) a fixed bias voltage. Initially, when the antenna electrodes are externally biased using the biased lines, then a static field gets built up inside the bulk semiconductor. The electrons and holes start flowing in a certain direction under the influence of a fixed DC bias voltage which results in the formation of the current. The generation of electron–hole pairs occurs when the input energy inside the photoconductive substrate becomes larger than that of its band gap energy. Moreover, this input energy varies rapidly on applying a short period optical pulse on the photoconductive gap. It results in the variation of the carrier density across the photoconductive gap, and due to different current intensity, there is emergence of the time-varying electric field, which leads to the generation of electromagnetic waves [35]. The DC field presents an initial force to drive the photoexcited carriers toward the antenna electrodes. However, the carrier dynamics which develops initially is evaluated using the Poisson's equation as:

$$\nabla^2 V = \frac{q}{\varepsilon}\left(n - p - N_D + N_A\right)$$

where V, q, ε, n, p, N_D, and N_A are voltage distribution, electric charge, permittivity of semiconductor, density of electrons, density of holes, impurity concentration due to donor ions, and impurity concentration due to acceptor ions, respectively. With the photoillumination of small-gap photoconductive dipole antenna, there is a generation of the time-dependent carrier density $n(t)$, which is determined using the equation of continuity as:

$$\frac{dn(t)}{dt} = \frac{\eta_l \alpha}{hf_l w_G w_e} P_{av} - \frac{n(t)}{\tau_c}$$

where η_l, f_l, h, α, $(w_G \times w_e)$, P_{av}, and τ_c are the photoconductor external quantum efficiency, laser frequency, Planck's constant, optical absorption coefficient, effective optically illuminated area with w_G width of the gap between the electrodes and w_e width of the photoconductor contact electrodes, average optical power, and carrier lifetime, respectively. The above equation can further be simplified and expressed in terms of laser beam intensity $I_l(r,t)$, which depends on the radial distance from the center axis of the beam, r, and time instant, t, as:

$$\frac{dn(t)}{dt} = -\frac{n(t)}{\tau_c} + \frac{\alpha}{hf_l} I_l(r,t) \tag{5.1}$$

where $I_l(r,t) = \frac{\eta_l}{w_g w_e} P_{av}$. The general expression of the laser intensity having electric field distribution of Gaussian laser pulses along the axis of propagation is:

$$I_l(r,t) = I_l(1-R)e^{\left(\frac{-2r^2}{w_0^2}\right)} e^{\left(\frac{-2t^2}{\tau_l^2}\right)} \tag{5.2}$$

where I_l, R, w_0, and τ_l are the peak laser intensity, power reflection coefficient, optical beam waist radius at which the field amplitude and intensity drop to $(1/e)$ of the axial value, and laser pulse duration, respectively. Moreover, it is implicit that at $z = 0$ the beam waist of laser pulse is onto the photoconductive gap (G) of the THz antenna. On putting the value of $I_l(r,t)$ from (5.2) in (5.1), an expression for time-dependent carrier density $n(t)$ becomes:

$$\frac{dn(t)}{dt} = -\frac{n(t)}{\tau_c} + \frac{\alpha}{hf_l}\left(I_l(1-R)e^{\left(\frac{-2r^2}{w_0^2}\right)} e^{\left(\frac{-2t^2}{\tau_l^2}\right)}\right)$$

This differential equation for $n(t)$ is solved using the following integral identity:

$$\int_{-\infty}^{a} e^{-u^2/T^2}\, du = \frac{\sqrt{\pi}}{2} T\left(1 + erf(a)\right) \text{ and } erf(x) = \frac{2}{\sqrt{\pi}} \int_0^x e^{-t^2}\, dt.$$

Hence, the final expression for the time-dependent carrier density for $r = w_0$ is:

$$n(\mathrm{t}) = \left[\sqrt{2\pi}\ \tau_l \alpha I_l / 4\mathrm{h}f_l\right](1-\mathrm{R})\mathrm{e}^{(-2)}\mathrm{e}^{\left[\left(\tau_l^2/8\tau_c^2\right)-\left(t/\tau_c\right)\right]} \left\{erf\left[\left(\sqrt{2}\,t/\tau_l\right)-\left(\sqrt{2}\,\tau_l/4\tau_c\right)\right]+1\right\} \tag{5.3}$$

The separation of electron–hole pairs as well as their accumulation near the antenna electrodes contribute in the generation of THz radiation. The residual space–charge pairs lying in the vicinity of the metal contacts, which are not capable to find the opposite sign pair for the recombination, remain as such near the dipole antenna electrodes. This results in the formation of time-dependent capacitance $C(t)$ and gets influenced by the generated carrier density $n(t)$ in addition to the recombination time of photoconductive semiconductor material.

$$C(t) = (\tau_r / Z_a)\left[1 + \left(e\mu_e Z_a S n(t)\right)/G\right] \tag{5.4}$$

where τ_r, Z_a, e, and μ_e are the recombination lifetime, antenna impedance, electron charge, and electron mobility in thin-layer photoconductive LT-GaAs superstrate, respectively. This time-varying capacitance $C(t)$ and time-dependent voltage controlled source $\beta(t)V_C(t)$, which is controlled by the voltage across capacitance $V_C(t)$, considers the screening effect across the photoconductive gap of dipole antenna. However, it is essential to consider this parameter in the equivalent circuit model as the screening effect limits the radiated THz power from the photoconductive antenna. An expression of $\beta(t)$ (described as the reverse voltage coefficient concerning the external bias voltage on the carrier density as well as the recombination lifetime) is expressed as:

$$\beta(t) = \left(e\mu_e n(t)\tau_r / \epsilon\xi\right) \tag{5.5}$$

where ϵ and ξ are the permittivity of LT-GaAs superstrate and geometrical factor of the substrate which signify the screening effect, respectively. Moreover, the expressions for time-varying capacitance, $C(t)$, and reverse voltage coefficient, $\beta(t)$, are computed by: (1) comparing the first-order differential equation of voltage across the antenna gap $V_C(t)$ as presented in [36] using the circuit analysis and (2) the expression of electric field in the photoconductive gap considering the screening effect formulated in [37]. In the photoconductive antenna, the input signal for the antenna is generated with the illumination of femtosecond laser pulses and its interaction with the photoconductive material. Therefore, the photoconductive gap act as a current source for the antenna with its resistance operating as the source resistance R_S (t) of the antenna [38]. The time-dependent source conductance $G_S(t)$ can be determined using the expression for time-dependent source resistance of the active volume as mentioned.

$$R_S(t) = G_S(t)^{-1} = \rho(t)\frac{G}{S} = \frac{G}{\sigma(t)S}$$

where G, $\sigma(t)$, and S are the gap length between the antenna electrodes, time-dependent bulk conductivity of photoconductive substrate material, and cross-sectional area of the active volume that is normal to the DC bias electric field, respectively. In view of the laser penetration depth into the photoconductive material, the time-dependent conductance $G_S(t)$ of the active volume is expressed as:

$$G_S(t) = \int_0^{T_{\text{LT-GaAs}}} \frac{\sigma(t)e^{-\alpha z}W\,dz}{G}$$

where W and $T_{\text{LT - GaAs}}$ are the width of dipole antenna electrodes and laser skin depth at the excitation region, respectively. When the integral for $G_S(t)$ is further solved, then the time-dependent conductance of the active volume becomes:

$$G_S(t) = \frac{W}{\alpha G}\sigma(t)\left[1 - e^{(-\alpha T_{\text{LT-GaAs}})}\right]$$

However, the time-dependent bulk conductivity $\sigma(t)$ of photoconductive substrate material is expressed in terms of carrier density $n(t)$ across the active volume as $\sigma(t) = e\,\mu_e n(t)$. On considering the expression of $\sigma(t)$ and $n(t)$, the expression for the source conductance becomes:

$$G_S(t) = \left(\frac{W}{G}\right)\left[1 - e^{(-\alpha T_{\text{LT-GaAs}})}\right] e\,\mu_e\,(1-R)e^{(-2)}e^{\left[\left(\tau_l^2/8\tau_c^2\right)-\left(t/\tau_c\right)\right]}$$
$$\left\{erf\left[\left(\sqrt{2}\,t/\tau_l\right)-\left(\sqrt{2}\,\tau_l/4\tau_c\right)\right]+1\right\} \tag{5.6}$$

It is significant to identify the exact source conductance $G_S(t)$ at the photoconductive gap because it provides a time-varying THz photocurrent to the antenna electrodes. Moreover, the time-dependent voltage across the capacitance of the antenna electrodes also depends on the source conductance and is determined from the Eq. (5.7):

$$\frac{dV_C(t)}{dt} + \left[\left(\frac{1+\beta(t)}{\mathrm{Z_a} \times \mathrm{C(t)}}\right) + \frac{\mathrm{G_S(t)}}{\mathrm{C(t)}} + \frac{1}{\mathrm{C(t)}}\frac{dC(t)}{dt}\right]V_C(t) = \frac{E_{\text{bias}}}{\mathrm{Z_a} \times \mathrm{C(t)}} \tag{5.7}$$

where E_{bias} is the constant biased voltage applied across the biased lines of photoconductive dipole antenna. The value of $V_C(t)$ is computed from Eq. 5.7 which is a first-order differential equation and is solved for $V_C(t)$ using Runge–Kutta numerical analysis method. To perform a consistency and stability analysis to the ordinary differential equations, the fourth-order Runge–Kutta method is most commonly used method. However, one of the benefits of this method is that the operational analysis is identical for both types of differential equation either linear or nonlinear. Moreover, the THz photocurrent, which is induced at the photoconductive gap (G), is calculated using Eq. 5.8, wherein the rise time of photocurrent is resolute by laser pulse duration and its decay time is influenced by the carrier lifetime of the photoconductive semiconductor material.

$$i_{\text{PC}}(t) = e\mu_e n(t) V_C(t) \frac{S}{G} \tag{5.8}$$

Besides these time-dependent lumped elements which are playing a considerable role in the antenna performance, the importance of biased lines which provides the initial biased potential from the fixed biased voltage (E_{bias}) cannot be ignored. Therefore, considering the equivalent circuit model based on the detailed numerical analysis [27], the contribution of constant valued half-wavelength biased lines parameters in the equivalent circuit model of small-gap photoconductive dipole antenna has been proposed. By considering the biased line distributed parameters along the length of line helps to improve the impedance matching efficiency of photoconductive dipole antenna. Moreover, the extension of these biased lines beyond the antenna electrodes also facilitates the implementation of linear array of photoconductive dipole antennas with half-wavelength spacing between the antenna elements on the common biased lines.

For THz imaging applications, there is a prospective demand of highly directive THz source to attain high depth-of-field to image the object under detection in addition to the high spatial resolution of imaging system. This can be realized by arranging the array configuration of photoconductive dipole antenna since a unit-cell photoconductive dipole antenna has a low directivity (5–7 dBi). However, for the array implementation of small-gap photoconductive dipole antenna, there is a necessity to optimize the unit cell of small-gap photoconductive dipole antenna while realizing the physical phenomenon using an appropriate equivalent circuit model occurring across it. In Fig. 5.1, a proposed unit-cell small-gap photoconductive dipole antenna is shown with its equivalent circuit. The physical parameters of the antenna configuration shown in Fig. 5.1a are as follows: (1) W is the width of antenna electrode, (2) G is photoconductive gap size, (3) L1 is width of the biased line, (4) L2 is length of antenna electrodes, (5) L is the separation between the biased lines, and (6) l_e is effective length of dipole antenna. In fabrication of photoconductive dipole antenna using biased lines, it is customary and convenient to describe the fixed length biased lines in terms of its line parameters such as its (1) resistance per unit length R, (2) inductance per unit length L_{ext}, (3) conductance per unit length G_{con}, and (4) capacitance per unit length C. For each one biased line, the conductors are characterized by σ_c, μ_c, and $\varepsilon_c = \varepsilon_0$ and the homogeneous dielectric separating the conductors is characterized by σ, μ, and ε, where these are the conductivity, permeability, and permittivity of the material, respectively. Moreover, to determine the values of the distributed parameters of half-wavelength biased lines, the following expressions are utilized.

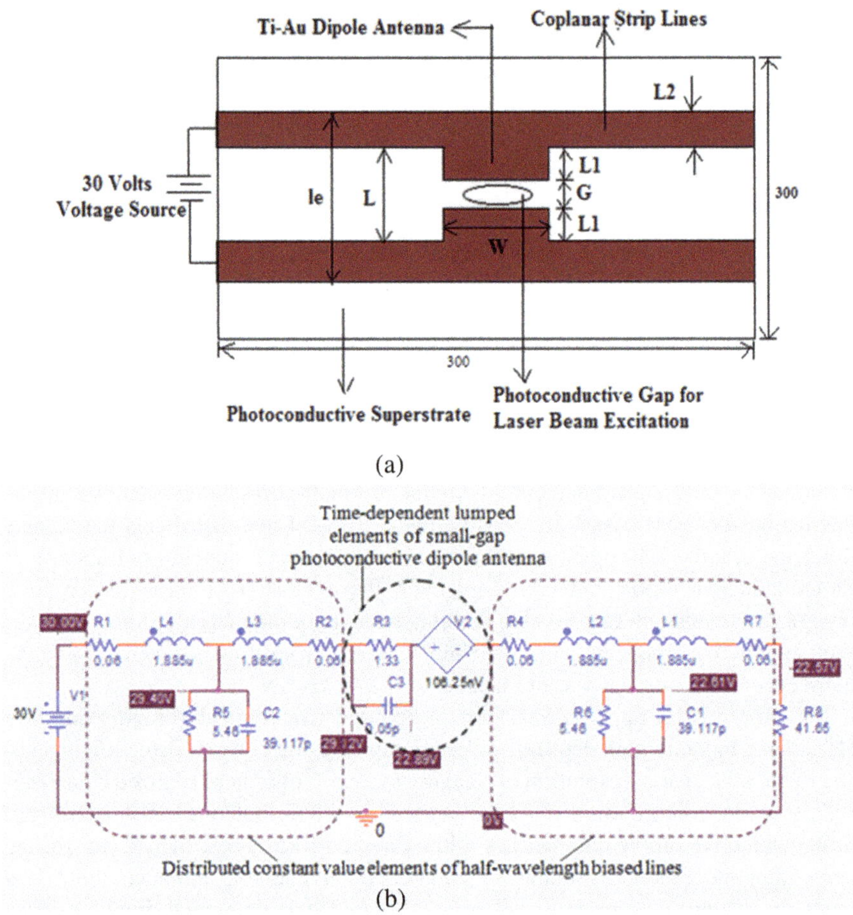

Fig. 5.1 The small-gap photoconductive dipole antenna (**a**) basic structure and (**b**) the equivalent circuit model with half-wavelength biased line designed in ORCAD PSPICE software, where *R*3, *C*3, and *V*2 are the source resistance determined from the time-dependent source conductance $G_S(t)$, time-dependent capacitance $C(t)$, and product of reverse voltage coefficient $\beta(t)$ and the voltage across the antenna gap $V_C(t)$, respectively. To determine the maximum power radiated from an antenna, the peak values of the time-dependent lumped elements of the small-gap photoconductive dipole antenna are taken [39]

$$C = \varepsilon' \frac{L_2}{L} = \varepsilon_r \varepsilon_0 \frac{L_2}{L}; \quad G_{\mathrm{con}} = \sigma \frac{L_2}{L}; \; L_{\mathrm{ext}} = \mu \frac{L}{L_2} = \mu_r \mu_0 \frac{L}{L_2}; \; \text{and} \quad R = \frac{2}{\sigma_c \delta L_2},$$

where $\delta = \frac{1}{\sqrt{\pi f \mu \sigma_c}}$ and f are the skin depth and operating frequency of the antenna, respectively. The resultant equivalent circuit is designed and simulated in ORCAD PSPICE software and is illustrated in Fig. 5.1b.

The radiated voltage across the antenna impedance R7 (taken as $Z_a = 41.65\ \Omega$, and the value is obtained from the simulation run of the antenna geometry in CST Microwave Studio) is attained through the PSPICE software. The voltage values at each node are also shown in Fig. 5.1b. However, for the circuit analysis, the peak conductance value in the photoconductive gap is considered for the reason that it corresponds to the maximum power generation in the antenna gap for maximum transformation of optical power to THz power. Mathematically, the time-dependent radiated voltage $V_{\text{rad}\,(t)}$ can be determined from Eq. 5.9 and the value depends on (1) antenna impedance, (2) carrier density, and (3) voltage across time-dependent capacitor developed across the antenna electrodes.

$$V_{\text{rad}(t)} = z_a e \mu_e n(t) V_c(t) \frac{S}{G} \tag{5.9}$$

Therefore, by selecting an appropriate material and using the earlier discussed structural framework based on the mathematical expressions to design THz small-gap photoconductive dipole antenna results in maximum radiated power from the antenna. Moreover, this analysis also helps to tune the THz imaging system with significantly improved efficiency.

5.4 Radiated Power and Total Efficiency

The THz photoconductive dipole antenna efficiency is defined as the ratio of radiated THz power to the laser power illuminated onto the photoconductive gap of antenna. In general, such efficiency is also identified as optical-to-THz power conversion efficiency. One of the major constraints of photoconductive antenna is the low efficiency of the antenna which obstructs its deployment in commercial applications of THz imaging system. Therefore, in this section, the parameters that influence the antenna efficiency are determined. From Eqs. 5.10 and 5.11, it is observed that the antenna radiated power, $P_{\text{THz}}(t)$, and the total efficiency of antenna, η_t, show their dependence on the radiated voltage, $V_{\text{rad}\,(t)}$, and optical laser source parameters.

$$P_{\text{THz}}(t) = \frac{V_{\text{rad}(t)}^{\ 2}}{Z_a} \tag{5.10}$$

$$\text{and } \eta_t = \frac{P_{\text{THz}}(t)_{\text{Peak}} \times \tau_l \times f_{\text{rep}}}{P_{av}} \tag{5.11}$$

where $P_{\text{THz}}(t)_{\text{Peak}}$ and f_{rep} are the peak THz radiated power and repetition frequency of laser pulse, respectively. Algorithm 5.1 describes the procedure to compute the radiated power and total efficiency for different values of average optical power of laser pulse across the gap of photoconductive dipole antenna [40]. Moreover, the total antenna efficiency η_t of the THz photoconductive antenna is determined from the multiplication of three different efficiencies: (1) the optical-to-electrical conversion efficiency (η_{LE}), (2) the impedance matching efficiency (η_m), and (3) the radiation efficiency (η_r).

Algorithm 5.1 Computation of Time-Varying Components of Equivalent Circuit of Photoconductive Dipole Antenna

---

Input $(\varepsilon, \varepsilon_0, \mu_0, \sigma_c, h, c, e, \mu_e, \tau_l, f_l, f_{rep}, \alpha, R, w_0, \tau_c, \tau_r, \xi, E_{bias}, W, G, S, T_{LT-GaAs}, Z_a, n, A(t), P_{av}, t, D)$

Output $(I(t), n(t), G_S(t), C(t), \beta(t), V_C(t), V_{rad}(t), P_{opt-peak}, \eta_t)$

BEGIN

{

Step 1: Variable declaration

ε = permittivity of thin – layer superstrate;

ε_0 = permittivity of free space;

μ_0 = permeability of free space;

σ_c = conductivity of metal;

h = *Planck's constant*;

c = speed of light in free space;

e = *electron charge*;

μ_e = *electron mobility in thin – layer superstrate*;

τ_l = *laser pulse duration*;

f_l = *light frequency*;

f_{rep} = *optical pulse repetition frequency*;

α = *optical absorption coefficient*;

R = *power reflection coefficient*;

w_0 = *beam waist*;

τ_c = *carrier lifetime*;

τ_r = *recombination lifetime*;

ξ = *geometrical factor of the substrate which denotes the screening factor*;

E_{bias} = *biased voltage*;

W = *width of antenna electrodes*;

G = *photoconductive gap*;

S = *photoconductive gap area*;

$T_{LT-GaAs}$ = *laser skin depth at the excitation region*;

Z_a = *antenna impedance*;

n = *refractive index of air*;

$A(t)$ = *temporal amplitude of Gaussian pulse laser*;

P_{av} = *average optical power*;

I(t) = temporal intensity function;

n(t) = time-varying carrier density;

$G_S(t)$ = *time – varying source conductance*;

C(t) = time-varying capacitance;

$\beta(t)$ = *reverse voltage coefficient*;

$V_C(t)$ = *voltage at the antenna gap*;

$V_{rad}(t)$ = *voltage across the antenna gap or votage across the time – varying capacitor*;

$P_{opt-peak}$ = *peak laser poer*;

η_t = *optical – to – THz conversion effeciency*;

t = varying time;

D = diameter of the laser beam falling on the lens;

Step 2: Computation of I(t) and then n(t)

$I(t) = (\varepsilon_0 \times c/2)\left[e^{(-2\log_e 2\,(t/\tau_l)^2)}\right]^2$

Peak of I(t) = I_l;

Compute the expression of n(t) in terms of t, by considering the peak value of I(t) for fixed value of P_{av}

$n(t) = I_l(1-R)\exp(-2r_0/w_0)\left[\sqrt{2\pi}\,\tau_l\,\alpha/4hf_l\right]\exp[(\tau_l^2/8\tau_c^2) - (t/\tau_c)]\{erf\,[(\sqrt{2}\,t/\tau_l) - (\sqrt{2}\,\tau_l/4\,\tau_c)\,] + 1\}$

Step 3: Computation of C(t) and $\beta(t)$

Using the expression of n(t), C(t) and $\beta(t)$ are determined using the following expressions,

$C(t) = (\tau_r/Z_a)\left[1 + \left(e\,\mu_e Z_a S\,n(t)\right)/G\right]$, and

$\beta(t) = \left(e\,\mu_e n(t)\tau_r/\epsilon\xi\right)$

Step 4: Calculation of source conductance $G_S(t)$

$$G_S(t) = (W/G)I_l e\,\mu_e(1 - R)\exp(-2r_0/w_0)\left[\sqrt{2\pi}\,\tau_l\,/4hf_l\right][1 - \exp(-\alpha T_{LT-GaAs})]\exp[(\tau_l^2/8\tau_c^2) - (t/\tau_c)]\{erf\,[(\sqrt{2}\,t/\tau_l) - (\sqrt{2}\,\tau_l/4\,\tau_c)\,] + 1\}$$

Step 5: Computation of voltage across the antenna gap or voltage across capacitor $V_c(t)$

Using Runga-kutta numerical analysis method for fourth order, the first order differential equation of $V_c(t)$ is computed to determine the value of $V_c(t)$.

$$\frac{dV_c(t)}{dt} + \left\{\left(\frac{1+\beta(t)}{Z_a \times C(t)}\right) + \frac{G_S(t)}{C(t)} + \frac{1}{C(t)}\frac{dC(t)}{dt}\right\}V_c(t) = \frac{E_{bias}}{Z_a \times C(t)}$$

Step 6: $V_{rad}(t)$ *calculations for time varying expressions of* n(t) *and* $V_c(t)$

$V_{rad\,(t)} = z_a e\mu_e n(t) V_c(t) S/G$

Step 7: Computation of radiated power $P_{THz}(t)$

$P_{THz} = V_{rad}^2(t)/Z_a$

Step 8: For different values of average optical power compute peak laser power

For defined optically illuminated area, Area= πr^2,

Where $r = \frac{4}{2\pi} \times (\text{laser wavelength}) \times \left(\frac{\text{lens focal length}}{D}\right)$

$P_{av} = I(t) \times Area$

$$P_{opt-peak} = \frac{P_{av}}{f_{rep} \times \tau_l}$$

Step 9: Computation of optical-to-THz conversion efficiency, η_t

$$\eta_t = \frac{P_{THz}}{P_{opt-peak}}$$

}

END

The laser-to-electrical power conversion efficiency [34] is determined as:

$$\eta_{\mathrm{LE}} = \frac{P_E}{P_L} \approx \frac{eE_{\mathrm{bias}}^{\;2}\mu_e\tau^2\eta_l f_{\mathrm{rep}}}{hf_l G^2} \tag{5.12}$$

where τ is the photocurrent decay time, and from Eq. 5.12, it is apparent that besides the laser source (τ, η_l, f_{rep}, f_l), this efficiency factor has also included another source which is the bias voltage (E_{bias}). Therefore, the electric power to the photoconductive dipole antenna is not exclusively provided by the laser source. However, the illumination efficiency η_l takes into account the reflection of laser on surface of substrate and the quantum efficiency. When the laser beam is focused onto the feeding gap (photoconductive gap) with the photon energy of laser equal to or slightly greater than that of the band gap of the semiconductor substrate, then the free electrons are efficiently created. This electro-optical operation converts the laser power P_L to the electrical power P_E. Next, the impedance matching efficiency η_{m} is determined from the antenna impedance Z_a and the source impedance Z_s using the expression as shown in Eq. 5.13.

$$\eta_m = 1 - \left(\frac{Z_a - Z_s}{Z_a + Z_s}\right)^2 \tag{5.13}$$

The source impedance is realized using the equivalent circuit as shown in Fig. 5.1b, wherein the computation of constant valued biased line components is performed by considering the Algorithm 5.2. The radiation efficiency η_{r} of photoconductive dipole antenna is defined as the ratio of gain and directivity of antenna at the frequency of operation which is obtained using an electromagnetic simulation tool CST Microwave Studio. The low radiation efficiency is the key challenge to the present photoconductive dipole antenna. It happens because of the occurrence of the excessive ohmic losses at THz frequencies. On the other hand, the impedance matching efficiency of antenna must be taken into account for proper impedance matching of laser source with photoconductive material of antenna.

Algorithm 5.2 Computation of Constant Valued Biased Line Parameters of Equivalent Circuit of Photoconductive Dipole Antenna

---

Input $(\varepsilon_r, \varepsilon_0, \mu_0, \mu_r, \sigma_c, \sigma, f, \delta, L_2, L)$

Output (C, G, L_{ext}, R)

BEGIN

{

Step 1: Variable declaration

ε_r = relative permittivity of photoconductive substrate
ε_0 = permittivity of free space
μ_0 = permeability of free space
μ_r = relative permeability
σ_c = conductivity of antenna
σ = conductivity of photoconductive material
f = operating frequency of antenna
δ = skin depth or depth of penetration
L_2 = width of biased line
L = separation between biase lines
C = line capacitance
G_{con} = line conductance
L_{ext} = line inductance
R = line resistance

Step 2: Computation of C

$$C = \varepsilon_r \varepsilon_0 \frac{L_2}{L}$$

Step 3: Computation of G_{con}

$$G_{con} = \sigma \frac{L_2}{L}$$

Step 4: Computation of L_{ext}

$$L_{ext} = \mu_r \mu_0 \frac{L}{L_2};$$

Step 5: Computation of R

For $\delta = \frac{1}{\sqrt{\pi f \mu \sigma_c}}$;

$$R = \frac{2}{\sigma_c \delta L_2}$$

}

END

5.5 Simulation Results and Discussions

In the equivalent circuit model, the photoconductive gap source components are having time-dependent features. Therefore, by considering the parameters for Eqs. 5.3 and 5.4 from Table 5.1, the numerically computed dynamics of $n(t)$ and $C(t)$ are shown in Fig. 5.2a, b, respectively. The capacitance between the antenna electrodes that has the time dependent behavior depends on the recombination lifetime of the carriers and the carrier density of the generated carriers. Therefore, the nature of two graphs is analogous and is illustrated in Fig. 5.2.

Table 5.1 Physical parameters used in the proposed photoconductive dipole antenna simulation

Parameter	Value
Light frequency f_l, (THz)	375
Laser pulse duration τ_l, (fs)	100
Power reflection coefficient, R	0.318
Optical absorption coefficient, α (/m)	6×10^6
Electron mobility in LT-GaAs, μ_e (m^2/V. sec)	0.1
Laser skin depth at the excitation region, $T_{LT-GaAs}$ (m)	10^{-6}
Recombination lifetime, τ_r (ps)	100
Geometrical factor of the substrate, ξ	900
Photoconductive area for laser illumination, S (μm^2)	50×10^{-12}
Laser repetition frequency, f_{rep} (MHz)	80

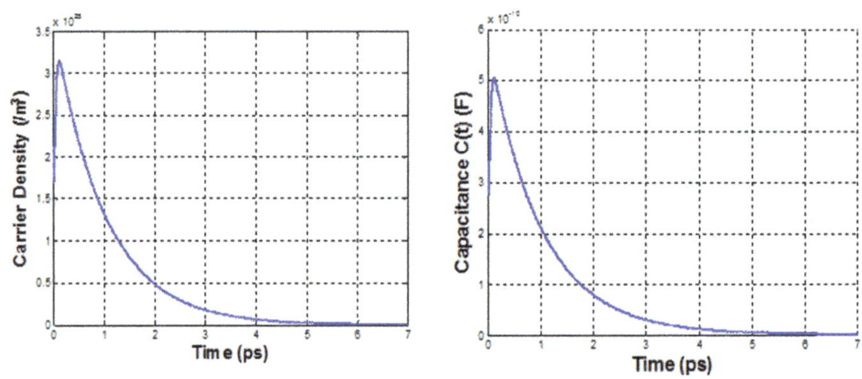

(a) (b)

Fig. 5.2 The response of time (ps) of optical illumination on photoconductive gap with single pulse of femtosecond laser beam over the photoconductive dipole antenna on the (**a**) carrier density and (**b**) capacitance across the antenna electrodes

Moreover, the influence of antenna gap size on the capacitance is investigated with fixed value of average optical power $P_{av} = 1W$. Figure 5.3 demonstrates that for small gap size of the capacitance is very large since it is inversely proportional to the gap size as shown in Eq. 5.4. On the other hand, it also means that the opposing effect of the field screening is smaller in large-gap antennas.

Now, the subsequent step is the calculation of gap conductance $G_S(t)$ through the behavior of generated carrier density $n(t)$. The $G_S(t)$ depends on parameters of (1) the optical source such as its power, pulse duration, and frequency, (2) the photoconductive material such as carrier lifetime and mobility, and (3) the antenna configuration which includes gap length and width of antenna electrodes. Therefore, under the peak laser power illumination, the source resistance $R_S(t)$ corresponding

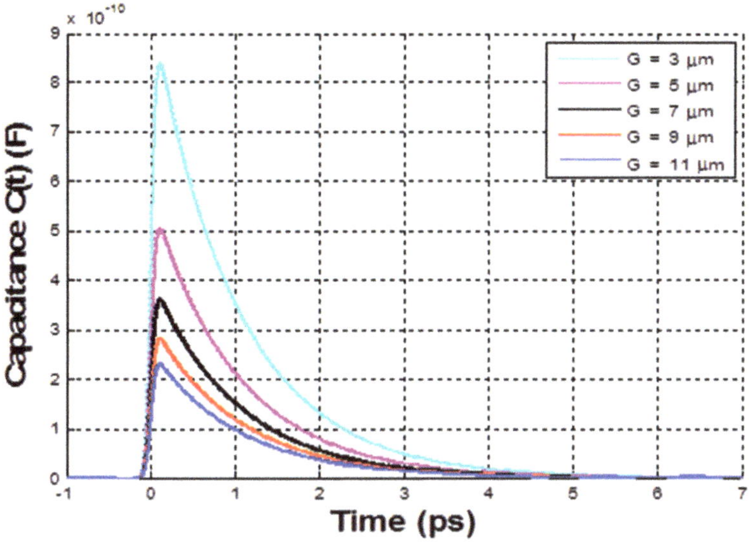

Fig. 5.3 The effect of change in photoconductive gap size (G) on the time-dependent capacitance $C(t)$

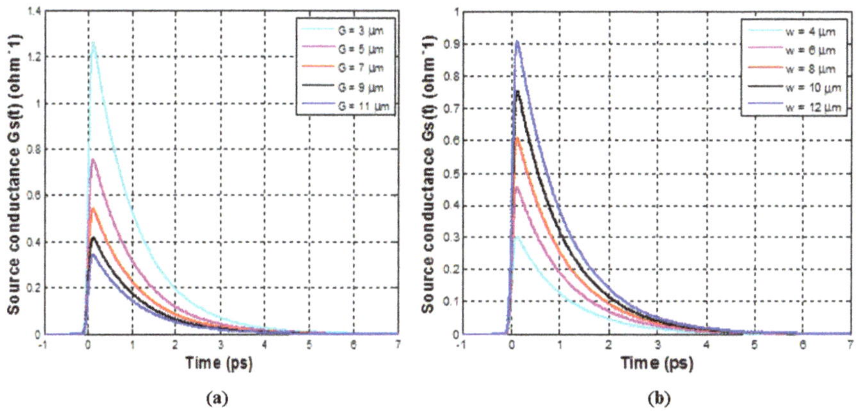

Fig. 5.4 Variation in time-dependent source conductance $G_S(t)$ with change in (**a**) photoconductive gap size (G) and (**b**) width of antenna electrodes (W) for constant average optical power $P_{av} = 1W$

to the maximum source conductivity $G_S(t)$ of the photoconductive semiconductor material is calculated as 1.33 Ω from Fig. 5.4a for $G = 5$ μm. Moreover, the change in source conductance with photoconductive gap size and width of antenna electrodes is illustrated in Fig. 5.4. It is observed from the graphs that with the increase of photoconductive gap, the peak value of source conductance $G_S(t)$ decreases. This occurs for the reason that at constant optical power, an increase in the gap size results in an increase in the distance between the generated photocarriers, which further results in decrease in the carrier density in the photoconductive antenna gap. The decrease in carrier density marks significant decrease in source conductance due to decrease in time-dependent bulk conductivity $\sigma(t)$. Likewise, on increasing the width of the antenna electrodes, the source conductance increases in view of the fact that time-dependent conductivity $\sigma(t)$ increases at the electrodes.

In Fig. 5.5, the variation in time-dependent radiated voltage $V_{\mathrm{rad}}(t)$ with change in photoconductive gap size is shown. Moreover, it is observed that with an increase in gap size there is a decrease in source conductance as well as carrier density and it results in the decrease in the voltage across the time-dependent capacitor. Consequently, the values of the radiated voltage decrease.

To support the proposed analytical analysis, it is worth to emphasize here that the value of the radiated voltage shown in Fig. 5.5 for gap size 5 μm is almost same as the voltage value obtained across antenna load resistance R7 = Z_a = 41.65 Ω from the equivalent circuit model realization using ORCAD PSPICE software as shown in Fig. 5.1b. Further, Fig. 5.6 represents the variation in the values of the radiated power and the total efficiency with respect to an increase in the average optical power. For small values of average optical power P_{av}, the radiated power as well as the total efficiency increases with an approximately linear dependence. However, for higher intensities of the laser pulse, more photogenerated carriers are generated in the photoconductive gap due to which a screening effect emerges near the electrodes because of the accumulation of the majority carriers. As an outcome, there is an inner field which opposes the incident field resulting in the saturation of the radiated power for high values of average optical power. Moreover, the effect of gap size

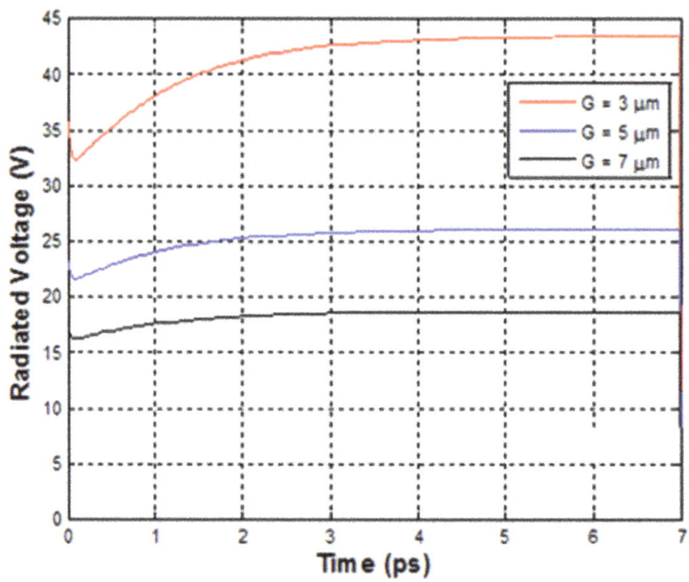

Fig. 5.5 Variation in radiated voltage of photoconductive antenna with gap size (G)

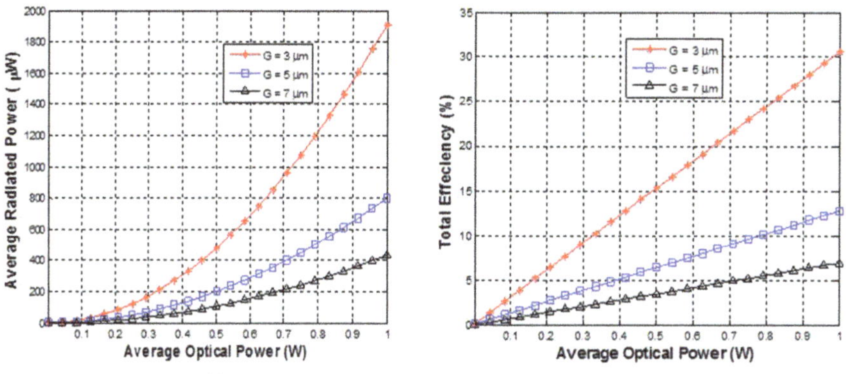

Fig. 5.6 Variation in (**a**) average radiated power and (**b**) total antenna efficiency with respect to average optical power for different values of photoconductive gap size (G)

on these antenna parameters is also shown in Fig. 5.6. It is observed that smaller the gap size of small-gap photoconductive dipole antenna, such as for $G = 3$ μm, higher is the radiated power as well as the total antenna efficiency. This happens because the generated electron–holes are nearly situated due to the smaller illuminated area. Further, the tight focusing of optical pump power across the small photoconductive gap can enhance the radiated power at low average optical power levels by reducing the carrier transport path to anode electrode. However, in such case, at high average optical power the carrier screening effect as well as the thermal breakdown severely limits the optical-to-terahertz conversion efficiency. For this reason, the photoconductive antennas with large values of gap size can withstand large optical power as well as bias voltage with smaller likelihood of saturation effect and device breakdown.

In real-time imaging system, the selection of THz photoconductive emitter also depends on the available optical source. When the laser sources with low optical powers are accessible for imaging system, then the use of a small-gap antenna marks significantly greater radiated power and optical-to-THz conversion efficiency in comparison to that of the large-gap antennas. Therefore, the optimum photoconductive gap size of 5 μm for small-gap photoconductive dipole antenna is useful as further increase in gap size results in the following: (1) small source conductance, (2) low down radiated voltage, (3) small average radiated power, and (4) low total efficiency of the antenna. Moreover, the investigations on gap size of the small-gap photoconductive antenna reveals that for the large transmitted THz signal, the antenna gap size as well as the antenna electrode width should be kept small and optimized. Further, by calculating the total efficiency based on gap size of 5 μm and the antenna parameters mentioned in Tables 5.1 and 5.2, the maximum achievable total efficiency is 0.135. The consideration of biased line distributed elements with time-dependent antenna elements in the equivalent circuit model improves the

Table 5.2 The structure parameters for proposed small-gap photoconductive dipole antenna considered for equivalent circuit realization

Parameter	Value
Biased lines (Ti-Au)	
Conductivity (S/m)	1.6×10^7
Width of line, L_2 (μm)	10
Length of line (μm)	200
Separation between the lines, L (μm)	30
Dipole antenna (Ti-Au) and ground	
Conductivity (S/m)	1.6×10^7
Width of antenna, W (μm)	10
Photoconductive gap, G (μm)	5
Length of antenna electrodes, L_1 (μm)	12.5
Ground thickness (μm)	0.35
Superstrate (LT-GaAs)	
Carrier lifetime, majority carriers (psec)	0.1
Mobility (cm²/Vs)	Greater than 200
Dielectric permittivity	13.26
Magnetic permeability	1
Loss tangent (S/m)	0.006
Electric breakdown field (V/cm)	Greater than 5×10^5
Conductivity, σ (W/cm°C)	0.55
Thickness (μm)	1
Substrate (GaAs)	
Carrier lifetime, majority carriers (psec)	0.25
Mobility (cm²/Vs)	200
Dielectric permittivity	12.9
Magnetic permeability	1
Loss tangent(S/m)	0.006
Electric breakdown field (V/cm)	Nearly 4×10^5
Thickness (μm)	10
Voltage source	
DC voltage, E_{bias} (V)	30

matching efficiency and so the total efficiency of antenna increases for the antenna impedance of 41.65 Ω.

Moreover, in free space, from a small dipole antenna, the radiation field $E_{rad}(t)$ at a distance r (which is much greater than the wavelength of the radiation) and time t is defined as:

$$E_{\text{rad}}(t) = e\mu_e \frac{(1-R)}{hf_l} \frac{\tau_l E_{\text{bias}}}{G^2} P_{av} \tag{5.14}$$

From Eq. 5.14, it is also clear that the size of the photoconductive gap has its influence on the radiated field for constant average optical power. As the simulation technique provides a powerful tool to study the detailed features of THz electric transients on the antenna and in the space, from near to far field, thus enabling one to optimize the antenna geometries such as the design, dimension, and dielectric constant of substrate. The transient solver of CST Microwave Studio is used which is based on the finite integration technique (FIT) and applies direct time-domain analysis and broadband computation of S parameters from one single calculation run by applying DFTs to time signals. Therefore, the simulation software is used to determine the effect of gap size on the radiation characteristics of the proposed antenna by considering Table 5.2 in which the structure parameters for the proposed small-gap photoconductive dipole antenna are mentioned. The effect of gap size (G) of the photoconductive dipole antenna on the gain and directivity values observed from the simulation results are listed in Table 5.3 for quick comparison.

From Table 5.3, it is concluded that with photoconductive gap size of 5 μm an improved radiation characteristic in terms of gain and directivity are achievable in comparison to 3 μm and 7 μm gap sizes. Moreover, using the simulation software, the computed radiation efficiency of photoconductive dipole antenna with photoconductive gap sizes 3 μm, 5 μm, and 7 μm is 76%, 81.2%, and 80% along with power flow of 0.0566 VA/μm^2, 0.0624 VA/μm^2, and 0.0651 VA/μm^2, respectively. In Fig. 5.7, the radiation characteristics of the photoconductive dipole antenna at 1.5 THz with optimized photoconductive gap size (5 μm) are shown.

5.6 Summary

A pragmatic description of the small-gap photoconductive dipole antenna, which includes the relationship among the parameters of equivalent circuit model, photoconductive material properties, and the dimensions of the photoconductive dipole antenna electrodes, is presented. For this, using equivalent circuit model, an analytical approach is used employing both optoelectronic and biased-line properties of the THz photoconductive dipole antenna. The consequence of photoconductive gap size on physical phenomenon such as time-dependent capacitance, radiated voltage, THz radiated power, and total antenna efficiency across the antenna electrodes are investigated. Moreover, the realization of distributed components for half-wavelength planar biased lines assists to improve the impedance matching efficiency of the antenna resulting in an increase in the total antenna efficiency. The total antenna efficiency 0.135 is achieved for the analyzed photoconductive dipole antenna gap size of 5 μm which is supported by the simulation results obtained for the antenna geometry designed using CST Microwave Studio. Therefore, the THz antenna performance and effect of different parameters can be accurately examined by considering the proposed analytical approach of equivalent circuit model before its fabrication. Further, this analytical approach also provides a more precise efficiency assessment of a small-gap photoconductive dipole antenna.

Table 5.3 The gain (dB) and directivity (dBi) for different values of gap size (G) of photoconductive dipole antenna

Principle plane	Photoconductive gap size (μm)	Gain (dB)				Directivity (dBi)			
		Main lobe magnitude (dB)	Main lobe direction (degree)	3 dB angular width (degree)	Side lobe level (dB)	Main lobe magnitude (dB)	Main lobe direction (degree)	3 dB angular width (degree)	Side lobe level (dB)
E (phi = 0°)	3	4.71	0	39.6	−2.9	5.75	0	39.6	−2.9
	5	4.85	0	39.6	−3.0	5.78	0	39.6	−3.0
	7	4.78	0	40.3	−2.7	5.65	0	40.3	−2.7
H (phi = 90°)	3	7.16	320	18.5	−2.4	8.20	320	18.5	−2.4
	5	7.40	40	18.5	−2.5	8.27	40	18.5	−2.5
	7	7.30	40	18.6	−2.6	8.24	40	18.6	−2.6

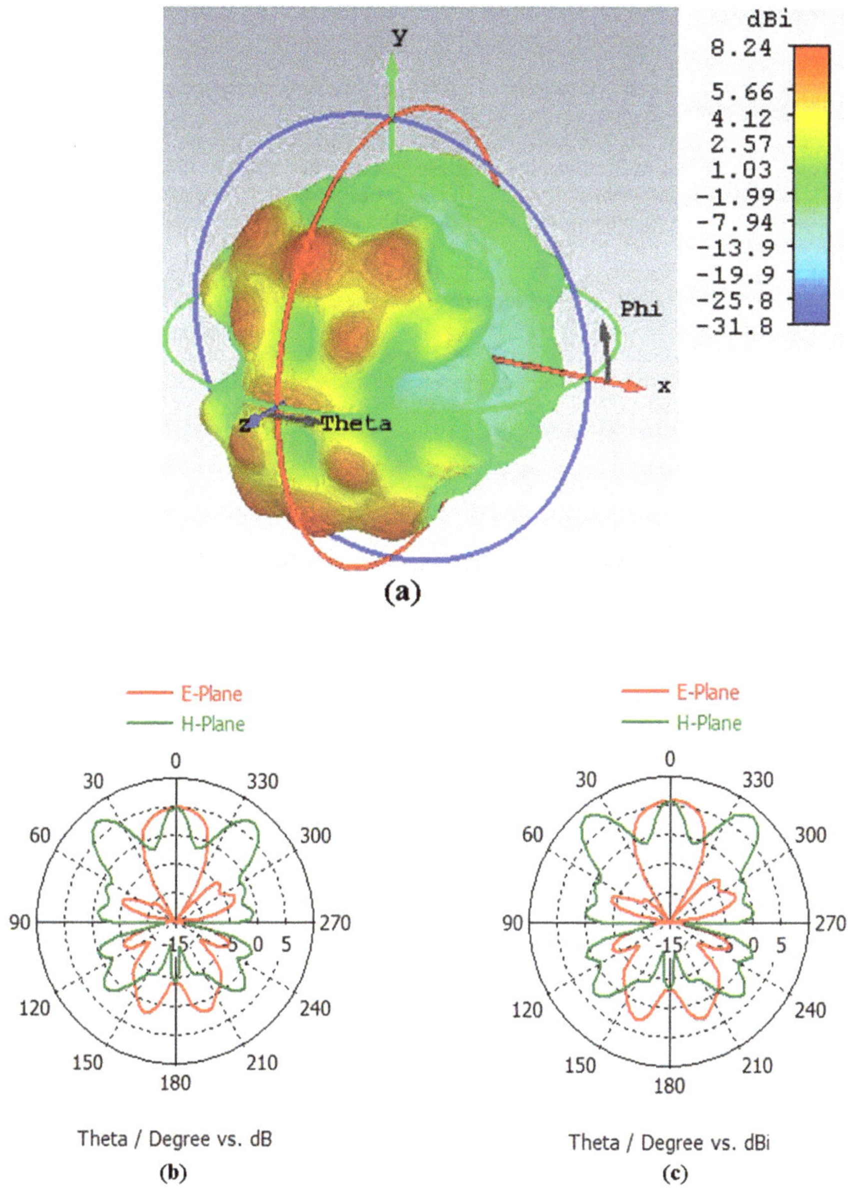

Fig. 5.7 (**a**) A far-field radiation pattern (3D view), (**b**) the gain characteristics of the photoconductive dipole antenna with gap size 5 μm in both principle planes E and H, and (**c**) the directivity characteristics of the photoconductive dipole antenna with gap size 5 μm in both principle planes E and H

References

1. C.W. Berry, M. Jarrahi, Principles of impedance matching in photoconductive antennas. J. Infrared Millim. Terahertz Waves **33**(12), 1182–1189 (2012)
2. J. Diao, F. Yang, L. Du, J.O. Yang, P. Yang, Enhancing terahertz radiation from dipole photoconductive antenna by blending tips. Prog. Electromagn. Res. Lett. **25**, 127–134 (2011)
3. H. Jafari, H. Heidarzadeh, G. Rostami, M. Dolatyari, A. Rostami, Continuous terahertz wave generation based on photomixers coupled to Fibonacci fractal tree antennas. Opt. Quant. Electron. **48**(12), 534/1–14 (2016)
4. M.R. Stone, M. Naftaly, R.E. Miles, J.R. Fletcher, D.P. Steenson, Electrical and radiation characteristics of semi-large photoconductive terahertz emitters. IEEE Trans. Microw. Theory Techn. **52**(10), 2420–2429 (2004)
5. M. Awad, M. Nagel, H. Kurz, J. Herfort, K. Ploog, Characterization of low temperature GaAs antenna array terahertz emitters. Appl. Phys. Lett. **91**(18), 181124/1–181124/4 (2007)
6. E. Moreno, M. Pantoja, A.R. Bretones, M. Ruiz-Cabello, S.G. Garcia, A comparison of the performance of THz photoconductive antennas. IEEE Antennas Wirel. Propag. Lett. **13**, 682–685 (2014)
7. Y. Zhang, X. Zhang, S. Li, J. Gu, Y. Li, Z. Tian, C. Ouyang, M. He, J. Han, W. Zhang, A broadband THz-TDS system based on DSTMS emitter and LTG InGaAs/InAlAs photoconductive antenna detector. Sci. Rep. **6**, 26949/1–26949/6 (2016)
8. S.B. Qadri, D.H. Wu, B.D. Graber, N.A. Mahadik, A. Garzarella, Failure mechanism of THz GaAs photoconductive antenna. Appl. Phys. Lett. **101**(1), 011910/1–011910/4 (2012)
9. R. Emadi, N. Barani, R. Safian, A.Z. Nezhad, Hybrid computational simulation and study of terahertz pulsed photoconductive antennas. J. Infrared Millim. Terahertz Waves **37**(11), 1069–1085 (2016)
10. F. Miyamaru, Y. Saito, K. Yamamoto, T. Furuya, S. Nishizawa, M. Tani, Dependence of emission of terahertz radiation on geometrical parameters of dipole photoconductive antennas. Appl. Phys. Lett. **96**(21), 211104/1–211104/3 (2010)
11. M.M. Seyedsharbaty, R.A. Sadeghzadeh, Antenna gain enhancement by using metamaterial radome at THz band with reconfigurable characteristics based on graphene load. Opt. Quant. Electron. **49**(6), 221/1–13 (2017)
12. Z. Piao, M. Tani, K. Sakai, Carrier dynamics and terahertz radiation in photoconductive antennas. Jpn. J. Appl. Phys. **39**(1R), 96–100 (2000)
13. K. Ezdi, B. Heinen, C. Jördens, N. Vieweg, N. Krumbholz, R. Wilk, M. Mikulics, M. Koch, A hybrid time-domain model for pulsed terahertz dipole antennas. J. Eur. Opt. Soc. Rapid Publ. **4**, 09001/1–09001/7 (2009)
14. J. Zhang, Characterization of the terahertz photoconductive antenna by three-dimensional finite-difference time-domain method. *Physics Optics*, arXiv preprint arXiv:1406.3872, pp. 1–21 (2014)
15. E. Moreno, M.F. Pantoja, S.G. Garcia, A.R. Bretones, R.G. Martin, Time-domain numerical modeling of THz photoconductive antennas. IEEE Trans. Terahertz Sci. Technol. **4**(4), 490–500 (2014)
16. Y.-S. Chung, C. Cheon, J.-H. Son, S.-Y. Hahn, FDTD analysis of propagation characteristics of terahertz electromagnetic pulses. IEEE Trans. Magn. **36**(4), 951–955 (2000)
17. K. Moon, H. Han, I. Park, Terahertz folded half-wavelength dipole antenna for high output power. *International Topical Meeting on Microwave Photonics, MWP 2005,* pp. 301–304, 12–14 October 2005, Seoul, South Korea
18. C.W. Berry, M. Jarrahi, Principles of impedance matching in photoconductive antennas. J. Infrared Millim. Terahertz Waves **33**(12), 1182–1189 (2012)
19. J. Suen, W. Li, Z. Taylor, E. Brown, Characterization and modeling of a terahertz photoconductive switch. Appl. Phys. Lett. **96**(14), 141103 1–141103 4 (2010)

20. G.C. Loata, M.D. Thomson, T. Löffler, H.G. Roskos, Radiation field screening in photoconductive antennae studied via pulsed terahertz emission spectroscopy. Appl. Phys. Lett. **91**(23), 232506 1–232506 4 (2007)
21. M. Teich, Field-theoretical treatment of photomixing. Appl. Phys. Lett. **14**(6), 201–203 (1969)
22. G. Slekas, Z. Kancleris, A. Urbanowicz, R. Ciegis, Comparison of full-wave models of terahertz photoconductive antenna based on ordinary differential equation and Monte Carlo method. Eur. Phys. J. Plus **135**(85), 1–16 (2020)
23. D. Saeedkia, S. Safavi-Naeini, A comprehensive model for photomixing in ultrafast photoconductors. IEEE Photon. Technol. Lett. **18**(13), 1457–1459 (2006)
24. J.H. Kim, A. Polley, S.E. Ralph, Efficient photoconductive terahertz source using line excitation. Opt. Lett. **30**(18), 2490–2492 (2005)
25. E. Castro-Camus, J. Lloyd-Hughes, M. Johnston, Three-dimensional carrier-dynamics simulation of terahertz emission from photoconductive switches. Phys. Rev. B **71**(19), 195301 1–195301 7 (2005)
26. M. Khorshidi, G. Dadashzadeh, Dielectric structure with periodic strips for increasing radiation power of photoconductive antennas: theoretical analysis. J. Infrared Millim. Terahertz Waves **38**(5), 609–629 (2017)
27. N. Khiabani, Y. Huang, Y. -C. Shen, and S. Boyes, "Time varying conductance in THz photoconductive antennas," Terahertz Sci. Technol., vol. 4, no. 3, pp. 116–122, 2011
28. J. Prajapati, M. Bharadwaj, A. Chatterjee, R. Bhattacharjee, Circuit modeling and performance analysis of photoconductive antenna. Opt. Commun. **394**, 69–79 (2017)
29. E. Moreno, R. Sohrabi, G. Klochok, E.A. Michael, Vertical versus planar pulsed photoconductive antennas that emit in the terahertz regime. Optix **166**, 257–269 (2018)
30. A. Singh, A. Pashkin, S. Winnerl, M. Welsch, C. Beckh, P. Sulzer, A. Leitenstorfer, M. Helm, H. Schneider, Upto 70 THz bandwidth from an implanted Ge photoconductive antenna excited by a femtosecond Er:fiber laser. Light Sci. Appl. **9**, 1–7 (2020)
31. I. Malhotra, K.R. Jha, G. Singh, Analysis of highly directive photoconductive dipole antenna at terahertz frequency for sensing and imaging applications. Opt. Commun. **397**, 129–139 (2017)
32. Z. Cong, S.U. Bo, Z.H. Fei, W.Y. Xiong, H.E. Jing-suo, Z.C. Lin, Study of low temperature gallium arsenide thin film photoconductive antenna in THz on-chip system. Spectrosc. Spectr. Anal. **39**(10), 3308–3312 (2019)
33. X. Zhonggang, D. Hu, X. Liang, Y. Jieping, S. Liping, Research on terahertz radiation peak control of photoconductive antenna. High Laser Part. Beams **32**(3), 033102 (2020)
34. R. Emadi, R. Safian, A.Z. Nezhad, Theoretical modeling of terahertz pulsed photoconductive antennas based on hot-carriers effect. IEEE J. Sel. Top. Quantum Electron. **23**(4), 1–9 (2017)
35. N. Zhu, R.W. Ziolkowski, Photoconductive THz antenna designs with high radiation efficiency, high directivity, and high aperture efficiency. IEEE Trans. Terahertz Sci. Technol. **3**(6), 721–730 (2013)
36. D. Li, Y. Huang, Y.-C. Shen, N. Khiabani, Effects of substrate on the performance of photoconductive THz antennas. *International Workshop on Antenna Technology (iWAT), 2010*, pp. 1–4, 1–3 March 2010, Lisbon, Portugal
37. N. Khiabani, Y. Huang, Y.-C. Shen, S. Boyes, Theoretical modeling of a photoconductive antenna in a terahertz pulsed system. IEEE Trans. Antennas Propag. **61**(4), 1538–1546 (2013)
38. G.C. Loata, Investigation of low-temperature-grown GaAs photoconductive antennae for continuous-wave and pulsed terahertz generation, *Ph.D. Thesis*, Goethe Universitat, Frankfurt, 2007
39. I. Malhotra, P. Thakur, S. Pandit, K.R. Jha, G. Singh, Analytical framework of small-gap photoconductive dipole antenna using equivalent circuit model. Opt. Quant. Electron. **49**, 334/1–23 (2017)
40. P.R. Smith, D.H. Auston, M.C. Nuss, Sub-picosecond photo-conducting dipole antennas. IEEE J. Quantum Electron. **24**(2), 255–260 (1988)

Chapter 6
Directivity Enhancement of Terahertz Photoconductive Dipole Antenna: Approach of Frequency Selective Surface

6.1 Introduction

For a photoconductive antenna emitter–detector configuration of THz pulsed imaging system, the main demerit to antenna is its low directivity along with low gain values. This indicates that the directivity enhancement methods of THz photoconducting dipole antenna need the fair dealing; therefore, in this chapter, efforts are made for a high-efficiency antenna design and subsequently expecting for an increased directivity of the THz photoconductive dipole antenna for imaging applications, keeping the antenna configuration planar and compact. In the photoconductive dipole antenna, the radiation efficiency is also an imperative concern because of the reduced conductivity of the metal at the THz frequencies and, particularly, the resonance effects of photoconductive antenna structure [1]. However, an array configuration [2] can be used to increase the gain or the radiation efficiency of the antenna as presented in next chapter of the thesis. The large size array configuration necessitates a higher illumination power from several separate femtosecond lasers. Moreover, the use of lens substrate to increase the directivity limits the number of antenna elements in a specified area, which is required in an imaging array.

Various researchers have accounted different directivity and gain enhancement techniques that are based on Fabry–Perot cavity (FPC) resonator, in general, which operates as a partially reflecting surface (PRS) [3, 4]. Similarly, artificial electromagnetic materials such as electromagnetic band gap (EBG), left-handed metamaterial (LHM), and frequency selective surface (FSS) are also explored to design high-gain microwave antennas [5, 6]. Moreover, these materials have attracted significant research interest because of their special electromagnetic possessions, which are applicable to a wide range of electromagnetic devices [7, 8]. Using Jerusalem Cross FSS structure, the radiation efficiency, as well as the bandwidth of the microstrip patch antenna, is enhanced and also results in considerable improvement in the microstrip patch antenna performance parameters [9]. Likewise, an EBG resonator antenna with strip dipole FSS array as a superstrate and tapered

I. Malhotra, G. Singh, *Terahertz Antenna Technology for Imaging and Sensing Applications*, https://doi.org/10.1007/978-3-030-68960-5_6

AMC ground substrate is presented in [10]. This results in an increase of gain with 2–3 dB in addition to 2.5 times more bandwidth in comparison to an EBG resonator antenna with AMC ground substrate. The bandwidth improvement of the FPC resonator antenna using FSS, which behaves as PRS, is examined in [11] by changing the distance amid the FSS elements in PRS. Moreover, a considerable advancement in the transmission characteristics of cascaded configuration of two-layered FSSs is revealed in [12] by introducing the resonators on one side of each substrate layer, which is positioned in a way that the resonators face each other with an air gap separating them. Such technique of using FSS with resonating antenna results in an improvement of the transmission response by bringing in an extra transmission zero at a frequency location that is lower than the resonant frequency. This helps to attain high roll-off rate for lower side of the stop band. Further, it also helps to suppress the unwanted resonances and so increasing the rejection/transmission bandwidth of the FSS filter. In [13], on applying metamaterials such as LHM, EBG, and FSS, the directivity enhancement of the patch antenna operating in the microwave frequency regime of electromagnetic spectrum is presented.

Moreover, the use of FSS structures at THz frequencies offers numerous interesting applications such as in imaging systems, laser cavities, Fabry–Perot interferometer, filter components, sensing systems, spectroscopy, and nondestructive testing and inspection. Furthermore, because of the simple geometry along with planar structure and compact size, the utilization of FSS with photoconductive dipole antenna offers an ease of deployment of the THz source with superior performance parameters for sensing and imaging application. Such antenna configuration (PCA with FSS) also provides flexibility in portability due to small volumetric scale. In the THz regime of electromagnetic spectrum, there are two types of frequency selective practices, which are utilized in the design of the FSS structure, such as: (1) The interference of electromagnetic waves reflected from the cascaded partially transmitting boundaries, and (2) The resonant interaction of electromagnetic waves with the segments of conductor (generally periodic arrays of the conducting elements in the dielectric) or the slots in the conducting screens for reflection/transmission of electromagnetic wave, respectively. In this chapter, we have explored the process of resonant interaction of wave with array of bandpass FSS for the gain and directivity enhancement of photoconductive dipole antenna in the THz frequency regime of the electromagnetic spectrum for the detection of hidden explosives and explosive-related compounds. In the next section, more discussion on the related work and our problem formulation has been compiled.

6.2 Related Work and Problem Formulation

In imaging applications, researchers are working to develop different techniques to detect the hidden explosives and illicit drugs using THz radiations [14]. Appleby et al. [15] have examined different THz techniques in conjunction with atmospheric, material, and component issues associated with the model design that provides potential applications in detection of weapons and contraband concealed on persons beneath clothing. Likewise, the detection of RDX by means of THz spectroscopic imaging using specular reflection has been established by Shen et al. [16]. In the real-world state of affairs, the targets generally have uneven surfaces; moreover, the surfaces are not aligned normal to the THz beam; therefore, the direction of the specular reflection is difficult to resolve. Therefore, it is more practical to detect and identify explosives using diffusely reflected THz waves [17]. In contrast, the transmission of THz wave through the object having flat spectra or the object that shows invariable background power spectrum helps to detect the type of object. In such THz spectroscopic system, the amplitude reduction in the THz wave due to the absorption of wave by the object facilitates to conclude the type of material. Moreover, the time shift of THz pulse also assists to discriminate the objects made of different explosives in powder form [18]. The angular frequency is an important parameter for designing a THz radiation source. If the source is capable of transmitting a THz pulse at the spectral fingerprint frequency of the material to be detected, only then it would be absorbed by the material under detection; therefore, the presence of such material is detected at the detector side using THz pulsed imaging system. The photoconductive antenna is a THz source, which absorbs power of incident laser and creates a number of carriers, and the applied bias voltage across the electrodes (acting as radiating source) accelerates the carriers and as a result a photocurrent flows through the radiation element. The antenna electrodes convert generated photocurrent to THz wave radiation. However, low efficiency in optical-to-THz conversion is an important drawback of photoconductive antenna. To enhance the performance parameters of photoconductive antenna, several methods have been adopted by researchers such as modifying the antenna layout [19, 20], the antenna contacts [21], or its structure and materials [22–24]. Singh et al. [19] have investigated the role of stripline photoconductive THz antenna in a regime, wherein both the direct emission of accelerated carriers in the semiconductor, as well as the antenna-mediated emission from the stripline, is considerable. They have observed the effect of varying the widths of two electrodes on the THz emission efficiency because an enhanced efficiency of THz emission will progress the data quality for many applications such as chemical identification, material characterization, and imaging. For enhancing the THz pulse emission, Park et al. [20] have presented a nanoplasmonic photoconductive antenna with metal nano-islands. These metal nano-islands serve up as plasmonic nano-antennas to locally increase the electric field of an ultrafast pulsed pump beam for the generation of large photocarriers. Their reported results show higher enhancement for THz pulse emission power by two times in comparison to the conventional photoconductive antenna. However, the use of plasmonic nanostructures within the photoconductive gap of photoconductive antenna is constrained by the factor of high cost involved in the nanofabrication of metal nanostructures using the technique such as e-beam lithography. Optimization of the device contacts further enhances the performance of photoconductive antenna. Vieweg et al. [21] have shown the higher output power using a lower contact resistance of the AuGe ohmic metal for making contacts for photoconductive THz antenna with low temperature grown GaAs photoconductive substrate, which leads to increased current flow. They have also reported that Ti/Au ohmic metal contacts are also useful for systems with high stability requirement due to higher thermal stability shown by Ti/Au contacts with low temperature grown GaAs. The use of plasmonic structures does not carry out well in the case of laser pulse with short duration time, and hence, its operation is limited to only transverse magnetic-polarized incident laser. Khorshidi et al. [22] have proposed the use of periodic dielectric structure inside the gap between the electrodes of photoconductive antenna to enhance the THz radiation power. In photoconductive dipole antenna, the material inhomogeneities arise during the growth leads to varying the emission strength. Abdulmunem et al. [23] have shown the relationship between THz emission strength and the surface properties of low temperature grown photoconductive antenna. Such correlation also helps to establish the factors to enhance the antenna performance parameters. Yardimci et al. [24] have analyzed that the choice of the substrate composition in addition to its growth process having short carrier lifetimes for photoconductive antenna (which are essential in determining substrate resistivity, carrier drift velocity, and carrier lifetime) has a direct impact on (1) optical-to-terahertz conversion efficiency, (2) radiation power, (3) radiation bandwidth, and (4) reliability of photoconductive emitters. Apart from abovementioned performance parameters of photoconductive antenna, for THz imaging application there is a need for highly directive THz source with optimum radiation efficiency. A highly directive photoconductive antenna is constructive to enhance the imaging capabilities of the THz imaging system to address the considerations such as limited depth-of-field (DoF), which is determined as the distance over which an object is considered in focus. When the THz source is deployed for short-range imaging, then a THz imaging system needs a narrow DoF so as to make the ability of the system to identify the hidden explosive carried by an individual moving toward the imaging system only for the concise moment. Moreover, such scanning over an extended volume possibly will provide security such as in a public place where the security is an essential concern; however, the visible display is not so much significant. Further, the size-weight-and-power (SWaP) of THz source for imaging application is another vital consideration. The use of compact and planar geometry of THz photoconductive dipole antenna with feasibility to have above-chip implementation supports the small SWaP values of THz source for imaging application. Moreover, high front-to-back ratio of THz antenna is advantageous in such applications so as to have interference free detection of hidden explosives.

Although the high directivity as required in THz imaging systems may be achieved by using THz photomixer as reported in [25] by using FSS, the photoconductive antenna [26–29] in pulsed THz imaging system is reasonably attractive because of the system-on-chip compatibility, its ability to work at room temperature, relatively small size, low power consumption, low manufacturing cost, and availability of compact lasers. Further, to preserve the purity of incoming wave on the receiving antenna side where the electromagnetic interference (EMI) is the major obstruction [30], the use of FSS structure with antenna geometry in THz region is beneficial to enhance the antenna performance parameters. To increase the directivity of the elementary dipole, as well as the planar antennas in the microwave frequency regime of the spectrum, highly reflective surfaces as superstrate is extensively used and it signifies that such type of arrangement is equally appropriate at the THz frequency [31]. In general, the directivity of antenna using FSS is envisaged by either using the transverse-equivalent-network model or the resonance-estimation ray tracing technique [6]. Although these techniques can only predict the peak directivity of antenna at resonance frequency, in case of FSS, array size and the ground plane size are adequately large or semiinfinite. When the FSS array or ground plane is curtailed, the directivity of cavity resonant antenna gets reduced and so in such conditions these techniques are not capable to predict the directivity of antenna accurately [32]. In such situations, the directivity of antenna array can be predicted using a comparison method at the THz frequency using FSS structure. In this chapter, a straightforward technique to improve the gain and directivity of the photoconductive dipole antenna using bandpass FSS as a superstrate at terahertz frequency for imaging and sensing applications is considered. Moreover, the THz absorption spectra has been well thought-out to design the THz photoconductive dipole antenna for commonly used explosives such as RDX, HMX, PETN, TNT, and other explosive-related compounds (ERCs), which illustrates their significant spectral absorption peaks in the range of 1–2 THz as accounted in [14, 33]. In this frequency regime of the spectrum (1–2 THz), the RDX shows spectral fingerprints with absorption at 1.05, 1.30, 1.5, and 1.9 THz. However, the HMX shows absorption at 1.58, 1.84, and 1.91 THz. Further, the PETN and TNT also have one absorption fingerprint nearly at 2.0 THz and 1.66 THz, respectively.

6.3 Theory of Operation

The frequency selective surfaces (FSSs) are described as the resonant periodic arrangement of conducting materials in one or two dimensions, which exhibit selectivity in the frequency and polarization [34, 35]. Several researchers have used different FSS structures such as dipole, tripod, cross dipole, Jerusalem cross, ring shaped, and square loop. From the literature, this has also been distinguished that the square loop shape of FSS presents superior performance in terms of bandwidth, band separation, and angular stability [36].

6.3.1 Analysis Procedure of Frequency Selective Surface

Various numerical techniques have been developed to analyze the frequency selective surface (FSS), where each one is connected with its own merits and demerits [37]. Moreover, the FSS structure can be designed as bandpass and bandstop spatial filters, which are complementary to each other [38]. Among the analytical procedures to analyze the FSS, the equivalent circuit model is very well accepted. This is due to its simplicity in which the equivalent lumped parameters of FSS are acquired from the inductive and capacitive behavior of the conductive sheet placed over a dielectric. As in the proposed antenna design, the FSS structure along with photoconductive dipole antenna has been used to increase the gain and directivity; therefore, the FSS as bandpass spatial filter is utilized. The unit-cell design of FSS bandpass and s-parameter response of unit-cell FSS are shown in Fig. 6.1. The metallic elements symbolize inductive screens, which provide rise to total reflection, while apertures in a metallic sheet correspond to capacitive screens and offer rise to total transmission [39, 40]. The analysis and design of bandpass FSS depends on the physical parameters such as the periodicity (*P*) of the unit-cell FSS, length of the slot (*d*), slot width (*s*) in addition to gap between two slots (*g*). The synthesis process of unit-cell FSS is explained in [38] to determine the parameters of the square-shaped bandpass FSS. Therefore, using the same synthesis technique with some more specific use of that technique in designing bandpass square-shaped FSS has been presented.

Further, in the proposed antenna design, the dielectric loaded FSS bandpass is used instead of using free standing since the use of dielectric substrate with FSS provides physical integrity with the FSS structure. Moreover, it also offers stable reflection and transmission characteristics and can also change the fundamental resonance frequency [41]. The unit-cell configuration of bandpass FSS is shown in Fig. 6.1a, which is created using aluminum for metallic patch with conductivity $\sigma = 3.5 \times 10^7$ S/m and is placed over the dielectric substrate, which is selected as thermocol having relative dielectric permittivity $\varepsilon_r = 1.05$. Moreover, the use of dielectric material transforms the performance of the FSS structure [41, 42]. It happens because the characteristic impedance above and below the FSS structure gets changed due to the presence of dielectric material.

Moreover, in order to avoid the evanescent waves to propagate all the way through the FSS periodic structure, the thickness of the dielectric substrate must follow the relation, $h < \lambda_0/20$, where h is the thickness of dielectric substrate of FSS and λ_0 is the wavelength in free-space; as in thicker dielectric substrates, the

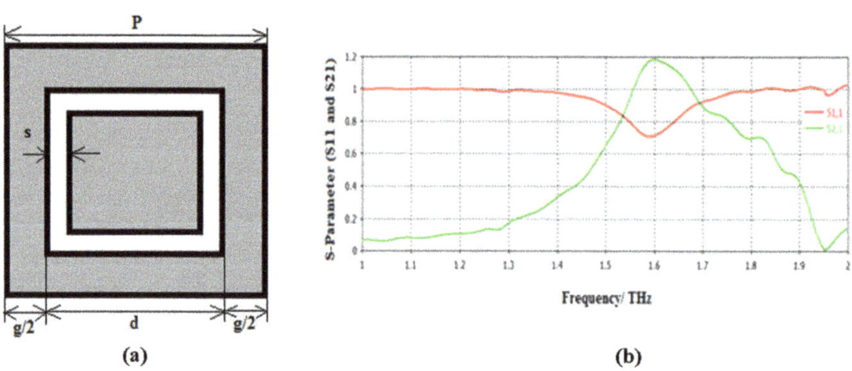

Fig. 6.1 The single square loop FSS (**a**) the unit-cell configuration with bandpass characteristics and (**b**) S-parameter of unit-cell with $P = 75$ μm, $d = 74.83$ μm, $g = 0.17$ μm and $s = 8$ μm

evanescent waves become stronger and affect the performance of the periodic structure [35]. Therefore, the thickness of substrate taken for bandpass FSS is 5 μm. The equivalent circuit elements are determined for the transverse electric (TE) polarized wave, using [40] as

$$\frac{\omega_r L}{Z_0} = \frac{d}{P}\cos\theta \times F(P,s,\lambda,\theta) \tag{6.1}$$

where

$$F(P,s,\lambda,\theta) = \frac{P}{\lambda}\left[\ln\csc\left(\frac{\pi s}{2P}\right) + G(P,s,\lambda,\theta)\right] \tag{6.2}$$

and

$$\frac{\omega_r C}{Y_0} = 4\frac{d}{\lambda}\sec\theta \times F(P,g,\lambda,\theta) \times \varepsilon_{\text{eff}} \tag{6.3}$$

where

$$F(P,g,\lambda,\theta) = \frac{P}{\lambda}\left[\ln\csc\left(\frac{\pi g}{2P}\right) + G(P,g,\lambda,\theta)\right] \tag{6.4}$$

In Eqs. (6.1)–(6.4), ε_{eff}, $G(P, s, \lambda, \theta)$, $G(P, g, \lambda, \theta)$, P, d, s, g, Z_0, and Y_0 are the effective dielectric permittivity of the media, the correction factors for the associated inductance and capacitance, respectively, periodicity of FSS unit-cell, length of the slot, slot width, gap between two FSS slots, characteristic impedance, and characteristic admittance, respectively. In case, the correction factors are neglected at the cost of a minor deviation in the results, Eqs. (6.1) and (6.3) can be rewritten as

$$\frac{\omega_r L}{Z_0} = \frac{d}{P}\cos\theta \times \frac{P}{\lambda}\ln\left[\csc\left(\frac{\pi s}{2P}\right)\right] \tag{6.5}$$

$$\frac{\omega_r C}{Y_0} = 4\frac{d}{\lambda}\sec\theta \times \frac{P}{\lambda}\ln\left[\csc\left(\frac{\pi g}{2P}\right)\right] \times \varepsilon_{\text{eff}} \tag{6.6}$$

With thermocol as the substrate ($\varepsilon_{\text{eff}} = 1.05$), the multiplication of Eqs. (6.5) and (6.6) gives

$$\omega_r^2 LC = 4.2\left(\frac{d}{P}\right)^2\left(\frac{P}{\lambda}\right)^2 \times \ln\left[\csc\left(\frac{\pi s}{2P}\right) + \csc\left(\frac{\pi g}{2P}\right)\right] \tag{6.7}$$

In Eq. (6.7), the left-hand side of the equation indicates the condition for resonance or antiresonance. Therefore, for the resonance condition of the FSS, the value of left-hand-side quantities must be 1 because $\omega_r^2 = \frac{1}{LC}$. Therefore, Eq. (6.7) can be rewritten as

$$1 = 4\left(\frac{d}{P}\right)^2 \left(\frac{P}{\lambda}\right)^2 \times \ln\left[\csc\left(\frac{\pi s}{2P}\right) + \csc\left(\frac{\pi g}{2P}\right)\right] \tag{6.8}$$

Further Eq. (6.8) is made simpler as

$$1 = 4\left(\frac{d}{P}\right)^2 \left(\frac{P}{\lambda}\right)^2 \times \ln\left[\frac{1}{\sin\left(\frac{\pi s}{2P}\right)} + \frac{1}{\sin\left(\frac{\pi g}{2P}\right)}\right] \tag{6.9}$$

For $s \ll 2P$ and $g \ll 2P$, Eq. (6.9) is inscribed as

$$1 = 4\left(\frac{d}{P}\right)^2 \left(\frac{P}{\lambda}\right)^2 \times \ln\left[\frac{2P}{\pi s} + \frac{2P}{\pi g}\right] \tag{6.10}$$

For the bandpass FSS, from the geometry it is clear that the value of g is greater than s; therefore, the ratio $\frac{2P}{\pi s}$ dominates over the ratio of $\frac{2P}{\pi g}$, so Eq. (6.10) is then simplified as shown in Eq. (6.11).

$$1 = 4\left(\frac{d}{P}\right)^2 \left(\frac{P}{\lambda}\right)^2 \times \ln\left[\frac{2P}{\pi s}\right] \tag{6.11}$$

Moreover, it is also recognized that for a specified FSS structure, the response changes with the variation in the angle of incidence of the electromagnetic wave θ, as well as with the period P of the FSS. Further, to circumvent the grating lobes, the wavelength λ is related to P and θ by the following relation from [39, 42] as mentioned in Eq. (6.12).

$$P\left(1 + \sin\theta\right) < \lambda \tag{6.12}$$

From Eq. (6.12), it is clear that for maximum incident angle (θ), a mathematical relationship between P and λ gets established on condition that the inequality is satisfied such as if a case is considered such that the structure must function satisfactorily over the angle θ_1. Then choose the value of $\theta = \theta_2$ in Eq. (6.12) where $\theta_2 > \theta_1$. Therefore, the inequality of Eq. (6.12) is contented, and the value of P is fixed as

$$P = M\lambda \tag{6.13}$$

In Eq. (6.13), M is a constant and its value lies in-between 0 and 1. On substituting Eq. (6.13) in Eq. (6.11), the expression of (6.11) becomes

$$1 = 4\left(\frac{d}{P}\right)^2 \left(\frac{P}{\lambda}\right)^2 \times \ln\left[\frac{2M\lambda}{\pi s}\right] \quad (6.14)$$

From Eqs. (6.12) and (6.14), it is concluded that the length of the FSS loop can be determined and can further be optimized using (1) the known operating frequency, (2) the desired slot size as the fraction of wavelength $\left(\frac{s}{\lambda}\right)$, and (3) the maximum accepted angle of the incident wave. From Fig. 6.1b, the S-parameters of unit cell of bandpass FSS with physical parameters P = 75 μm, d = 74.83 μm, g = 0.17 μm and s = 8 μm, the magnitude of reflection coefficient, $|S_{11}| = 0.895$ and $|S_{21}| = 0.65$, are obtained at 1.5 THz using CST Microwave Studio, and ideally at this operating frequency of designed photoconductive dipole antenna, the value of $|S_{11}|$ should be equal to 1.0 for the lossless condition. The slight difference of the value of $|S_{11}|$ from the ideal value is due to the losses and the potential challenges that are experienced while operating the FSS structures at THz regime of electromagnetic spectrum. Moreover, in the lossless condition, ideally $|S_{21}|^2 + |S_{11}|^2 = 1$, and this means that where S_{11} is minimum the value of S_{21} reaches to maximum. Moreover, the flatness of the transmission property of FSS gets enlarged. The reason for having the value greater than 1 is due to the presence of voltage source across the biased lines of the photoconductive dipole antenna. In the THz regime of the electromagnetic spectrum, the performance of the FSSs is largely influenced by the ohmic losses [43], surface roughness [44], and dispersion effect [45]. The ohmic loss gets considerably increased at THz regime because of the compact size of FSS, and in such situation, the metal no longer can be approximated as the perfect electric conductor (PEC) [46]. Moreover, owing to the resonant characteristics, the FSSs exhibit more losses than that of a metallic sheet. Therefore, the ohmic losses, which occur from the finite conductivity of the metallic elements, boost extensively and dominate the cause of heat dissipation in the FSSs. The surface roughness or deformities of FSS are also accountable for scattering in the THz region, which together with the surface plasmons gives rise to a supplementary drop in the power reflectance [43]. Moreover, the effect of dispersion loss in FSS at THz frequency is because of the relative dielectric permittivity of the dielectric substrate, which is generally a complex frequency-dependent quantity. The real part of the relative dielectric permittivity corresponds to the energy stored, and the imaginary part is associated with the dissipation. Therefore, for the dispersion less dielectric, the resonant frequency decreases as square root of the relative dielectric permittivity. Moreover, the dispersion becomes major concern at higher frequencies, which results in modified reflection characteristics through contraction/expansion of bandwidth. In the THz regime of electromagnetic spectrum, the skin depth is much smaller than the conductor cross-section due to which the current flows mainly at the surface of the conductor and such occurrence further add to the dispersion loss and results in a substantial decrease in the performance of the FSSs [47]. In general, the aforementioned analysis procedure is equally suitable for the TM polarized wave to determine the physical parameters of square-shaped bandpass FSS. The square loop FSS is polarization independent, therefore the formulae described for the TE mode are also applicable to TM mode of operation.

6.3.2 Modeling of FSS Bandpass Structure

The choice of dielectric in addition to conducting material is an imperative aspect to design the FSS, which also depends on the application in particular, still there are vital concerns related to the material selection and are required to be contemplate [48]. The material to be employed as a dielectric substrate must be transparent enough to demonstrate adequate absorption along with significant thickness so as to reduce the dielectric losses, dispersion and also evade the onset of the grating lobes over the frequency band of interest [44, 49]. On the other hand, the selection of conducting material is influenced by the value of conductivity of the material as well as the width of the material that considerably influenced the performance of FSS. Mainly metals such as aluminum or copper are utilized for the construction of FSS because of low conductor losses associated with such materials due to the high conductivity. Moreover, these metals are easily available with fiscal in use. However, aluminum is preferred above copper since the microfabrication processes are employed to fabricate the FSSs in terahertz region; consequently from the fabrication perspective, the dielectric substrate must endure the chemical reactions involved in common microfabrication processes and then again, aluminum is easy to use and provides ease in fabrication [50].

To analyze the theory discussed in the previous subsection about the FSS bandpass unit cell, the physical parameters of FSS unit cell at 1.5 THz have been computed. The physical parameters of FSS bandpass unit cell as shown in Fig. 6.1 are calculated for the normal incident wave (where $\Theta = 0°$) because the transient solver of CST Microwave Studio supports normal incident of the electromagnetic wave on to the structure, and it is also presumed that THz radiations generated from PCA strikes the lower FSSs array with normal incidence only. Using Eq. (6.12), the relation between periodicity (P) and length of the slot (d) of the unit cell of FSS at normal incidence, that is, $\Theta = 0°$of the plane wave, can be determined as $[P(1 + \sin 0°)] < \lambda$ and $P < \lambda$. The operating frequency of designed PCA is 1.5 THz, and the wavelength in free space becomes 200 μm. The periodicity of FSS unit cell is kept below 200 μm. The substrate size of designed PCA is $300 \times 300\ \mu m^2$ and for using FSS bandpass unit-cell array with the designed antenna with overall compact configuration, the size of the substrate for FSSs is also taken as $300 \times 300\ \mu m^2$. With this, the periodicity P for different size of arrays are taken as $P = 150$ μm for 2 × 2 FSSs, $P = 100$ μm for 3 × 3 FSSs, and $P = 75$ μm for 4 × 4 FSSs. In order to satisfy (Eq. 6.13), for different values of P (150, 100, and 75 μm), the value of M is computed as 0.75, 0.5, and 0.375, respectively. When the value of M is set, then for different value of s/λ, the value for the length of slot is computed using Eq. (6.14) and are presented in Tables 6.1, 6.2, and 6.3,

Table 6.1 FSS bandpass unit-cell physical parameters for 2 × 2 FSS array

s/λ	s (μm) for λ = 200 μm	P (μm)	d (μm)	$g = P - d$ (μm)	$g/2$ (μm)
0.01	2	150	49.64	100.36	50.18
0.02	4	150	54.77	95.23	47.62
0.03	6	150	58.69	91.31	45.66
0.04	8	150	61.97	88.03	44.015

Table 6.2 FSS bandpass unit-cell physical parameters for 3 × 3 FSS array

s/λ	s (μm) for λ = 200 μm	P (μm)	d (μm)	$g = P - d$ (μm)	$g/2$ (μm)
0.01	2	100	53.76	46.24	23.12
0.02	4	100	60.12	39.88	19.94
0.03	6	100	65.07	34.93	17.47
0.04	8	100	69.44	30.56	15.28

Table 6.3 FSS bandpass unit-cell physical parameters for 4 × 4 FSS array

s/λ	s (μm) for λ = 200 μm	P (μm)	d (μm)	$g = P - d$ (μm)	$g/2$ (μm)
0.01	2	75	56.14	18.86	9.43
0.02	4	75	63.5	11.5	5.75
0.03	6	75	69.57	5.43	2.715
0.04	8	75	74.83	0.17	0.085

for three types of array setups with photoconductive dipole antenna corresponding to (2 × 2) FSSs, (3 × 3) FSSs, and (4 × 4) FSSs, respectively. The numerical value for gap between two slots is also computed with the help of geometry of the FSS bandpass unit cell shown in Fig. 6.1.

However, it is observed from the simulation results presented in the section of numerical analysis and simulation results that for the fixed value of P and an increase in the value of s/λ, there is an increase in the deviation of the resonant frequency from the intended frequency. It is also being observed from the simulation results that with an increase in value of the slot size, the capacitive effect of FSS array gets increased, which causes the bandwidth enhancement.

6.4 Design of FSS-PCA

Three geometrical configurations of a photoconductive dipole antenna are presented in Fig. 6.2. The front view and top view of basic structure of photoconductive dipole antenna are shown in Fig. 6.2a, b, respectively. It consists of an antenna structure, a thin layer of photoconductive superstrate (1 μm thick) followed by a photoconductive substrate (10 μm thick), and a ground plane (0.35 μm thick) is placed under the substrate. The antenna has a photoconductive gap of 5 μm at the center of the dipole antenna with distance between the coplanar striplines of 30 μm. The length, width,

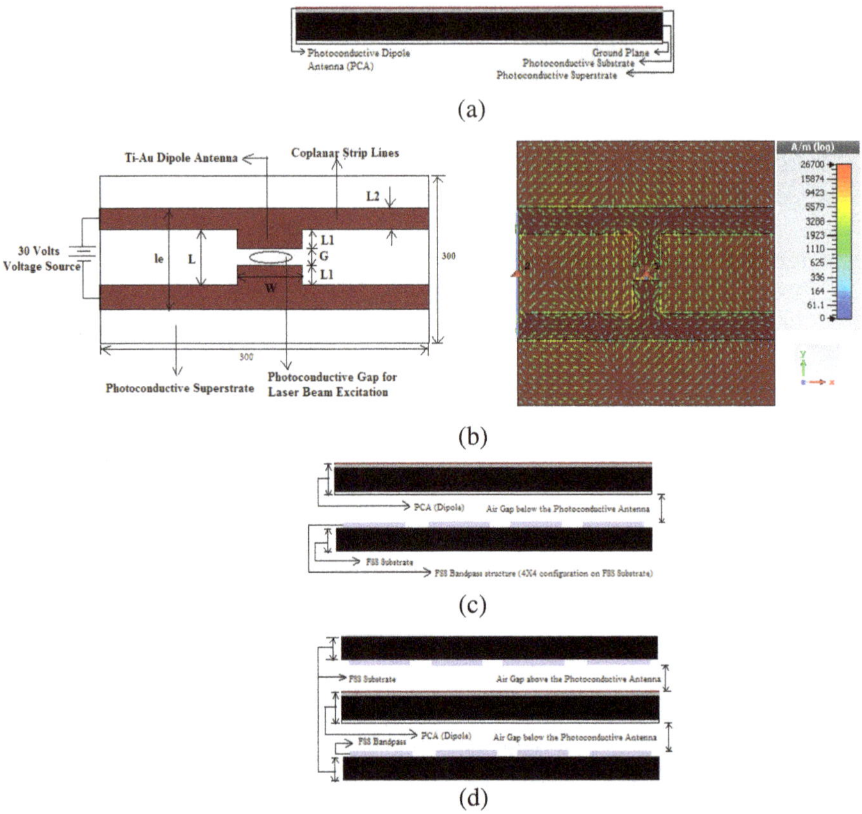

Fig. 6.2 The photoconductive dipole antenna (**a**) front-view, (**b**) top-view, (**c**) the surface current distribution on the planar surface using CST Microwave Studio (**d**) with 4 × 4 FSS bandpass substrate placed below the ground plane of photoconductive dipole antenna (PCA) at 8.65 μm with air gap, and (**e**) with two 4 × 4 FSS bandpass substrate placed above and below of the photoconductive dipole antenna (PCA) at 8.65 μm with air gap distance on both sides of antenna [54]

and thickness of the antenna coplanar striplines are 150, 10, and 0.35 μm. The coplanar striplines are biased with a voltage source of 30 volts. In the photoconductive dipole antenna, the optical source is a femtosecond laser pulse, which has a Gaussian beam distribution at its output; therefore, in the CST simulation software, we have applied the Gaussian beam excitation with pulse duration less than 100 fs on to the superstrate from the gap of the photoconductive dipole antenna. The antenna length (L) is computed at operating frequency 1.5 THz, $f = c/2$ nL, where c is the speed of light in vacuum, L is the separation distance between the two coplanar striplines (length of the dipoles), and n is the refractive index of the material. This value of frequency has been chosen so that the designed antenna must show high gain and directivity at this frequency to sense the RDX explosive, which shows its spectral fingerprint by absorbing the frequency at 1.5 THz. For the resonance condition, $L = m \times \lambda_n/2$ where, $m = 1, 2, 3, \ldots$, and the wavelength λ_n in the material depends on the refractive index n, which is given by $\lambda_n = \lambda/n$. If we take $m = 1$, then $L = \lambda/2n$. The refractive index (n) of semiconductor antenna material for low temperature Gallium Arsenide (LT-GaAs) at THz frequencies is 3.4 and for frequency 1.5 THz, the value of L (length of dipole) is obtained as $L = \dfrac{c}{2nf_r} = 30\,\mu\text{m}$ and $n = [(1 + \varepsilon_d)/2]^{1/2}$, where $\varepsilon_d = 12.9$ is the dielectric permittivity of GaAs substrate used for photoconductive dipole antenna.

An increase of the effective refractive index of the substrate (GaAs) results in the shift of resonant frequency toward the lower value. Moreover, an accurate assessment of the refractive index with change of the substrate in the present case is not straight forward since its local value depends on the local THz electric field in a nonlinear manner. Further, the use of material with high dielectric permittivity results in the emergence of shock wave at the air–substrate interface in the THz regime of the electromagnetic spectrum [51]. In addition, the application of thick substrate leads to surface wave loss because of the trapping of energy within the substrate [52], and reduction in the substrate thickness reduces the performance, as well as mechanical strength of the antenna. In order to reduce the surface wave loss, either the effective dielectric permittivity or thickness of material has to be trimmed down. However, the effective dielectric permittivity can be reduced by artificially transforming the property of material by using the periodic air-defects in the homogeneous host material, which generally acts as photonic bandgap material. In the proposed antenna model, the thickness of photoconductive substrate GaAs is taken as 10 μm, which is small enough to minimize the surface wave loss while providing sufficient strength to the THz antenna structure. Figure 6.2c shows the surface current distribution on the planar surface of photoconductive dipole antenna when an optical beam of short duration (approximately 100 fs duration) is allowed to excite the photoconductive gap of the antenna. From the Maxwell's equations, the THz electric field in the far-field region is proportional to the time derivative of the transient current [53] and is expressed as $E(t) \propto dJ/dt \propto E_b\,(dn(t)/dt)$, where J, E_b, and n are the current density, applied bias field, and current density, respectively. In Fig. 6.2d, the FSS bandpass array is placed below the PCA with air gap distance of 8.65 μm from the ground plane of PCA. This value is computed using parameter sweep of transient solver of CST Microwave Studio to optimize the height of PCA from FSS bandpass array for desired increase in the performance parameters of PCA. In Fig. 6.2e, an additional FSS bandpass array is used as superstrate (placed at 8.65 μm above the structure of PCA with air gap in between) to PCA so as to observe its effect on the complete antenna configuration performance in terms of resonance frequency, bandwidth, gain, directivity, radiation efficiency, and front-to-back ratio.

All THz-PCA antenna configurations are designed using the CST Microwave Studio, which is one of the simulation modules of CST Studio Suite. It provides a simulation platform for all types of electromagnetic field applications. Moreover, in CST Microwave Studio multiplicities of different solvers operating in time and frequency domains are included. The transient solver of CST Microwave Studio is

Table 6.4 The parameters used in simulation performed in transient solver of CST Microwave Studio

Parameter	Value
Dipole antenna (Ti-Au) and ground	
Conductivity (S/m)	1.6×10^7
Superstrate (LT-GaAs)	
Carrier lifetime, majority carriers (ps)	1
Mobility (cm^2/Vs)	Greater than 200
Permittivity	13.26
Permeability	1
Loss tangent (S/m)	0.006
Electric breakdown field (V/cm)	Greater than 5×10^5
Surface recombination velocity(m/s)	5000
Substrate (GaAs)	
Carrier lifetime, majority carriers (ps)	20
Mobility (cm^2/Vs)	200
Permittivity	12.9
Permeability	1
Loss tangent (S/m)	0.006
Electric breakdown field (V/cm)	Nearly 4×10^5
FSS bandpass unit cell (aluminum)	
Conductivity (S/m)	3.5×10^7 S/m
FSS substrate	
Permittivity	1.05
Permeability	1

used to obtain the simulation results. It works on finite integration technique (FIT) in which DFTs to time signals are applied for the direct time-domain analysis and broadband calculation of S-parameters using single calculation run. The adaptive mesh refinement in 3D using S-parameters is also used in the simulation run, which is supported by the transient solver of CST Microwave Studio.

Table 6.4 presents the parameters in terms of material conductivity, dielectric permittivity, magnetic permeability, and loss tangent considered in CST Microwave Studio for designing the three geometrical configurations of photoconductive dipole antenna as shown in Fig. 6.2.

6.5 Numerical Analysis and Simulation Results

The designed photoconductive dipole antenna at 1.5 THz frequency is simulated using CST Microwave Studio, and then three resonance frequencies have been observed as shown in Fig. 6.3, at 1.3, 1.58, and 1.87 THz. Mathematically, these frequencies are computed using the expression $f_r = c/(2l_e[(1 + \varepsilon_d)/2]^{1/2})$, where f_r is the resonance frequency, c is the speed of light in vacuum, l_e is the effective length of dipole antenna, and ε_d is the dielectric permittivity of LT-GaAs superstrate of photoconductive dipole antenna. With the computation of effective length of dipole antenna, the width of the coplanar stripline is also considered as shown in Fig. 6.2a

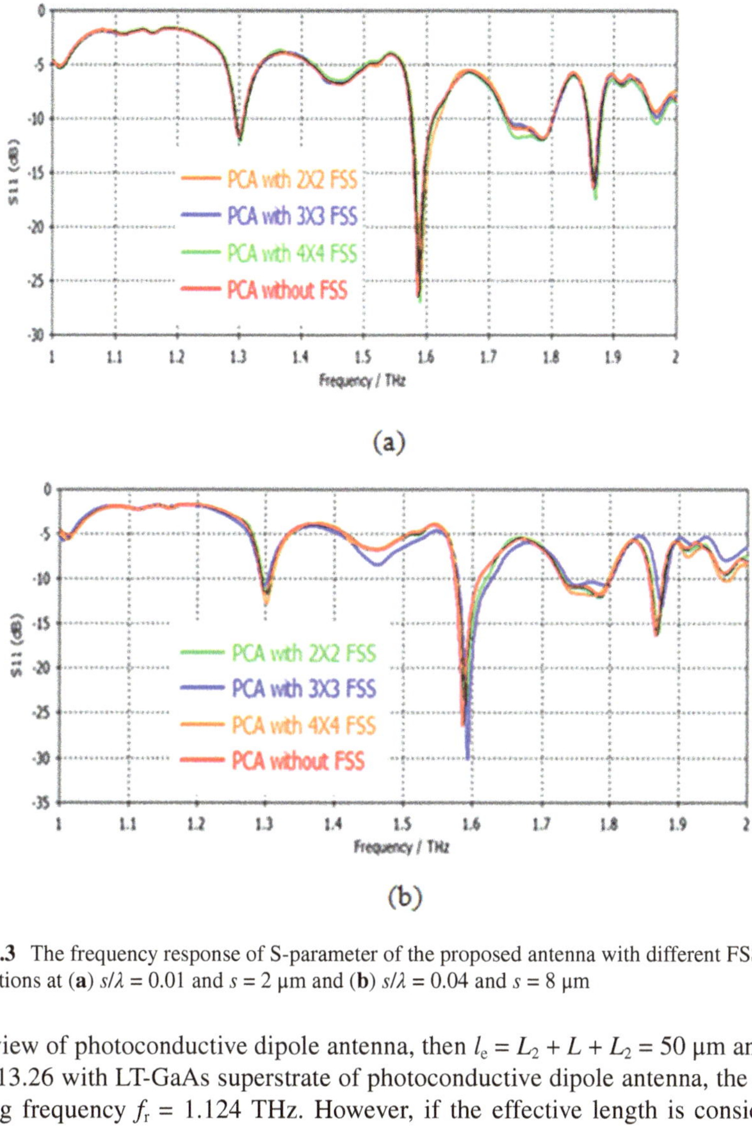

Fig. 6.3 The frequency response of S-parameter of the proposed antenna with different FSS configurations at (**a**) $s/\lambda = 0.01$ and $s = 2$ µm and (**b**) $s/\lambda = 0.04$ and $s = 8$ µm

top-view of photoconductive dipole antenna, then $l_e = L_2 + L + L_2 = 50$ µm and for $\varepsilon_d = 13.26$ with LT-GaAs superstrate of photoconductive dipole antenna, the resonating frequency $f_r = 1.124$ THz. However, if the effective length is considered equal to the separation between the coplanar striplines, $l_e = L = 30$ µm, the resonating frequency $f_r = 1.87$ THz. However, there is a small deviation seen from the simulated results of return loss in the lower resonant frequency values, which may be caused by the effect of relative dielectric permittivity of the GaAs substrate of thickness 10 µm placed below the LT-GaAs superstrate. The return loss values at 1.3, 1.58, and 1.87 THz are −11.54, −22.26, and −15 dB with 10-dB impedance bandwidth of 13.9 GHz, 35.8 GHz, and 20.1 GHz, respectively. Further, in order to achieve an antenna configuration with improved antenna performance, it is important to study the effect of the slot size of FSS unit-cell and the effect of using variable size bandpass FSS array on the performance parameters of photoconductive dipole antenna.

6.5.1 Effect of Slot Size on Antenna Performance Parameters

When a bandpass FSS is placed below the photoconductive dipole antenna, first, the effect of varying the slot size and s/λ ratio of FSS bandpass unit cell with different array sizes in the values of S-parameters of antenna configurations are observed as mentioned in Tables 6.1, 6.2, and 6.3, respectively. For the comparison, the simulated results for the smallest value of $s/\lambda = 0.01$ with 2 μm slot size and for largest value of $s/\lambda = 0.04$ with 8 μm slot size corresponding to 300 × 300 μm^2 FSS substrate area are shown in Fig. 6.3.

For the small slot size, $s = 2$ μm, at 1.3 THz, the return loss of the proposed antenna with 2 × 2, 3 × 3, and 4 × 4 FSSs configuration are −11.76 dB, −11.97 dB, and −12.23 dB, respectively, and the corresponding 10-dB impedance bandwidth are 13.8, 14.2, and 14.6 GHz. However, at 1.58 THz, the return loss of the proposed antenna with 2 × 2, 3 × 3, and 4 × 4 FSSs configuration are −25.40, −24.34, and −27.08 with 10-dB impedance bandwidth of 42.2 GHz, 36.4 GHz, and 35.7 GHz, respectively. Similarly, at 1.87 THz, the values are −16.07, −16.13, and −17.51 with 10-dB impedance bandwidth of 19.4 GHz, 20 GHz and 21 GHz, respectively. Here, it can be concluded from the observed results that for all the three resonant frequency values, the return loss for photoconductive dipole antenna with FSS below the antenna is showing significant increase in its respective values in comparison to that of the photoconductive dipole antenna without FSS. With increased slot size, $s = 8$ μm, at 1.3 THz, the return loss values are −11.38, −10.79, and −12.68 dB for 2 × 2 FSSs, 3 × 3 FSSs, and 4 × 4 FSSs, respectively, and the corresponding 10-dB impedance bandwidth are 13.7, 10.9, and 16.5 GHz. However, at 1.58 THz, with these FSS configurations, the corresponding S-parameters are −21.02, −17.44, and −19.58 dB with 10-dB impedance bandwidth of 42.5 GHz, 49.5 GHz, and 33.3 GHz, respectively. Similarly, at 1.87 THz, the values are −16.24, −11.65, and −14.93 dB with 10-dB impedance bandwidth of 18.8 GHz, 15 GHz, and 18.4 GHz, respectively. The values of return loss and 10-dB bandwidth are presented in Table 6.5 for quick comparison. It is observed from the obtained results that for all the three resonant frequency values, the return loss for photoconductive dipole antenna having 4 × 4 FSSs structure below the antenna is illustrating significantly improved values. Moreover, at 1.58 THz 3 × 3 FSSs with slot size 8 μm, a bandwidth with almost 50 GHz is obtained.

Furthermore, with an increase in the value of slot size, the capacitive effect of FSS array gets increased and this causes the increase in the width of the scattering parameter at resonating frequency 1.58 THz. However, the simulation results for directivity and gain values remain almost same at the resonant frequencies when these three types of FSS bandpass array structure (2 × 2, 3 × 3, and 4 × 4) are individually placed below the photoconductive dipole antenna, as shown in Figs. 6.4 and 6.5.

The values of gain and directivity at operating frequency of our designed photoconductive dipole antenna are shown in Table 6.6.

As discussed in preceding sections, for THz sensing and imaging application, there is a requirement of highly directive photoconductive dipole antenna at THz

Table 6.5 The return loss (dB) and 10-dB impedance bandwidth (GHz) comparison of various FSS array structure used below the PCA

			FSS placed below the ground plane of PCA with 8.65 μm air gap distance and slot size $s = 2$ μm			FSS placed below the ground plane of PCA with 8.65 μm air gap distance and slot size $s = 8$ μm		
Frequency (THz)	Parameter	PCA without FSS	PCA with (2 × 2) FSS	PCA with (3 × 3) FSS	PCA with (4 × 4) FSS	PCA with (2 × 2) FSS	PCA with (3 × 3) FSS	PCA with (4 × 4) FSS
1.3	Return loss (dB)	−11.54	−11.76	−11.97	−12.23	−11.38	−10.79	−12.68
	10 dB bandwidth (GHz)	13.9	13.8	14.2	14.6	13.7	10.9	16.5
1.58	Return loss (dB)	−22.26	−25.40	−24.34	−27.08	−21.02	−17.44	−19.58
	10 dB bandwidth (GHz)	35.8	42.2	36.4	35.7	42.5	49.5	33.3
1.87	Return loss (dB)	−15	−16.07	−16.13	−17.51	−16.24	−11.65	−14.93
	10 dB bandwidth (GHz)	20.1	19.4	20	21	18.8	15	18.4

frequencies to obtain a high-resolution image of detected hidden explosive material. Therefore, to enhance the gain and directivity of photoconductive dipole antenna, an array of FSS bandpass structure is placed above the photoconductive dipole antenna at the air gap distance of 8.65 μm from the Ti-Au dipole antenna structure of photoconductive dipole antenna, and the observed effect on the antenna performance parameters is discussed below.

6.5.2 *Effect of FSS as Superstrate*

With the use of bandpass FSS above and below the photoconductive dipole antenna, the resonance peaks shift to higher frequencies and their number has also been increased as shown in Fig. 6.6. It has been observed from the simulated results that the FSS has the ability to enhance the performance of photoconductive dipole antenna at 1.5 THz because of the electromagnetic wave collimation effect of the effectively low refractive indexed metamaterial and is prominent while using FSSs as superstrate.

For the small slot size, $s = 2$ μm, at 1.3 THz, only 4 × 4 FSSs bandpass structure shows the resonant peak with return loss −11.66 dB with 10-dB impedance bandwidth of 14.5 GHz. At 1.59 THz, 2 × 2 FSSs bandpass structure shows the resonant peak of −25 dB with 27.3 GHz impedance bandwidth along with two more shifted

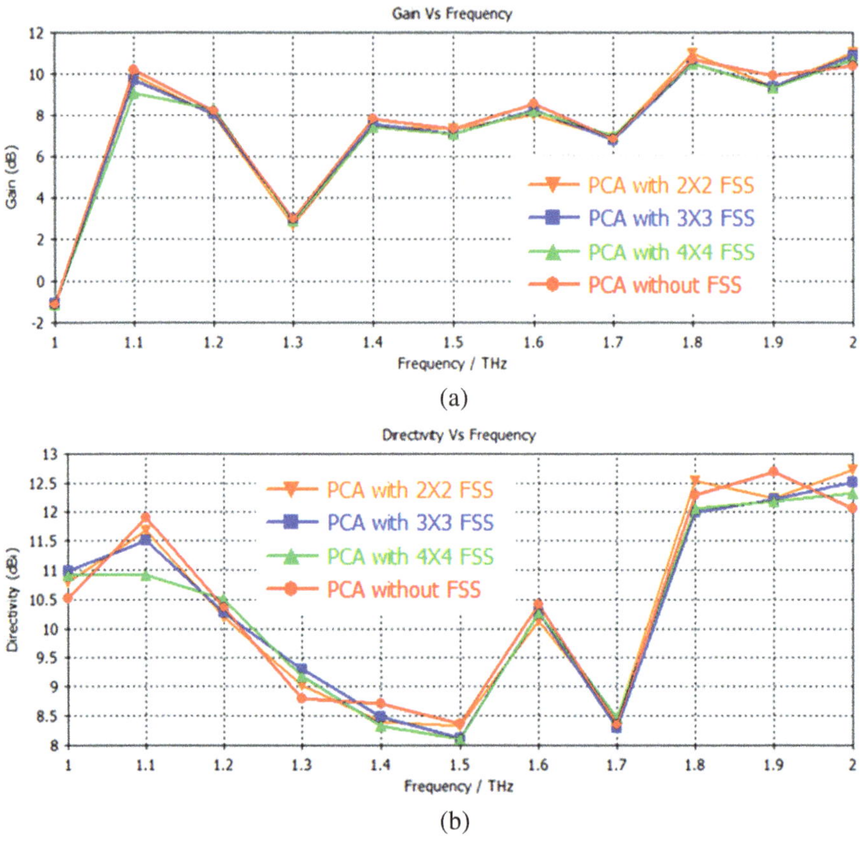

Fig. 6.4 The frequency response with different FSS configuration of the proposed antenna over the (**a**) gain and (**b**) directivity for the chosen values of $s/\lambda = 0.01$ and $s = 2$ μm

resonant peaks at 1.673 and 1.722 THz having return loss values −15.41 dB and −16.74 dB, respectively. The 10-dB impedance bandwidth at these resonant peaks are 27.7 GHz and 35.4 GHz, respectively. Similarly, for 3 × 3 FSSs bandpass structure, the resonance peak is at 1.744 THz with return loss −36.45 dB and 42.7 GHz 10-dB impedance bandwidth. The 3 × 3 FSSs bandpass structure is also showing a resonant peaks at 1.87 and 1.94 THz with return loss −17.73 and −17.76 dB with 10-dB impedance bandwidth 26 GHz and 25 GHz, respectively. In case of 4 × 4 FSSs bandpass structure, the resonant peaks are observed at 1.24, 1.3, 1.61, and 1.785 THz with −12.81, −11.66, −14.46, and −19.04 dB and 9, 14.5, 68, and 62.8 GHz 10-dB impedance bandwidth, respectively. Moreover, at 1.87 THz, only 4 × 4 FSSs bandpass structure is resonating with return loss −10.57 dB and 8 GHz 10-dB impedance bandwidth. At 1.93 THz, there is one more resonant peak being observed for 4 × 4 FSSs bandpass structure with return loss −12.15 dB and 28 GHz 10-dB impedance bandwidth. From these results, it is easy to say that the use of 4 × 4 FSSs structure as superstrate, as well as same structure below the photoconductive dipole antenna, not only shows the resonant peaks at same frequency as

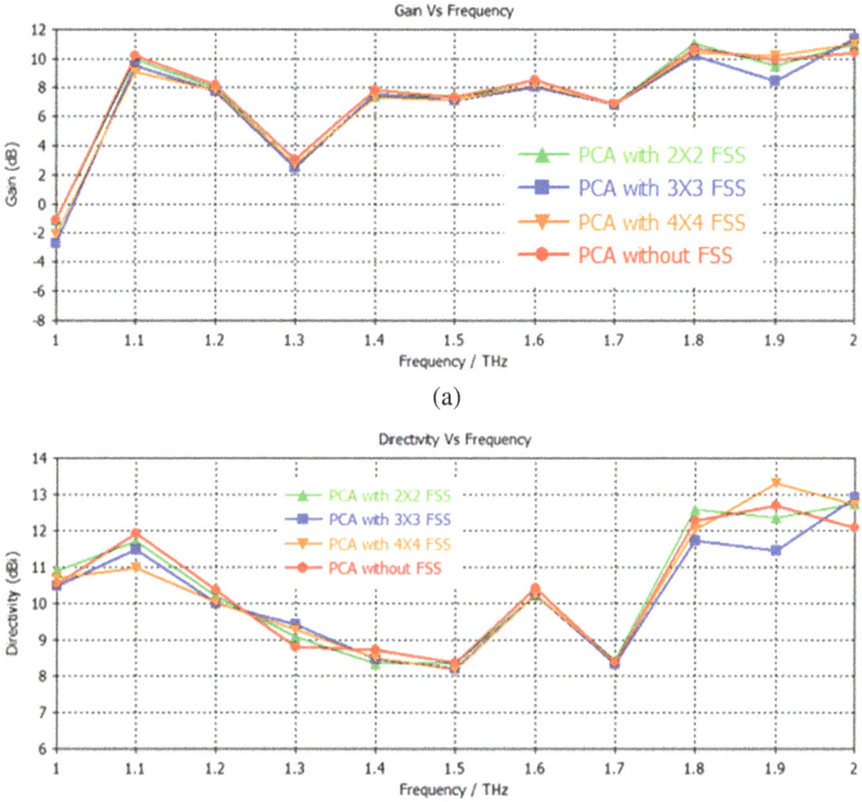

Fig. 6.5 The frequency response with different FSS configuration of the proposed antenna over the (**a**) gain and (**b**) directivity for chosen values of $s/\lambda = 0.04$ and $s = 8$ μm

Table 6.6 The gain (dB) and directivity (dBi) comparison of various FSS array structure used below the photoconductive dipole antenna (PCA)

Antenna performance parameter	PCA without FSS	PCA with 2×2 FSS	PCA with 3×3 FSS	PCA with 4×4 FSS
Gain (dB) at 1.5 THz with $s = 2$ μm	7.29	7.34	7.08	7.07
Gain (dB) at 1.5 THz with $s = 8$ μm	7.29	7.38	7.09	7.11
Directivity (dBi) at 1.5 THz with $s = 2$ μm	8.36	8.32	8.12	8.09
Directivity (dBi) at 1.5 THz with $s = 8$ μm	8.24	8.36	8.18	8.15

shown by the photoconductive dipole antenna without FSS structure but also more resonant peaks with significantly increased 10-dB impedance bandwidth are observed.

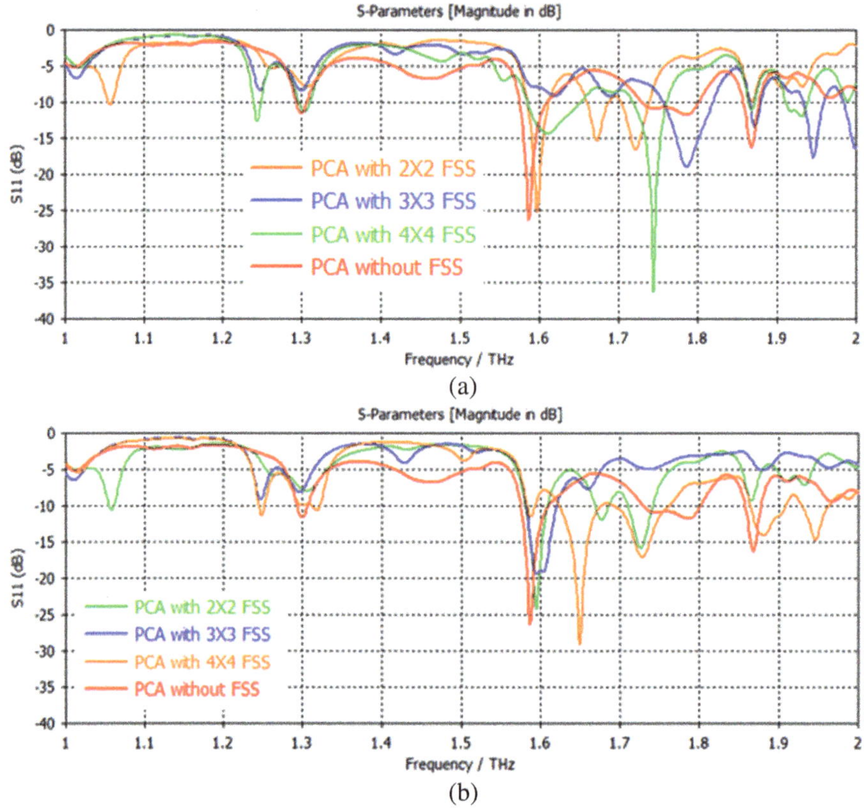

Fig. 6.6 The S-parameters for different configuration of FSS over (**a**) $s/\lambda = 0.01$ and $s = 2$ μm and (**b**) $s/\lambda = 0.04$ and $s = 8$ μm

Further, to check the performance of 4 × 4 FSSs bandpass structure with photoconductive dipole antenna, the antenna design with FSS unit-cell slot size of 8 μm has also been simulated. With increased slot size, $s = 8$ μm, at 1.3 THz, again only 4 × 4 FSSs bandpass structure shows the resonant peak with return loss −10.72 dB and 11 GHz 10-dB impedance bandwidth. Along with this, it also shows one more resonant peak at 1.249 THz with return loss −11.36 dB and 7.5 GHz 10-dB impedance bandwidth. The 2 × 2 FSSs bandpass structure shows two shifted resonant peaks at 1.6 and 1.72 THz having return loss values −12.05 dB and −15.99 dB, respectively. The 10-dB impedance bandwidth corresponding to these frequencies are 17.6 GHz and 31.2 GHz, respectively. For 3 × 3 FSSs bandpass structure, the resonance peak is at 1.58 THz with return loss −19.49 dB and 38.4 GHz 10-dB impedance bandwidth. Moreover, in case of 4 × 4 FSSs bandpass structure, the resonant peak is observed at 1.65 THz with −25.6 dB and 54.2 GHz 10-dB impedance bandwidth. These results can be used for the detection of RDX explosive, which shows its absorption spectral fingerprint at this frequency [6]. The other resonant band shown by 4 × 4 FSSs is at 1.728 THz with return loss −17.19 dB and

71 GHz 10-dB impedance bandwidth, which can be used to detect the HMX explosive whose spectral fingerprint of absorption is at 1.78 THz for the range 1–2 THz [6]. At 1.87 THz, again only 4 × 4 FSSs bandpass structure is resonating with return loss −14.16 dB and 45.5 GHz 10-dB impedance bandwidth. One more resonating band is shown by 4 × 4 FSSs, and it is located at 1.946 THz with return loss −14.85 dB and 33.8 GHz 10-dB impedance bandwidth. This result is again useful to detect the RDX explosive, which has one more spectral fingerprint of absorption at 1.9 THz [6]. Further with this analysis, it is depicted that by varying the value of s/λ, the desired band of the transmission of signal is attainable. Moreover, from the simulated results it is observed that when the length of slot (d) is reduced and the slot width (s) is increased, the transmission bandwidth of FSS also gets increased and with the enhancement in the length of the slot (d), the bandwidth decreased in case of bandpass FSS at frequency of operation 1.5 THz. This means the bandwidth is inversely related to the length of slot and therefore, with the periodicity of FSS. In this study, we have chosen two values of slot size $s = 2$ μm and $s = 8$ μm for comparison of each case for (2 × 2), (3 × 3), and (4 × 4) FSS of Tables 6.1, 6.2, and 6.3. From the simulation results, it is observed that with an increase in the value of length of the slot, the resonance frequency shifts downwards. This happens because of the reduction in the quality factor which gets reduced due to change in the capacitive value (where the capacitance is related to the slot width of FSS). The return loss and 10-dB bandwidth comparison of various FSS array structures used as superstrates above and below the PCA are shown in Table 6.7.

In addition to bandwidth and return loss, the other performance parameters like gain and directivity are also important. Therefore, we have illustrated the effect of size of FSS array on the gain and directivity of the photoconductive dipole antenna. Figures 6.7 and 6.8 show the values of gain and directivity for different values of slot size of FSS unit cell in the frequency range 1–2 THz. Further, with this simulation results, it is observed that there is a significant change in the values of gain, as well as directivity of an antenna with the use of FSS superstrate with the photoconductive dipole antenna. Using FSS as superstrate along with the FSS structure of same size below the photoconductive dipole antenna having equal air gaps from the antenna, results into the formation of cavity-resonator-type antenna at 1.5 THz. The enhancement in gain and directivity at the frequency of operation of the resulting antenna configuration occurs because of the electromagnetic wave collimation effect of the effectively low refractive indexed metamaterial inside the cavity. Moreover, with the reduction in periodicity of FSS, the transmission bandwidth gets increased and the directivity gets decreased. This is the reason due to which at some values of frequency in the range 1–2 THz, (2 × 2) FSS shows better gain and directivity in comparison to that of (4 × 4) FSS.

As the proposed antenna configuration has been designed to use for detection of explosive particular for which has its spectral fingerprint at 1.5 THz, therefore the values of gain and directivity are tabulated in Table 6.8. It is clearly observed that by using 4 × 4 FSSs bandpass structure with slot size 8 μm, the value of gain and directivity has been increased.

Table 6.7 The return loss (dB) and 10-dB bandwidth (GHz) comparison of various FSS array structures used as superstrates above and below the PCA

			FSS placed above and below the PCA with 8.65 μm air gap distance on both sides and slot size $s = 2$ μm			FSS placed above and below the PCA with 8.65 μm air gap distance on both sides and slot size $s = 8$ μm		
Frequency (THz)	Parameter	PCA without FSS	PCA with (2 × 2) FSS	PCA with (3 × 3) FSS	PCA with (4 × 4) FSS	PCA with (2 × 2) FSS	PCA with (3 × 3) FSS	PCA with (4 × 4) FSS
1.3	Return loss (dB)	−11.54	No resonant peak at 1.3 THz	No resonant peak at 1.3 THz	−11.66	No resonant peak at 1.3 THz	No resonant peak at 1.3 THz	−10.72
	10 dB Bandwidth (GHz)	13.9			14.5			11
1.58	Return loss (dB)	−22.26	−25	Shifted to 1.744 THz with −36.45 dB return loss and 42.7 GHz 10 dB bandwidth	Shows multiple resonances at 1.61 THz and 1.785 THz with −14.46 and −19.04 dB return loss & 68 and 62.8 GHz 10 dB bandwidth	Shifted to 1.6 THz with −12.05 dB return loss and 17.6 GHz 10 dB bandwidth	−19.49	Shows multiple resonance at 1.65 and 1.728 THz with −25.6, −17.19 dB return loss and 54.2, 71 GHz 10 dB bandwidth, respectively
	10 dB Bandwidth (GHz)	35.8	27.3				38.4	
1.87	Return loss (dB)	−15	Resonates at 1.673 and 1.722 THz with −15.41 and −16.74 dB return loss and 27.7 and 35.4 GHz 10 dB bandwidth, respectively	−17.73	−10.57	Shifted to 1.72 THz with −15.99 dB return loss and 31.2 GHz 10 dB bandwidth	No resonant peak at 1.87 THz	−14.16
	10 dB Bandwidth (GHz)	20.1		26	8.0			45.5
				Also resonates at 1.94 THz with −17.76 dB return loss and 25 GHz 10 dB bandwidth	Also resonates at 1.93 THz with −12.15 dB return loss and 28 GHz 10 dB bandwidth			Also resonates at 1.946 THz with −14.85 dB return loss and 33.8 GHz 10 dB bandwidth

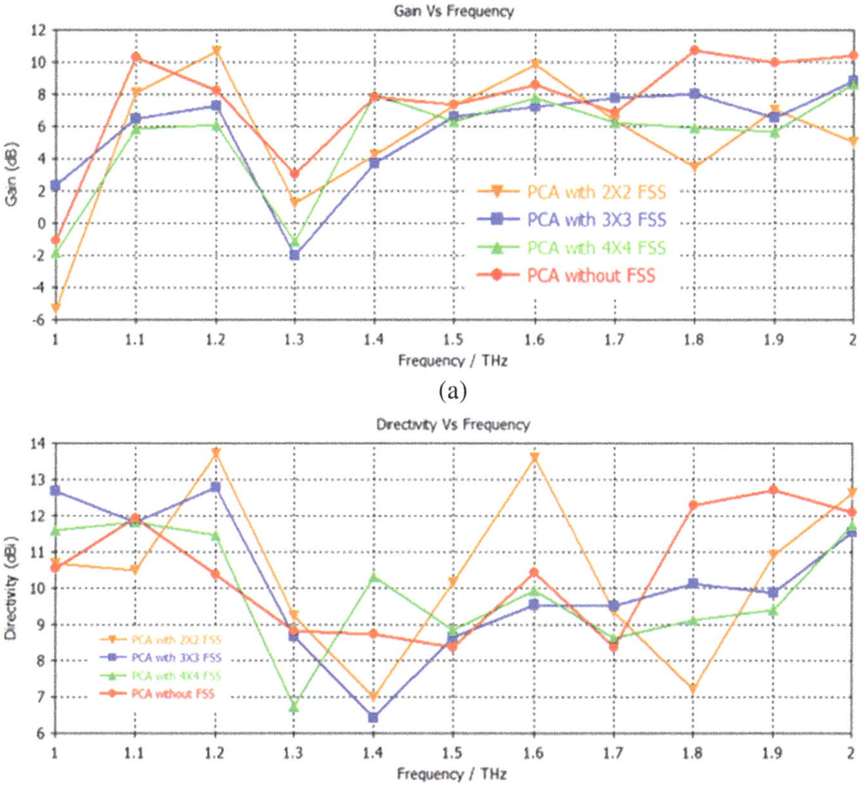

Fig. 6.7 The frequency response of the proposed antenna for different FSS configuration over (**a**) gain and (**b**) directivity for chosen value of s/λ =0.01 and s = 2 μm

The other important antenna performance parameter has been considered for highly directive antenna is the front-to-back ratio (FBR). Therefore, a comparison of FBR for all configurations of photoconductive dipole antenna with different sized FSS array is shown in Fig. 6.9. It is illustrated that at 1.5 THz, the FBR ratio has also been improved significantly with the use of 4 × 4 FSSs bandpass structure with photoconductive dipole antenna.

In general, with various array configurations of FSS used with photoconductive dipole antenna and simulated using CST Microwave Studio reveals that when the periodicity, P is close to the wavelength (λ = 200 μm, as in proposed antenna design), multiple resonances are achieved. Therefore, the analysis procedure perhaps be used in the design of multiresonance FSS structure [55]. It is also observed from the simulation results that for wideband selection smaller slot size is required. However, for the high directivity and less return loss, the periodicity of FSS structure has to be very small in comparison to that of the wavelength, and the slot size must be large enough to transmit the selected frequency. Therefore, the band of operation of FSS bandpass array is mainly influenced by the slot length (d) and slot

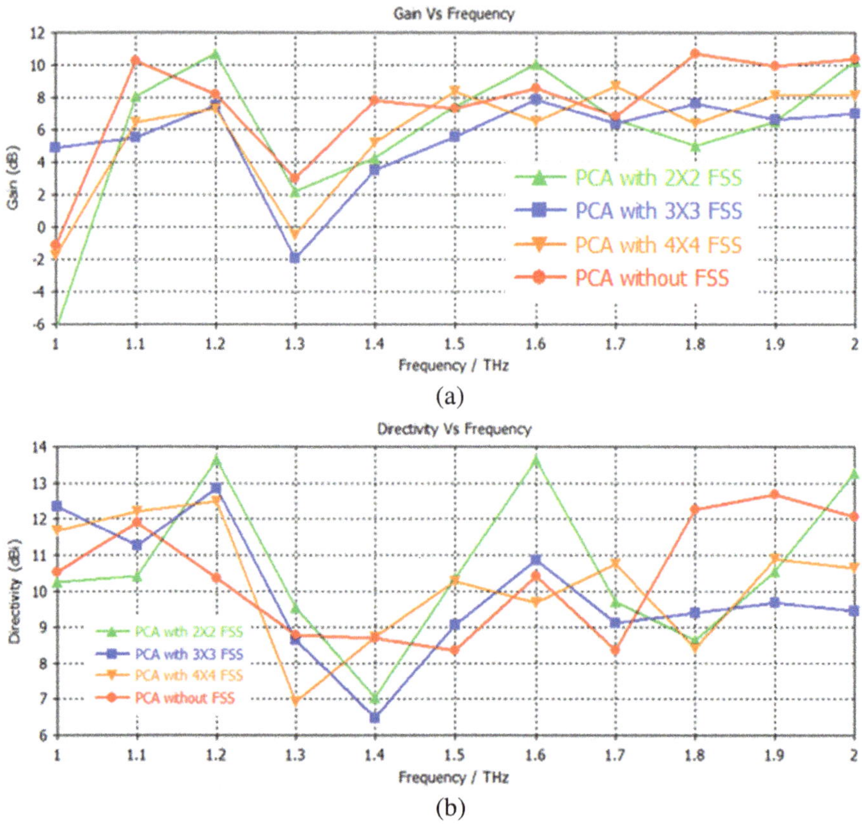

Fig. 6.8 The frequency response for different configuration of FSS over the (**a**) gain and (**b**) directivity for chosen values of s/λ =0.04 and s = 8 μm

Table 6.8 The gain (dB) and directivity (dBi) comparison of various FSS array structure used as superstrates with photoconductive dipole antenna (PCA)

Antenna performance parameter	PCA with 2 × 2 FSS	PCA with 3 × 3 FSS	PCA with 4 × 4 FSS
Gain (dB) at 1.5 THz with s = 2 μm	7.26	6.55	6.26
Gain (dB) at 1.5 THz with s = 8 μm	7.41	5.56	8.38
Directivity (dBi) at 1.5 THz with s = 2 μm	10.12	8.63	8.81
Directivity (dBi) at 1.5 THz with s = 8 μm	10.36	9.07	10.27

size (s). When the slot length is reduced and slot size is increased, the transmission bandwidth of FSS gets increased. However, with an increase in the slot length, the bandwidth reduces with fixed value of periodicity (P) of FSS unit cell. Therefore, with the use of different sized FSS bandpass unit cell for array configuration, the

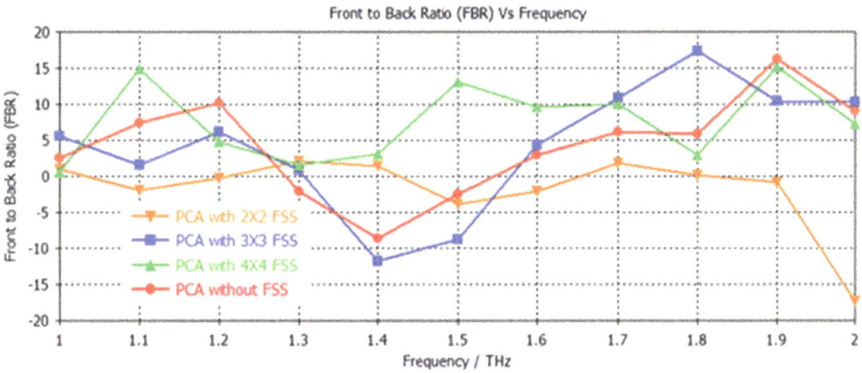

Fig. 6.9 The frequency response over front-to-back ratio (FBR) for chosen values of $s/\lambda = 0.04$ and $s = 8$ μm

resonance condition of the structure can be used to detect other explosives which shows their spectral fingerprints in the range of 1–2 THz.

6.6 Proposed Photoconductive Dipole Antenna with 4 × 4 FSS Bandpass Superstrate

On the basis of proposed simulated results as discussed in preceding sections on S-parameters, gain and directivity are related to three types of array configurations of FSS bandpass structure with photoconductive dipole antenna. For the detection of explosive corresponding to its prominent spectral fingerprint absorption peaks, it is useful to design photoconductive antenna with 4 × 4 FSSs bandpass structure placed above and below the antenna. The electric field pattern in the principle planes obtained from the simulation result is shown in Fig. 6.10. For both the principle planes, the maximum electric field is obtained in main lobe direction at 180° with main lobe magnitude of 20.8 dBV/m. The 3-dB angular width of 47.8° and 11.5° is obtained in E- and H-plane, respectively.

The directivity of an antenna depends on the electric field as:

$$D(\theta,\varphi) = \frac{4\pi}{\Omega_A} \left| E_{\text{norm}}(\theta,\varphi) \right|^2 \tag{6.15}$$

where $D(\theta, \varphi)$ is the directivity of the antenna configuration? $\Omega_A = \iint |E_{\text{norm}}(\theta, \varphi)|^2 \sin\theta \, d\theta d\varphi$ is the beam solid angle, and $|E_{\text{norm}}(\theta, \varphi)|$ is the normalized magnitude of the electric field pattern. The gain (G) of the antenna configuration is related to the directivity (D) of the antenna as $G = \eta D$, where η is the radiation effeciency of the antenna. The radiation patterns for gain (dB) and directivity (dBi) obtained in the principle planes for PCA without FSS structure and with

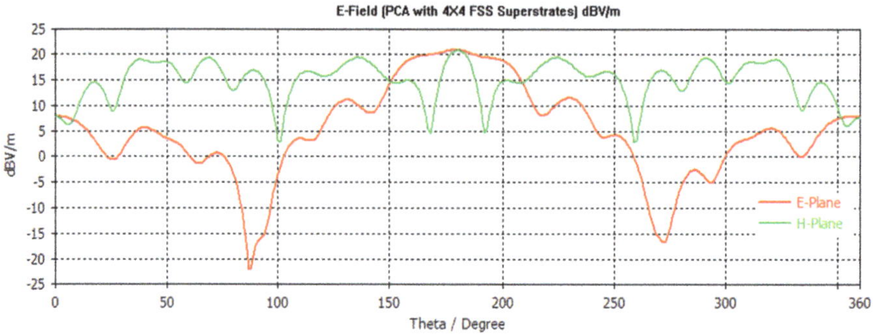

Fig. 6.10 The electric field of PCA in both principle planes with 4 × 4 FSS in E-plane and H-plane

4 × 4 FSSs structure are illustrated in Figs. 6.11 and 6.12. By means of 4 × 4 FSSs bandpass structure above and below the photoconductive dipole antenna, significantly high gain and directivity is achieved, which is required for the use of this antenna configuration for THz sensing and imaging of explosives in powder form.

For the antenna configuration without FSS, the gain of antenna is 4.85 and 7.3 dB in E-plane and H-plane, respectively. However, the gain is increased up to 8.38 dB in both E-plane and H-plane with the use of 4 × 4 FSSs bandpass structure (having slot size of unit cell 8 μm and $s/\lambda = 0.04$) with photoconductive dipole antenna. Similarly, the directivity of photoconductive dipole antenna without FSS is 5.78 and 8.24 dBi in E-plane and H-plane, respectively. With the use of 4 × 4 FSSs bandpass structure (having slot size of unit cell 8 μm and $s/\lambda = 0.04$), the directivity of photoconductive dipole antenna has been increased to 10.27 dBi in both the principle planes at 1.5 THz. The radiation efficiency of the proposed antenna configuration at 1.5 THz is 81.6%.

6.7 Summary

In this chapter, the significant improvement of gain and directivity of the photoconductive dipole antenna using bandpass FSSs as a superstrate at terahertz frequency for sensing and imaging is presented. Further, the effect of unit-cell periodicity of FSS on the resonance condition of antenna structure is studied. From the proposed analysis, it is illustrated that by varying the value of s/λ, the required band of transmission of the THz signal is achieved using FSS superstrate with photoconductive dipole antenna. The proposed photoconductive dipole antenna with 4 × 4 FSS bandpass structure is used above and below the antenna, which demonstrate significant improvement in the performance values of the return loss, gain, directivity, and front-to-back ratio. Therefore, the proposed antenna can be used to enhance the imaging capabilities of THz radiating source to address the

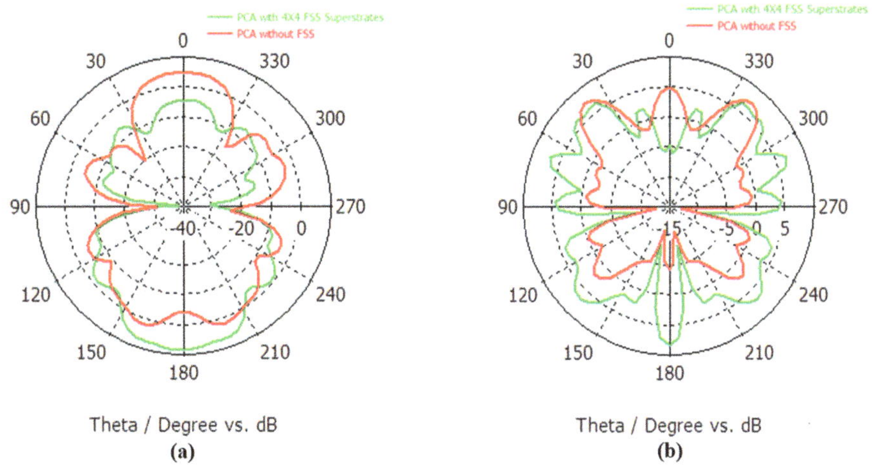

Fig. 6.11 The radiation characteristics such as gain (dB) for photoconductive dipole antenna without FSS and with FSS in (**a**) E-plane and (**b**) H-plane

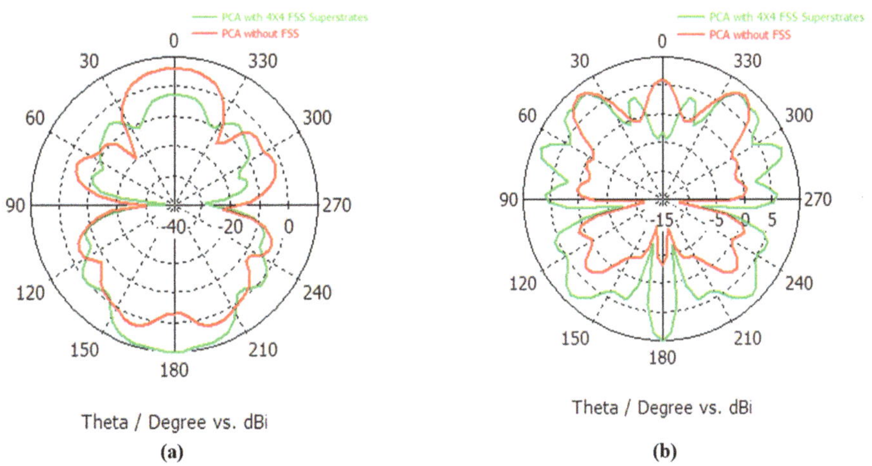

Fig. 6.12 The radiation characteristics such as directivity (dBi) for photoconductive dipole antenna without FSS and with FSS in (**a**) E-plane and (**b**) H-plane

considerations such as limited depth-of-field (DoF), which is the distance over which an object is considered in focus. However, the performance of the FSS is widely affected by the ohmic losses and various other factors such as surface roughness, dispersion through the conductor, and the skin depth, which become prominent in the THz regime of the electromagnetic spectrum. Further, the implementation of array structure of photoconductive dipole antenna with FSS bandpass superstrate is a challenging issue, which is discussed in detail in the next chapter.

References

1. I. Malhotra, P. Thakur, S. Pandit, K. R. Jha, G. Singh, Analytical framework of small-gap photoconductive dipole antenna using equivalent circuit model. Optical and Quantum Electronics **49**(10), 334/1–23 (2017)
2. N. Zhu, R.W. Ziolkowski, Photoconductive THz antenna designs with high radiation efficiency, high directivity, and high aperture efficiency. IEEE Trans. Terahertz Sci. Technol. **3**(6), 721–730 (2013)
3. F. Costa, A. Monorchio, Design of subwavelength tunable and steerable Fabry-Perot/leaky wave antennas. Prog. Electromagn. Res. **111**, 467–481 (2011)
4. G.V. Trentini, Partially reflecting sheet arrays. IRE Trans. Antennas Propag. **4**(4), 666–671 (1956)
5. A. Pirhadi, F. Keshmiri, M. Hakkak, M. Tayarani, Analysis and design of dual band high directive EBG resonator antenna using square loop FSS as superstrate layer. Prog. Electromagn. Res. **70**, 1–20 (2007)
6. D.H. Lee, Y.J. Lee, J. Yeo, R. Mittra, W.S. Park, Directivity enhancement of circular polarized patch antenna using ring-shaped frequency selective surface superstrate. Microw. Opt. Technol. Lett. **49**(1), 199–201 (2007)
7. A. Foroozesh, L. Shafai, Investigation into the effects of the patch-type FSS superstrate on the high-gain cavity resonance antenna design. IEEE Trans. Antennas Propag. **58**(2), 258–270 (2010)
8. H. Yang, N. Alexopoulos, Gain enhancement methods for printed circuit antennas through multiple superstrates. IEEE Trans. Antennas Propag. **35**(7), 860–863 (1987)
9. H. Gharbaninejad-Foumani, M. Khalaj-Amirhosseini, Compact spatial band-pass filters using FSS. Prog. Electromagn. Res. C **21**, 59–73 (2011)
10. J. Yeo, D. Kim, Novel design of a high-gain and wideband Fabry-Perot cavity antenna using a tapered AMC substrate. J. Infrared Millim. Terahertz Waves **30**(3), 217–224 (2009)
11. H. Boutayeb, T.A. Denidni, M. Nedil, Bandwidth widening technique for directive antenna based on partially reflecting surfaces. Prog. Electromagn. Res. **74**, 407–419 (2007)
12. S. Vegesna, Z. Yanhan, A. Bernussi, M. Saed, Terahertz two-layer FSSs with improved transmission characteristics. IEEE Trans. Terahertz Sci. Technol. **2**(4), 441–448 (2012)
13. H. Liu, S. Lei, X. Shi, L. Li, Study of antenna superstrates using metamaterials for directivity enhancement based on Fabry-Perot resonant cavity. Int. J. Antennas Propag. **13**, 1–10 (2013)
14. J. Liu, W.-H. Fan, X. Chen, J. Xie, Identification of high explosive RDX using terahertz imaging and spectral fingerprints. J. Phys. Conf. Ser. **680**(1), 012030/1–9 (2016)., IOP Publishing, Hangzhou, China
15. R. Appleby, H.B. Wallace, Standoff detection of weapons and contraband in the 100 GHz to 1 THz region. IEEE Trans. Antennas Propag. **55**(11), 2944–2956 (2007)
16. Y. Shen, A.T. Lo, P. Taday, B. Cole, W. Tribe, M. Kemp, Detection and identification of explosives using terahertz pulsed spectroscopic imaging. Appl. Phys. Lett. **86**(24), 241116/1–4 (2005)
17. H.-B. Liu, Y. Chen, G.J. Bastiaans, X.-C. Zhang, Detection and identification of explosive RDX by THz diffuse reflection spectroscopy. Opt. Express **14**(1), 415–423 (2006)
18. M. Tani, M. Herrmann, K. Sakai, Generation and detection of terahertz pulsed radiation with photoconductive antennas and its application to imaging. Meas. Sci. Technol. **13**(11), 1739–1745 (2002)
19. A. Singh, S. Winnerl, J.C. König-Otto, D.R. Stephan, M. Helm, H. Schneider, Plasmonic efficiency enhancement at the anode of strip line photoconductive terahertz emitters. Opt. Express **24**(20), 22628–22634 (2016)
20. S.-G. Park, Y. Choi, Y.-J. Oh, K.-H. Jeong, Terahertz photoconductive antenna with metal nanoislands. Opt. Express **20**(23), 25530–25535 (2012)

21. N. Vieweg, M. Mikulics, M. Scheller, K. Ezdi, R. Wilk, H.-W. Hübers, M. Koch, Impact of the contact metallization on the performance of photoconductive THz antennas. Opt. Express **16**(24), 19695–19705 (2008)
22. M. Khorshidi, G. Dadashzadeh, Dielectric structure with periodic strips for increasing radiation power of photoconductive antennas: Theoretical analysis. J. Infrared Millim. Terahertz Waves **38**(5), 609–629 (2017)
23. O. Abdulmunem, K. Hassoon, J. Völkner, M. Mikulics, K. Gries, J. Balzer, Photoconductive LT-GaAs terahertz antennas: Correlation between surface quality and emission strength. J. Infrared Millim. Terahertz Waves **38**(5), 574–582 (2017)
24. N.T. Yardimci, R. Salas, E.M. Krivoy, H.P. Nair, S.R. Bank, M. Jarrahi, Impact of substrate characteristics on performance of large area plasmonic photoconductive emitters. Opt. Express **23**(25), 32035–32043 (2015)
25. T.K. Nguyen, H. Han, I. Park, Frequency selective surface for directivity enhancement of a terahertz photomixer antenna, in *International Workshop on Antenna Technology (iWAT)*, Karlsruhe, Germany, 4–6 March 2013, pp. 100–103
26. I. Malhotra, K.R. Jha, G. Singh, Analysis of highly directive photoconductive dipole antenna at terahertz frequency for sensing and imaging applications. Opt. Commun. **397**, 129–139 (2017)
27. N.M. Burford, M.O. El-Shenawee, Review of terahertz photoconductive antenna technology. Opt. Eng. **56**(1), 010901 (2017)
28. S.I. Lepeshov, A.A. Gorodetsky, N.A. Toropov, T.A. Vartanyan, A.E. Krasnok, P.A. Belov, E.U. Rafailov, Novel optimized hybrid terahertz photoconductive antennas. J. Phys. Conf. Ser. **1092**, 012076 (2018)
29. X. Ropagnol, Intense sub-terahertz radiation from wide-bandgap semiconductor based large aperture photoconductive antennas pumped by UV lasers. New J. Phys. **21**, 113042/1–10 (2019)
30. I. Malhotra, K. R Jha, G.Singh, Design of highly directive lens-less photoconductive dipole antenna array with frequency selective surface for terahertz imaging applications. Optik: International Journal for Light and Electron Optics **173**, 206–219 (2018)
31. P. Carelli, F. Chiarello, S. Cibella, A.D. Gaspare, R. Leoni, M. Ortolani, G. Torrioli, A fast terahertz spectrometer based on frequency selective surface filters. J. Infrared Millim. Terahertz Waves **33**(5), 505–512 (2012)
32. K.R. Jha, G. Singh, Prediction of highly directive probe-fed microstrip antenna at terahertz frequency. Int. J. Numer. Modell. **25**(2), 175–191 (2012)
33. H.-B. Liu, H. Zhong, N. Karpowicz, Y. Chen, X.-C. Zhang, Terahertz spectroscopy and imaging for defense and security applications. Proc. IEEE **95**(8), 1514–1527 (2007)
34. T.-K. Wu, *Frequency Selective Surface and Grid Array*, 1st edn. (Wiley-Interscience, New York, 1995)
35. B.A. Munk, *Frequency Selective Surfaces: Theory and Design* (Wiley Online Library, New York, 2000)
36. F. Bayatpur, Metamaterial-inspired frequency-selective surfaces. PhD thesis, University of Michigan, 2009
37. R. Mittra, C.H. Chan, T. Cwik, Techniques for analyzing frequency selective surfaces-a review. Proc. IEEE **76**(12), 1593–1615 (1988)
38. K.R. Jha, G. Singh, R. Jyoti, A simple synthesis technique of single-square-loop frequency selective surface. Prog. Electromagn. Res. B **45**, 165–185 (2012)
39. J.A. Reed, Dissertation proposal: Frequency selective surfaces with multiple periodic elements. PhD thesis, University of Texas at Dallas, 1997
40. A.E. Yilmaz, M. Kuzuoglu, Design of the square loop frequency selective surfaces with particle swarm optimization via the equivalent circuit model. Radioengineering **18**(2), 95–102 (2009)
41. P. Callaghan, E.A. Parker, R.J. Langley, Influence of supporting dielectric layers on the transmission properties of frequency selective surfaces. IEEE Proc. Microw. Antennas Propag. **138**(5), 448–454 (1991)., IET Digital Library

42. N. Behdad, M. Al-Joumayly, M. Salehi, A low-profile third-order bandpass frequency selective surface. IEEE Trans. Antenna Propag. **57**(2), 460–466 (2009)
43. J.E. Raynolds, B.A. Mank, J.B. Pryor, R.J. Marhefka, Ohmic loss in frequency-selective surfaces. J. Appl. Phys. **93**(9), 5346–5358 (2003)
44. E.D. Palik, *Handbook of Optical Constants of Solids* (American Press, London, 1985)
45. P.W. Gilberd, The anomalous skin effect and the optical properties of metals. J. Phys. F **12**(8), 1845–1860 (1982)
46. J.S. Fernandez, G. Goussetis, R. Cheung, Tunable 2D electromagnetic bandgap (EBG) structures based on micro-electro-mechanical systems (MEMS) for THz frequencies, in *Proceedings of IEEE International Symposium on Antennas and Propagation Society (APSURSI)*, Toronto, 11–17 July 2010, pp. 1–4
47. T.K. Ostmann, K. Pierz, G. Hein, P. Dawson, M. Koch, Audio signal transmission over THz communication channel using semiconductor modulator. Electron. Lett. **40**(2), 124–126 (2004)
48. G. Bharti, K.R. Jha, G. Singh, Terahertz frequency selective surface for future wireless communication systems. Optik **126**(24), 5909–5917 (2015)
49. M.E. MacDonald, A. Alexanian, R.A. York, Z. Popovic, E.N. Grossman, Spectral transmittance of lossy printed resonant-grid terahertz bandpass filters. IEEE Trans. Microw. Theory Tech. **48**(4), 712–718 (2000)
50. N. Engheta, R. W. Ziolkowski (eds.), *Metamaterials: Physics and Engineering Explorations* (John Wiley & Sons, Hoboken, 2006)
51. D. Grischkowsky, I. Duling III, J. Chen, C.-C. Chi, Electromagnetic shock waves from transmission lines. Phys. Rev. Lett. **59**(15), 1663–1666 (1987)
52. A.K. Bhattacharya, Characteristics of space and surface waves in a multilayered structure. IEEE Trans. Antennas Propag. **38**(8), 1231–1238 (1990)
53. M.R. Stone, M. Naftaly, R.E. Miles, J.R. Fletcher, D.P. Steenson, Electrical and radiation characteristics of semilarge photoconductive terahertz emitters. IEEE Trans. Microw. Theory Tech. **52**(10), 2420–2429 (2004)
54. I. Malhotra, K.R. Jha, G. Singh, Design of highly directive terahertz photoconductive dipole antenna using frequency selective surface for sensing and imaging application. J. Comput. Electron. **17**, 1–20 (2018)
55. A. Pirahadi, F. Keshmiri, M. Hakkak, M. Tayarani, Analysis and design of dual band high directivity EBG resonator antenna using square loop FSS as superstrate layer. Prog. Electromagn. Res. **70**, 1–20 (2007)

Chapter 7
Highly Directive Lens-Less Photoconductive Dipole Antenna Array for Imaging Applications

7.1 Introduction

Among the concepts for THz imaging, several noncoherent techniques such as microbolometer arrays are presented [1], whereas to increase the measurement speeds, such approach provides only limited information owing to lack of information about the phase. However, the pulsed THz systems using coherent techniques offer potential advantage of broad spectral information with a single scan. However, the most basic coherent imaging can be attained by raster scanning a sample through the THz wave focus and creating complete spectroscopic information at each pixel corresponding to the antenna structure. On the other hand, based on the delay line concept, the required lock-in time constants to record a high-quality THz pulse takes a long time to trace out the full image [2] with single antenna. Therefore, to speed up the measurements, there is a requirement of making a THz system for parallel or accelerated measurements. One potential approach to such parallelization is the implementation of multiantenna setup, which reduces the measurement time through deployment of a number of THz emitter and detector pairs. On the other hand, the compact size arrays need to be used to counter the increase in system cost [3].

In broadband pulsed THz systems, the optical gating of photoconductive dipole antennas is one of the most accepted THz generation and detection techniques [4–6]. Using the photoconductive dipole antenna sources and detectors in the imaging system, the setup measures a first pulse with no sample in the THz beam path and a second pulse after passing through the sample to be detected which allows the concurrent extraction of the refractive index, the absorption coefficient in addition to the thickness of sample, and therefore, provides the detailed information about sample under investigation [7]. The visualization techniques for coherent THz image information comprise (1) peak-to-peak measurement, (2) time-of-flight measurement, (3) amplitude measurement with single frequency, (4) integrated signal amplitude over a specified frequency range, and (5) evaluation of the dielectric

I. Malhotra, G. Singh, *Terahertz Antenna Technology for Imaging and Sensing Applications*, https://doi.org/10.1007/978-3-030-68960-5_7

material properties such as the refractive index and absorption coefficient. Therefore, several images revealing different kinds of information can be produced from a single dataset corresponding to the phase and amplitude of the detected THz pulses. Likewise, the information such as variation in thickness, humidity, material composition, density, and inclusions of air or other impurities in the sample [8] is obtained.

For the THz imaging applications, the planar THz antennas, such as photoconductive antennas, have convincing advantages, such as: (a) low profile, (b) easy integration to other planar devices, and (c) easy fabrication [9]. Moreover, the fabrication of planar transmission line [10] at THz frequencies also proves a potential means of the fabrication of array of planar THz antennas. Therefore, a submillimeter-wave, highly directive photoconductive, dipole array antenna is presented in this chapter for terahertz (THz) imaging system operating in 1.4–2.2 THz frequency range. The array antenna is formed on a single photoconductive substrate to make it compact and suitable for THz imaging applications. The photoconductive dipole array antenna improves the gain and directivity, and therefore, is useful to enhance the imaging capabilities to address the considerations, such as limited depth-of-field (DoF), that is, the distance over which an object is well thought out in focus and size, weight, and power (SWaP) of THz source for imaging applications. These are important considerations for applications such as standoff imaging and surveillance of moving targets where the high angular resolution as well as extended depth-of-field is the key for successful detection of hidden explosives and illicit drugs. A technique of using frequency selective surface with photoconductive array antenna is also presented to further enhance the directivity from the radiating structure to yield high image resolution. Moreover, by means of this technique, radiation energy gets confined to the desired frequency band rather than spreading over a broad-spectrum range.

7.2 Related Work and Problem Formulation

For imaging application, both the photoconductive antenna and photomixer techniques are considered as THz sources by THz community. The THz antennas based upon the photoconduction method have several advantages as compared to that of the other THz pulsed sources, such as THz antennas work in the room temperature environment and compact, which can be used both at the emitter and at detector of pulsed imaging system. Moreover, the THz generation and detection using photoconductive antenna is useful for coherent THz imaging since this technique has the ability to record THz waves in the time domain together with both the intensity and phase information which can further be utilized to acquire more details of the target, for example, spectral and depth information. For photoconductive antenna emitter–detector configuration of THz pulsed imaging system, the potential efforts are made for a high-efficiency antenna design and consequently expecting an increased directivity of a THz photoconductive dipole antenna for sensing and imaging applications.

There are several ways to increase the directivity such as the use of array configuration [11] or a lens substrate [12]. Earlier, to achieve high image resolution, a lens-based system is generally used with photoconductive dipole antenna for diffraction-limited image extraction. Pradarutti et al. [13] have presented THz line detection with 16 channels using microlens array coupled photoconductive antenna array. However, using a large apertures and lens-based systems results in volumetric scale to the THz source and also presents challenges from a portability perspective. On the other hand, the array configuration requires a higher illumination power from multiple femtosecond lasers; however, the use of lens substrate limits the number of antenna elements in a specified area, which is required in an imaging array. Moreover, the advent of Er^{+}-doped femtosecond fiber lasers has replaced the high system cost which has involved with the use of titanium: sapphire lasers for pulsed THz systems based on photoconductive array antenna. Further, the use of defective photonic crystal (DPC) substrates in arrays of THz antenna reduces the amount of power coupled to the substrate surface waves, and it helps to decrease the mutual coupling between the elements of the antenna array, and therefore, increase the directivity in the boresight direction [14]. Gorodetsky et al. [15] have studied the behavior of photoexcited carriers in GaAs wafers, which contain InAs quantum dots. Such quantum dots help to shorten the overall carrier lifetime and therefore increases the THz intensity at higher frequencies. High-power and broadband photoconductive THz emitter, which operates at telecommunication optical wavelengths, is presented in [16]. The authors have used ErAs:InGaAs substrate for two-dimensional array of plasmonic nanoantennas to achieve high resistivity as well as short carrier lifetime characteristics, which are required for robust operation at telecommunication optical wavelengths. The plasmonic nanoantennas [17, 18] offer significantly higher optical-to-terahertz conversion efficiencies in comparison to the conventional photoconductive emitters while maintaining broad operation bandwidths [19].

In addition, the phenomenon of scanning through array implementation over an extended volume could offer security such as in a public marketplace where the security is essential, but a visible display is not so much important. Therefore, for this specific application of imaging, a compact array structure of photoconductive dipole antenna with pulsed optical beam from femtosecond laser pulse can be used with high directive ability of the radiating THz source. Even though in THz imaging, multipixel approaches provide an acceleration of THz measurements and therefore lead to attractive system architectures in future, however, stern problems in the alignment process are still major shortcomings which need to be addressed.

7.3 Unit-Cell Antenna Modeling

The designed photoconductive dipole antenna as shown in Fig. 7.1 consists of a dipole structure of Ti-Au metal with coplanar transmission line of length $\lambda/2$ on both sides of the dipole antenna. The complete metal structure is placed on a thin

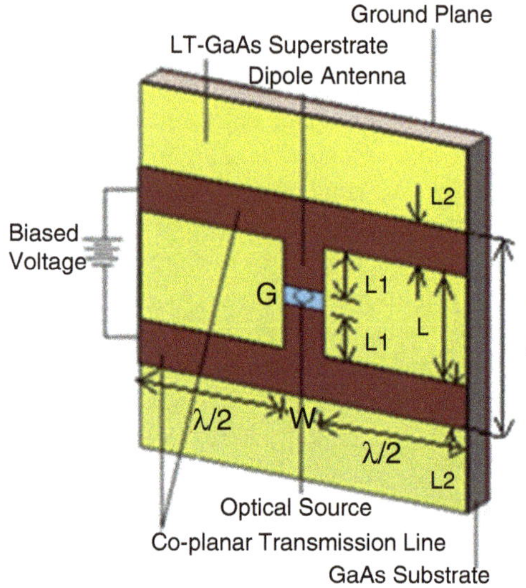

Fig. 7.1 The basic structure of THz photoconductive dipole antenna designed in CST Microwave Studio [20]

layer of LT-GaAs superstrate, which is stacked on GaAs photoconductive substrate. A ground plane with same material as of antenna is used below the substrate. The antenna has a photoconductive gap length (*G*) and width (*W*) at the center of dipole antenna of effective length ($L = l_e$), also considered as the distance between the transmission lines, which is biased with a voltage source and is illuminated at the center by means of femtosecond laser pulse. The distance between the transmission lines or the effective length is determined from the relation $L = c/2nf$, where c is the speed of light in free space and n is the refractive index of material.

At 1.5 THz, using the value of refractive index ($n = 3.4$), for GaAs photoconductive material, the effective length of dipole antenna l_e is 30 μm. The resonant frequency of the photoconductive dipole antenna is determined using the relation $f_r = c/(2l_e[(1 + \varepsilon_d)/2]^{1/2})$, where f_r is the resonating frequency, and ε_d is the dielectric permittivity of GaAs substrate and is taken as 12.9. By putting the values of c, ε_d, and l_e, the resonant frequency of the designed photoconductive dipole antenna is 1.95 THz [21]. The length of coplanar transmission line is generally set to be long enough to evade the reflection at the line end. In the designed photoconductive dipole antenna, the length of transmission lines is taken as 110 μm with width and thickness 10 μm and 0.35 μm, respectively. As the emission efficiency is inversely proportional to the photoconductive dipole antenna gap, the bias field and pump power are considered to be constant. For that reason, it is important to exploit a photoconductive gap as small as possible by focusing the laser beam (optical excitation) very closely to the gap. Therefore, the photoconductive gap is taken as 5 μm for small gap photoconductive dipole antenna in all antenna-designed configurations in this paper. Moreover, the dimensions of the superstrate, substrate and ground are taken as (110 × 110) μm^2, with thickness 1 μm, 10 μm, and 0.35 μm, respectively.

7.4 Design of Photoconductive Antenna Array

In THz imaging applications, there is a requirement of efficient THz antennas in terms of losses and cost both at transmitter and receiver side. Moreover, in case of active imaging systems, the image acquisition time is required to be small. Therefore, several types of illumination of the object with THz radiation under investigation are selected, such as single frequency, multifrequency, or ultra-wideband signals [22]. For such arrangement, highly directive and optimum radiation efficiency from the radiating source is essential. In THz imaging applications, to yield enhanced depth-of-field (DoF) for imaging purpose by using THz source, the high directivity of the radiating source is required. Various other antenna configurations, such as spiral-type, bow-tie shape, are used; however, the photoconductive dipole antenna is comparatively attractive as small gap dipole antenna because of its ease of fabrication and are also useful for system-on-chip compatibility in spite of its poor directivity. One straightforward solution to overcome this problem is to design an array of photoconductive dipole antenna [23]. Moreover, the noise equivalent power (NEP) of an imaging array depends on the antenna efficiency. Therefore, it is essential to facilitate high-efficiency on-chip antennas for low system NEP. Such on-chip antennas can be put into practice with different radiation phenomena such as above-chip radiation, above-chip with focusing techniques, below-chip radiation, and below-chip with a dielectric lens as discussed in [24]. Here, Uzunkol et al. [24] have recommended the use of above-chip radiation from the THz array source due to low-cost and being the most robust system as it works with low-cost carrier substrate. Moreover, use of THz system with above-chip also supports the small size, weight, and power (SWaP) values of THz source for imaging applications. Therefore, the above-chip radiation phenomena and above-chip with focusing technique such as use of frequency selective surface (FSS) to further increase the directivity of the THz array antenna are presented in this chapter. The array designs are shown in Fig. 7.2a, b.

First, a (1 × 2) photoconductive dipole linear array antenna has been designed to determine the radiation parameters. Then, (2 × 2) photoconductive dipole array antenna is designed to evaluate the radiation parameters, such as gain, directivity, and radiation efficiency. The spacing between photoconductive dipole antennas in an array arrangement is taken as $\lambda/2$, and for (2 × 2) photoconductive dipole array, the spacing between two sets of transmission line is also considered as $\lambda/2$. Both types of arrays designed are further considered for directivity enhancement using FSS as a superstrate and placed above the antenna array with air gap separation between the antenna structure and FSS structure [25]. The detailed discussion on making a choice of using FSS is presented in next section.

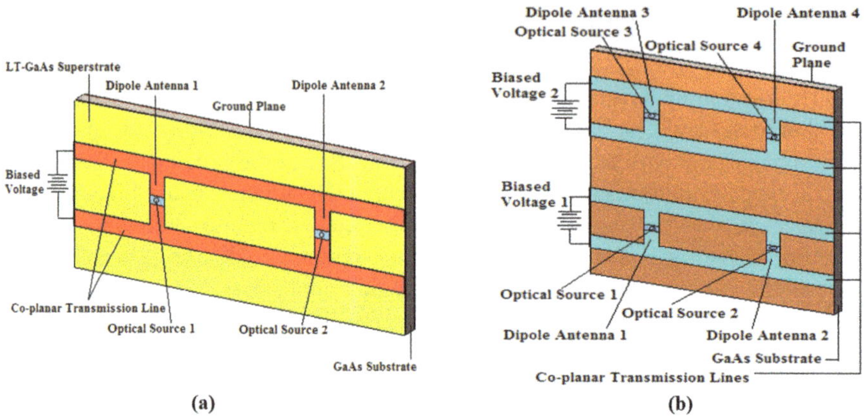

Fig. 7.2 Photoconductive dipole antenna array: (**a**) (1 × 2) photoconductive dipoles and (**b**) (2 × 2) photoconductive dipoles

7.5 Frequency Selective Surface for Photoconductive Antenna Array

To increase the directivity of the elementary dipole and planar antennas in the microwave frequency regime of the spectrum, extremely reflective surfaces as superstrate are extensively used [26–28] and it signify that such type of structure is equally appropriate at the terahertz frequency [29, 30]. In general, the directivity of antenna using FSS is calculated by using either the transverse equivalent network model or the resonance estimation ray tracing technique. However, these techniques can only predict the peak directivity of the antenna at the resonance frequency, while the FSS array and the ground plane size are suitably large or semi-infinite. When the FSS array or ground plane is truncated, the directivity of the cavity resonant antenna is reduced, and then these techniques are unable to predict the directivity of the antenna correctly [31]. In such case, using a comparison method, the directivity of antenna array can be predicted at the THz frequency using FSS structure.

7.5.1 Analysis of Unit-Cell Frequency Selective Surface

In the photoconductive dipole array antenna design, the FSS structure is used along with (1 × 2) and (2 × 2) antenna array configurations to enhance the gain and directivity, therefore, the FSS as bandpass spatial filter is used. The unit-cell design of FSS bandpass is shown in Fig. 7.3. The FSS structure made of copper with thickness 0.35 μm is followed by the polytetrafluoroethylene substrate. Moreover, in order to keep away from evanescent waves to propagate through the periodic

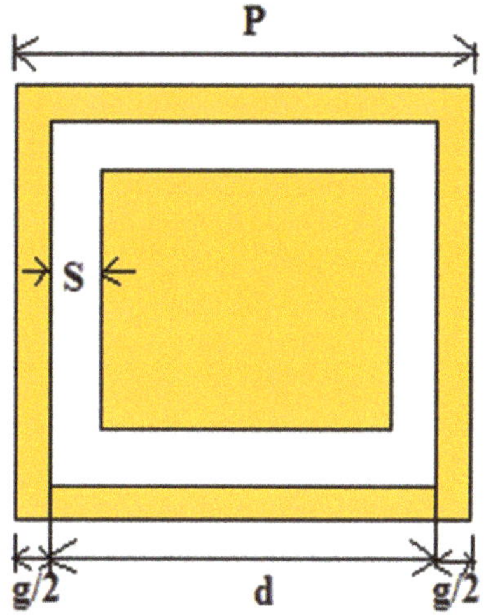

Fig. 7.3 The unit-cell configuration of bandpass FSS with FSS structure made of copper

structure, the thickness of dielectric substrate must follow the relation, $h < \lambda_0/20$, where h is the thickness of dielectric substrate of FSS and λ_0 is the wavelength in free space. Therefore, the thickness of substrate taken for bandpass FSS is 5 μm. The metallic elements signify inductive screens and give rise to total reflection, while apertures in a metallic sheet symbolize capacitive screens and give rise to total transmission. The performance of bandpass FSS depends on the physical parameters, such as the periodicity (P) of the unit cell, length of the slot (d), slot width (s), and gap between two slots (g) [25].

The values of the physical parameters of the FSS unit cell are determined using the synthesis technique of single square loop bandpass FSS at resonance frequency as mentioned in [32]. The relation between the physical parameters of unit-cell FSS is described as $4\ (d/\lambda)^2 \ln (2M\lambda/\pi s)\varepsilon_{\text{eff}} = w_r^2LC$, where M is a constant of equality that is related as $P = M\lambda$, in which λ is the operating wavelength, and ε_{eff} is the effective dielectric permittivity of the FSS substrate. From the geometry of the unit cell in Fig. 7.3, it is clear that $P - d = g$ and for fixed value of P, the value d can be evaluated for the resonance condition $w_r^2LC = 1$, using $4\ (d/\lambda)^2 \ln (2M\lambda/\pi s)\varepsilon_{\text{eff}} = 1$.

7.5.2 Estimation of Resonance Condition Using Ray Tracing

The ray tracing technique [33] is used to calculate the height (h) of the superstrate above the ground plane of the photoconductive dipole array antenna. When the unit-cell FSS surface, as in Fig. 7.3, is excited with a normal incident wave, whose electric field is aligned along the length of the surface, then bandpass unit-cell FSS shows the high-transmission property. Such surface property is achieved through simulating the unit-cell FSS in the transient solver of the CST microwave studio. The amplitude as well as the phase of the S-parameter at resonance frequency obtained are as follows: mag(S_{11}) = 0.845 and $\varnothing_{FSS} = 16.46^{\circ}$ at resonance frequency. The value of the ground plane phase is determined using the relation $\varnothing_g = \pi - 2\tan^{-1}(Z_d \tan(\beta d)/Z_0)$, where Z_d and Z_0 are the characteristic impedance of the dielectric substrate and air, respectively. The value obtained is $\varnothing_g = 49.66^{\circ}$. β represents the dielectric phase constant and d is the dielectric substrate thickness. The resonant air gap length (l) of 118.37 μm is obtained using the relation: $l = (N\lambda/2) + [(\varnothing_{FSS} + \varnothing_g)/\pi](\lambda/4)$, where N is an integer number. The schematic view of the designed photoconductive dipole array antennas with (2 × 2) FSS unit-cell structure across each dipole of the photoconductive array antenna is shown in Fig. 7.4.

However, Table 7.1 presents the parameters in terms of material conductivity, dielectric permittivity, magnetic permeability, and loss-tangent considered in CST Microwave Studio for simulating the antenna configurations. The results of single-unit, array antenna implementation of (1 × 2) and (2 × 2) modeling, design and simulation for photoconductive dipole antenna and implementation of highly reflective bandpass FSS superstrate are presented in the next section.

7.6 Numerical Analysis and Simulation Results

The designed single unit of photoconductive dipole antenna at 1.95 THz frequency is simulated using transient solver of CST Microwave Studio, and the 10-dB impedance bandwidth of 317.1 GHz is obtained as shown in Fig. 7.5, with S-parameter magnitude of −21.88 dB.

The radiation patterns for gain (dB) and directivity (dBi) obtained in the principle planes for photoconductive dipole antenna are illustrated in Fig. 7.6.

The simulated results in Fig. 7.6a show the value of gain 5.47 dB in E-plane with main lobe direction at 8.0°. The angular width (3-dB) is 72.3° with the side lobe

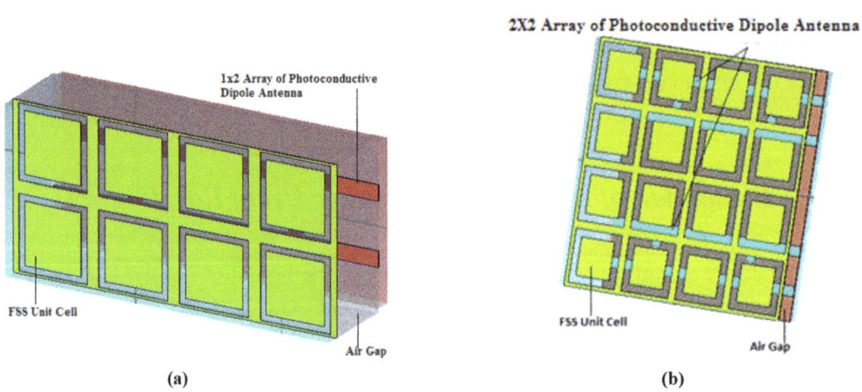

Fig. 7.4 Schematic of photoconductive dipole array antenna with (2 × 2) array of FSS structure across each dipole antenna with periodicity $P = 62.4$ μm of unit-cell FSS across (**a**) (1 × 2) photoconductive dipole linear array antenna and (**b**) (2 × 2) photoconductive dipole array antenna [21]

Table 7.1 The parameters used in simulation performed using the transient solver of CST Microwave Studio

Parameter	Value
Photoconductive dipole antenna (Ti-Au), transmission lines, and ground	
Conductivity (S/m)	1.6×10^7
Superstrate (LT-GaAs)	
Carrier lifetime, majority carriers (psec)	0.25
Mobility (cm^2/Vs)	Greater than 200
Permittivity	13.26
Permeability	1
Loss tangent (S/m)	0.006
Electric breakdown field (V/cm)	Greater than 5×10^5
Substrate (GaAs)	
Carrier lifetime, majority carriers (psec)	0.1
Mobility (cm^2/Vs)	200
Permittivity	12.9
Permeability	1
Loss tangent (S/m)	0.006
Electric breakdown field (V/cm)	Nearly 4×10^5
FSS bandpass unit cell (copper)	
Conductivity (S/m)	5.8×10^7 S/m
FSS substrate (polytetrafluoroethylene)	
Permittivity	2.08
Loss tangent(S/m)	0.0004

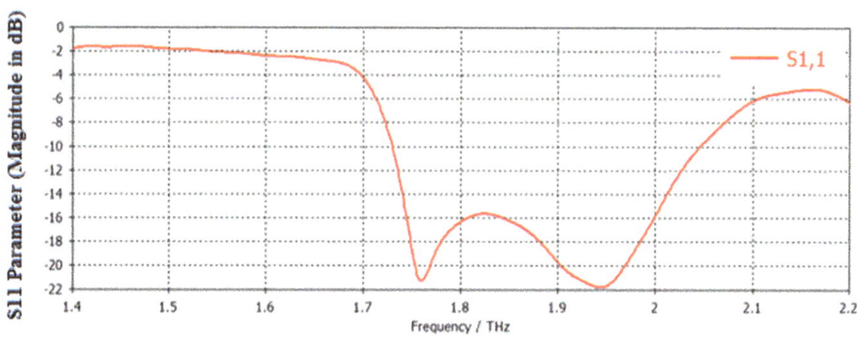

Fig. 7.5 The frequency response of S-parameter for simple photoconductive dipole antenna for 1.4–2.2 THz

level −16.4 dB. However, in H-plane, the gain is 5.35 dB with its main lobe direction 5.0°. The angular width (3-dB) is 82.2° having a side lobe level −22.3 dB. In Fig. 7.6b, the directivity in E-plane is 6.25 dBi with main lobe direction 8.0°. The angular width (3-dB) is 72.3° and the side lobe level is −16.4 dB and, in H-plane, the directivity is 6.14 dBi with its main lobe direction 5.0°. The angular width

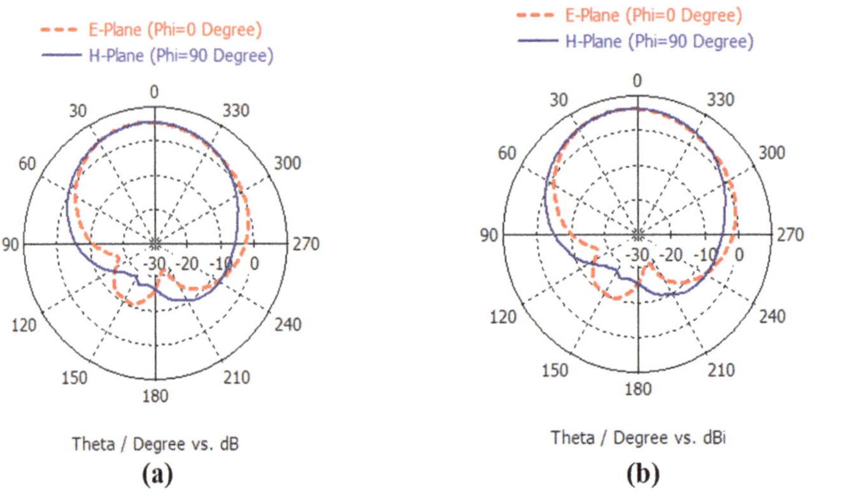

Fig. 7.6 The radiation characteristics of single photoconductive dipole antenna (**a**) gain (dB) in E- and H-plane, and (**b**) directivity (dBi) in E- and H-plane

(3-dB) is 82.2° having a side lobe level −22.3 dB. The radiation efficiency and total efficiency obtained are 0.84 and 0.83, respectively, with current density of 1.06×10^7 A/m^2. It is clearly observed from simulated results as shown in Fig. 7.6 and the theory of operation discussed in previous section for a single photoconductive dipole antenna that the gain and directivity values are very small in spite of its wideband of operation. Therefore, there is a requirement to use the array technique to increase the values of gain and directivity so that the simple THz photoconductive dipole antenna can be used in imaging applications. Initially, a linear array implementation of photoconductive dipole antenna is proposed. In this, the photoconductive dipoles are constructed with common bias and ground plane and are placed at $\lambda/2$ spacing, which makes the fabrication easy. The antenna geometry is placed on a 220×110 μm^2 area of photoconductive substrate. The simulation results obtained in terms of radiation patterns for gain and directivity are shown in Fig. 7.7. The simulated results in Fig. 7.7a show the radiation pattern of gain 5.86 dB in E-plane with main lobe direction at 1.0°. The angular width (3-dB) is 27.6° and the side lobe level is −3.3 dB. However, in H-plane, the gain is 5.83 dB with its main lobe direction 359.0°. The angular width (3-dB) is 82.2° having a side lobe level −17 dB.

Likewise, the simulated results as in Fig. 7.7b show the directivity in E-plane of 8.14 dBi with main lobe direction at 1.0°. The angular width (3-dB) is 27.6° and the side lobe level is −3.3 dB and, for H-plane, the directivity is 8.11 dBi with its main lobe direction 359.0°. The angular width (3-dB) is 82.2° having a side lobe level −17 dB. The radiation efficiency and total efficiency obtained are 0.72 and 0.59, respectively, with current density of 2.37×10^7 A/m^2. It is observed that with the increase in dipole antenna elements in linear form, the directivity is increased by 2 dBi but there is not much increase in the gain value. Moreover, the radiation efficiency of the (1×2) photoconductive linear dipole array antenna decreases and it

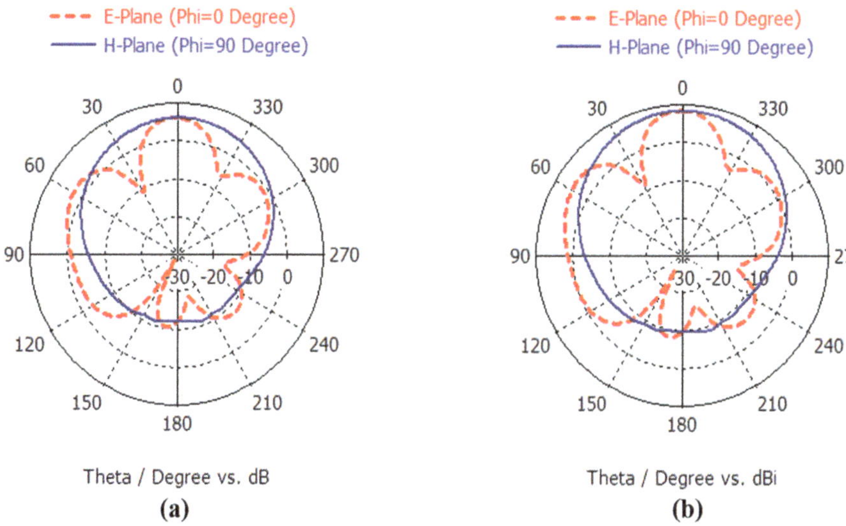

Fig. 7.7 The radiation characteristics of (1 × 2) photoconductive dipole array antenna (**a**) gain (dB) in E- and H-plane, and (**b**) directivity (dBi) in E- and H-plane

affects the total efficiency of the array antenna. The main reason for the enhancement in the directivity is the increase in the current density across the two dipoles of linear-array arrangement.

However, for imaging application, an optimum gain and radiation efficiency values are also required. Therefore, the FSS structure is used as superstrate with (1 × 2) array antenna, which is placed at a height of 118.37 μm from the ground plane of the photoconductive dipole array antenna. The radiation patterns obtained from the simulation result are shown in Fig. 7.8. Figure 7.8a shows the simulated results for gain with value 7.74 dB in E-plane and main lobe direction at 1.0°. The angular width (3-dB) is 31.1° and the side lobe level is −10.2 dB. However, in H-plane, the gain is 7.73 dB with its main lobe direction 359.0°. The angular width (3-dB) is 65.7° having a side lobe level −10.3 dB. The directivity in E-plane is 10.8 dBi with main lobe direction at 1.0° as shown in Fig. 7.8b. The angular width (3-dB) is 31.1° and the side lobe level is −10.2 dB and, in H-plane, the directivity is 10.8 dBi with its main lobe direction 359.0°.

The angular width (3-dB) is 65.7° having a side lobe level −10.3 dB. The radiation efficiency and total efficiency obtained are 0.72 and 0.59, respectively, with current density of 2.6×10^7 A/m^2. The use of FSS as superstrate with photoconductive dipole array antenna results in a further increase in the value of directivity by 2.66 dBi and gain by 2 dB. Furthermore, an array of four dipole antennas is designed by using two separate set of transmission lines separated by $\lambda/2$ distance. The complete (2 × 2) photoconductive dipole array antenna is placed on (220 × 220) μm^2 area of photoconductive substrate. The radiation parameters obtained from the simulation results are shown in Fig. 7.9.

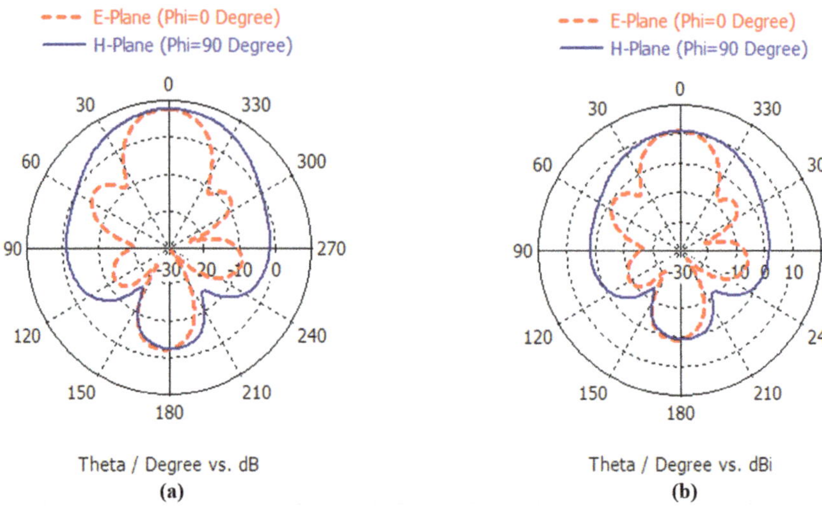

Fig. 7.8 The radiation characteristics of (1 × 2) photoconductive dipole array antenna with FSS superstrate (**a**) gain (dB) in E- and H-plane, and (**b**) directivity (dBi) in E- and H-plane

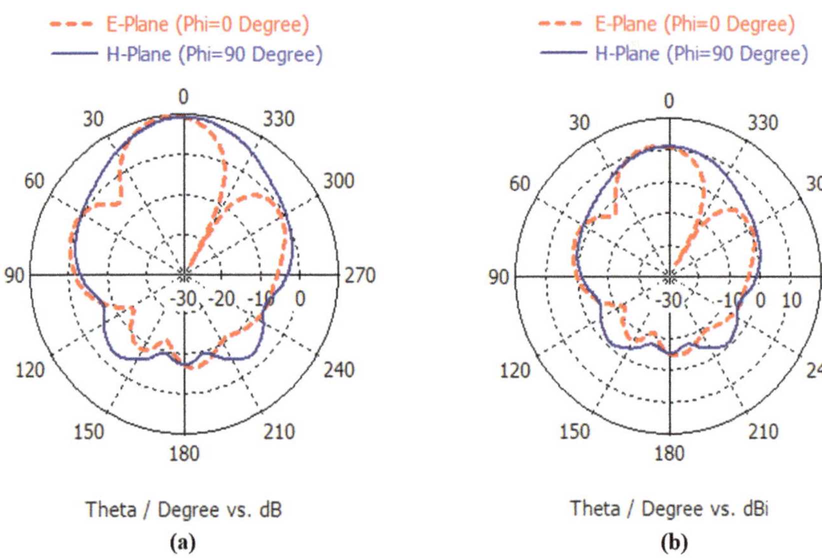

Fig. 7.9 The radiation characteristics of (2 × 2) photoconductive dipole array antenna (**a**) gain (dB) in E- and H-plane and (**b**) directivity (dBi) in E- and H-plane

The simulated results in Fig. 7.9a show the value of gain 9.58 dB in E-plane with main lobe direction at 6.0°. The angular width (3-dB) is 32.2° with the side lobe level −8.7 dB. However, in H-plane, the gain is 9.15 dB with its main lobe direction 0.0°. The angular width (3-dB) is 50.7° having a side lobe level −12.5 dB. In Fig. 7.9b, the directivity in E-plane is 11.4 dBi with main lobe direction at 6.0°. The angular width (3-dB) is 32.2° and the side lobe level is −8.7 dB and, in H-plane, the directivity of 10.9 dBi with its main lobe direction 0.0° is obtained. The angular width (3-dB) is 50.7° having a side lobe level −12.5 dB. The radiation efficiency and total efficiency are 0.84 and 0.67, respectively, with current density of 2.7×10^5 A/m^2. In comparison to (1 × 2) photoconductive dipole array antenna, the (2 × 2) photoconductive dipole array antenna shows high gain and directivity values in both principle planes. Moreover, the radiation efficiency and the total efficiency of the array configuration are also increased. To further enhance the radiation parameters of array antenna, the FSS structure is placed above the array configuration. The radiation patterns obtained from the simulation are shown in Fig. 7.10. The gain with value 10.4 dB in E-plane and main lobe direction at 3.0° having angular width (3-dB) 26.5° and the side lobe level −6.4 dB is shown in Fig. 7.10a. However, in H-plane, the gain is 10.3 dB with its main lobe direction 0.0°. The angular width (3-dB) is 27.0° having a side lobe level −9.2 dB. In Fig. 7.10b, the directivity obtained in E-plane is 13.2 dBi with main lobe direction at 3.0°. The angular width (3-dB) is 26.5° and the side lobe level is −6.4 dB and, in H-plane, the directivity is 13.1 dBi with its main lobe direction 0.0°. The angular width (3-dB) is 27.0° having a side lobe level −9.2 dB. The radiation efficiency and total efficiency obtained are 0.79 and 0.52, respectively, with current density of 2.74×10^5 A/m^2. Therefore, the use of FSS as superstrate with photoconductive dipole array antenna [34] results in further increase in the value of directivity by 3 dBi.

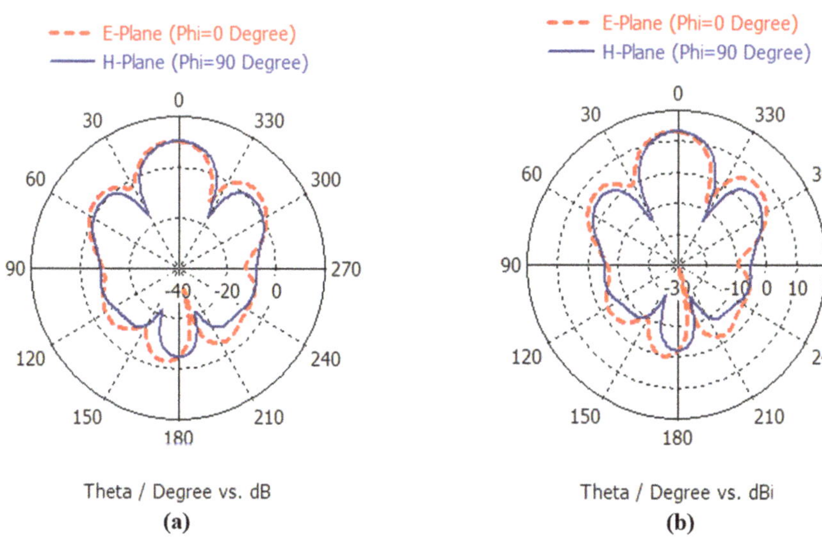

Fig. 7.10 The radiation characteristics of (2 × 2) photoconductive dipole array antenna with FSS superstrate (**a**) gain (dB) in E- and H-plane, and (**b**) directivity (dBi) in E- and H-plane

7.7 Summary

Because of the limitation of small gain and directivity values of single photoconductive dipole antenna, we have explored linear- and planar-array implementation of photoconductive dipole antennas for THz imaging application. However, for imaging applications, there is a requirement of planar and compact THz antenna sources, which provide high directivity so as to achieve large DoF for improved image resolution. In this chapter, a technique of using frequency selective surface to further enhance the radiation parameters of array configuration of photoconductive dipole antenna is discussed. The resulting array antenna configuration produced a radiation efficiency of 72% for (1 × 2) and 84% for (2 × 2) PCA array, respectively. The directivity of 10.8 dBi and 13.2 dBi is obtained using FSS superstrate with (1 × 2) and (2 × 2) photoconductive dipole array antenna in comparison with 6.25 dBi directivity of single photoconductive dipole antenna at the resonance frequency 1.95 THz. The proposed array antenna with FSS results in planar configuration of highly directive, radiating THz source for the imaging applications.

References

1. A.W. Lee, Q. Hu, Real-time, continuous-wave terahertz imaging by use of a microbolometer focal-plane array. Opt. Lett. **30**(19), 2563–2565 (2005)
2. C. Jansen, S. Wietzke, O. Peters, M. Scheller, N. Vieweg, M. Salhi, N. Krumbholz, C. Jordens, T. Hochrein, M. Koch, Terahertz imaging: Applications and perspectives. Appl. Opt. **49**(19), E48–E57 (2010)
3. B. Pradarutti, R. Müller, G. Matthäus, C. Brückner, S. Riehemann, G. Notni, S. Nolte, A. Tunnermann, Multichannel balanced electro-optic detection for terahertz imaging. Opt. Express **15**(26), 17652–17660 (2007)
4. E. Castro-Camus, J. Lloyd-Hughes, M. Johnston, Three-dimensional carrier-dynamics simulation of terahertz emission from photoconductive switches. Phys. Rev. B **71**(19), 195301 1–7 (2005)
5. X.-C. Zhang, D. Auston, Optically induced THz electromagnetic radiation from planar photoconducting structures. J. Electromagn. Waves Appl. **6**(1–4), 85–106 (1992)
6. S.E. Ralph, D. Grischkowsky, Trap-enhanced electric fields in semi-insulators: The role of electrical and optical carrier injection. Appl. Phys. Lett. **59**(16), 1972–1974 (1991)
7. M. Scheller, C. Jansen, M. Koch, Analyzing sub-100-μm samples with transmission terahertz time domain spectroscopy. Opt. Commun. **282**(7), 1304–1306 (2009)
8. M. Herrmann, M. Tani, K. Sakai, Display modes in time-resolved terahertz imaging. Jpn. J. Appl. Phys. **39**(11R), 6254 (2000)
9. K.R. Jha, G. Singh, Terahertz planar antennas for future wireless communication: A technical review. Infrared Phys. Technol. **60**, 71–80 (2013)
10. Y. Kadoya, THz wave propagation on strip lines: Devices, properties, and applications, in *19th International Conference on Applied Electromagnetics and Communications, ICECom*, (Dubrovnik, Croatia, 24–26 September 2007), pp. 1–4
11. A. Rivera-Lavado, L.-E. García-Muñoz, A. Generalov, D. Lioubtchenko, K.-A. Abdalmalak, S. Llorente-Romano, A. Garcia-Lamperez, D. Segovia-Vargas, A.V. Raisanen, Design of a dielectric rod waveguide antenna array for millimeter waves. J. Infrared Millimeter Terahertz Waves **38**(1), 33–46 (2017)

12. G.C. Trichopoulos, G. Mumcu, K. Sertel, H.L. Mosbacker, P. Smith, A novel approach for improving off-axis pixel performance of terahertz focal plane arrays. IEEE Trans. Microwave Theory Tech. **58**(7), 2014–2021 (2010)
13. B. Pradarutti, R. Müller, W. Freese, G. Matthäus, S. Riehemann, G. Notni, S. Nolte, A. Tunnermann, Terahertz line detection by a microlens array coupled photoconductive antenna array. Opt. Express **16**(22), 18443–18450 (2008)
14. E. Rahmati, M.A. Boroujeni, Design of terahertz photoconductive antenna arrays based on defective photonic crystal substrates. Opt. Laser Technol. **114**, 89–94 (2019)
15. A. Gorodetsky, N. Bazieva, E.U. Rafailov, Pump dependent carrier lifetimes in InAs/GaAs quantum dot photoconductive terahertz antenna structures. J. Appl. Phys. **125**, 151606 (2019)
16. N.T. Yardimci, H. Lu, M. Jarrahi, High power telecommunication- compatible photoconductive terahertz emitters based on plasmonic nano-antenna arrays. Appl. Phys. Lett. **109**, 191103 1–5 (2016)
17. S.I. Lepeshov, A.A. Gorodetsky, N.A. Toropov, T.A. Vartanyan, E.U. Rafailov, A.E. Krasnok, P.A. Belov, Optimization of nanoantenna- enhanced terahertz emission from photoconductive antennas. J. Phys. **917**, 062060 (2017)
18. S. Lepeshov, A. Gorodetsky, A. Krasnok, N. Toropov, T.A. Vartanyan, P. Belov, A. Alu, E.U. Rafailov, Boosting terahertz photoconductive antenna performance with optimized plasmonic nanostructures. Sci. Rep. **8**, 6624 1–7 (2018)
19. I. Malhotra, K.R. Jha, G. Singh, Terahertz antenna technology for imaging applications- a technical review. Int. J. Microw. Wirel. Technol. **10**(3), 271–290 (2018)
20. I. Malhotra, P. Thakur, S. Pandit, K.R. Jha, G. Singh, Analytical framework of small-gap photoconductive dipole antenna using equivalent circuit model. Opt. Quant. Electron. **49**(10), 334/1–23 (2017)
21. I. Malhotra, K. R. Jha, G. Singh, Design of highly directive lens-less photoconductive dipole antenna array with frequency selective surface for terahertz imaging applications. Optik: International Journal for Light and Electron Optics **173**, 206–219 (2018)
22. R. Camblor-Diaz, S. Ver-Hoeye, C. Vazquez-Antuna, G.R. Hotopan, M. Fernandez-Garcia, F. Las Heras Andres, Sub-millimeter wave frequency scanning 8 x 1 antenna array. Prog. Electromagn. Res. **132**, 215–232 (2012)
23. G.H. Döhler, L.E. Garcia-Muñoz, S. Preu, S. Malzer, S. Bauerschmidt, J. Montero-de-Paz, E. Ugarte-Munoz, A. Rivera-Lavado, V. Gonzalez-Posadas, D. Segovia-argas, From arrays of THz antennas to large-area emitters. IEEE Trans. Terahertz Sci. Technol. **3**(5), 532–544 (2013)
24. M. Uzunkol, O.D. Gurbuz, F. Golcuk, G.M. Rebeiz, A 0.32 THz SiGe 4x4 imaging array using high-efficiency on-chip antennas. IEEE J. Solid State Circuits **48**(9), 2056–2066 (2013)
25. I. Malhotra, K.R. Jha, G. Singh, Design of highly directive terahertz photoconductive dipole antenna using frequency selective surface for sensing and imaging application. J. Comput. Electron. **17**, 1–20 (2018)
26. G. Bharti, K.R. Jha, G. Singh, R. Jyoti, Azimuthally periodic wedge-shaped metal vane loaded circular ring frequency selective surface. Int. J. Microw. Wirel. Technol. **7**(1), 95–106 (2015)
27. G. Bharti, K.R. Jha, G. Singh, R. Jyoti, Design of dual-polarized and angular stable new bandpass frequency selective surface in X-band. Telecommun. Syst. **61**(3), 559–567 (2016)
28. K.R. Jha, G. Singh, Performance Analysis of an Open-Loop Resonator Loaded Terahertz Microstrip Antenna. Microelectronics Journal, **42**(7), 950–956 (2011)
29. P. Carelli, F. Chiarello, S. Cibella, A.D. Gaspare, R. Leoni, M. Ortolani, G. Torrioli, A fast terahertz spectrometer based on frequency selective surface filters. J. Infrared Millimeter Terahertz Waves **33**(5), 505–512 (2012)
30. S. Vegesna, Y. Zhu, A. Bernussi, M. Saed, Terahertz two-layer frequency selective surfaces with improved transmission characteristics. IEEE Trans. Terahertz Sci. Technol. **2**(4), 441–448 (2012)
31. K.R. Jha, G. Singh, Prediction of highly directive probe-fed microstrip antenna at terahertz frequency. Int. J. Numer. Modell. Electron. Networks Devices Fields **25**(2), 175–191 (2012)

32. G. Bharti, K.R. Jha, G. Singh, A synthesis technique of single square loop frequency selective surface at terahertz frequency. Optik-Int. J. Light Electron Opt. **125**(21), 6428–6435 (2014)
33. A. Foroozesh, L. Shafai, Investigation into the effects of the patch-type FSS superstrate on the high-gain cavity resonance antenna design. IEEE Trans. Antennas Propag. **58**(2), 258–270 (2010)
34. I. Malhotra, K.R. Jha, G. Singh, Analysis of highly directive photoconductive dipole antenna at terahertz frequency for sensing and imaging applications. Optics Communication, **397**, 129–139 (2017)

Chapter 8
Beam-Steering Characteristics of Highly Directive Photoconductive Dipole Phased Array Antenna

8.1 Introduction

An antenna array permits us to attain high gain with multiple radiating elements, and a phased array additionally put forwards the possibility to shape as well as steer the beam without varying the array geometry. Conventionally, phased arrays have been used in defense applications, mainly for radar, but their deployment is changing with an increase in demand for the commercial use, especially with the arrival of on-the-move access to high bandwidth data. However, the challenge lies in realizing the radiating characteristics using phased array antenna while keeping the undesired effects such as the mutual coupling between the elements because of their proximity to each other. The design of large phased arrays is characteristically accomplished in two steps where, first, the element design is obtained by optimizing the active element impedance and active element pattern over all frequencies along with scan angles. Second, the array design is authenticated by inspecting the real scanning behavior besides including the non-periodic structure effects. Each of aforementioned stages/steps entitles for an appropriate array design functionality that makes the design of phased arrays, at both the cell and full array level, substantially more efficient and powerful.

For commercial imaging applications, fast measurements are required when the THz imaging technique is applied especially to the scrutiny of large number of targets [1]. Conventionally, the THz images are obtained by moving the samples in the course of the focused beam or else by scanning the emitter and detector in the case of immobile samples. However, efficient methods of steering the THz beam are essential for practical applications as the measurement speed is primarily limited by the speed of the mechanical devices used in conventional imagers. Moreover, due to the lack of active system, components required to dynamically manage free-space THz wave propagation still obstruct the broad-scale accomplishment of THz imaging systems. The potential of dynamically controlling the propagation of free-space

I. Malhotra, G. Singh, *Terahertz Antenna Technology for Imaging and Sensing Applications*, https://doi.org/10.1007/978-3-030-68960-5_8

THz waves, such as by steering or focusing the THz beams, is the solution to a multiplicity of THz imaging applications [2]. For example, in THz sensing and imaging application, beam-steering devices can be used to substitute the mechanical scanning stages and therefore can results in a significant increase in the measurement speed as well as the level of system integration [3, 4]. Further, a high-speed beam-steering technique also provides new avenues to several THz measurement methodologies, for instance, single-pixel THz imaging depending on compressed sensing [5], imaging of moving objects through tracking the targets by means of THz beams, etc. Primarily, to achieve THz beam steering, the simple technique is to either use mechanical scanning device or deployment of spatial phase modulator that generally operates as phased arrays and is extensively used in microwave regime of electromagnetic spectrum. In Fig. 8.1, by combining the THz antenna with graphene THz phase shifters, a reconfigurable transmitter [6] with functionalities, such as beam forming and beam steering, is illustrated.

Initially, the functioning of THz phase shifters has been realized on free carrier excitation in semiconductors [7–9]. In this, the dielectric characteristics of the semiconductor material are modulated through the externally excited free charge carriers that results in phase shift in the THz wave while passing through the medium. However, according to the Kramers–Kronig relation, the real part and the imaginary part of the dielectric of the semiconductor material are dependent on

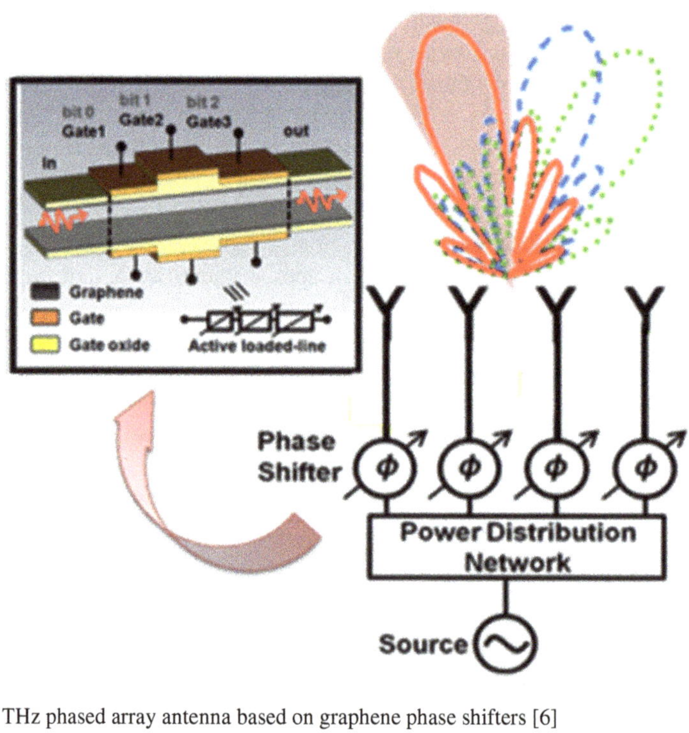

Fig. 8.1 THz phased array antenna based on graphene phase shifters [6]

each other, and therefore the phase modulation, as a consequence, generates insertion losses that are inseparable from the amplitude modulation [7–9]. Similarly, structures based on resonant metamaterial phase shifters are designed and reported in [10–12], in addition to the phase shifters using liquid crystals [13–15] which operate at high voltages and low speed. Another simple approach of beam steering which has been explored in our proposed antenna configuration is the phase-controlled optical excitation of arrays of small-gap photoconductive dipole antenna. Since this approach necessitates ultrashort optical laser pulse excitation, the number of potential applications is limited. However, as reported in [1, 16, 17], using the spatial dispersion of ultrafast laser pulse, the phenomena of beam steering from THz antenna array has been achieved. Moreover, in THz active imaging systems, the deployment of THz antenna array with beam-steering capability reduces the image acquisition time [18].

For millimeter-wave imaging systems, several measurement setups are anticipated which are either based on a rotary scanner with variable radius [19], or mechanically rotating mirrors with linear movement of the probe [20]. A THz imaging setup based on THz-frequency scanning antenna array is shown in Fig. 8.2, wherein a vector network analyzer (VNA) is used with auxiliary circuitry [21]. The object under inspection (placed on an azimuthal positioned) is illuminated with the help of frequency scanning antenna array whose main radiating lobe is steered in the vertical plane [22]. The beam is progressively rotated to evaluate the response for multiple incidence angles. To receive the reflected power from the object under inspection, which is illuminated from different vertical segments, a directive horn antenna is used. For the implementation of power divider as well as the phase shifting network, a rectangular waveguide structure is used, and open-ended waveguides are used as radiating elements.

Therefore, considering the requirement of scanning beam for the detection of hidden explosives carried by some moving object, there is a need for exploring the beam-steering characteristics of highly directive THz antenna array with features of compact size and planar profile.

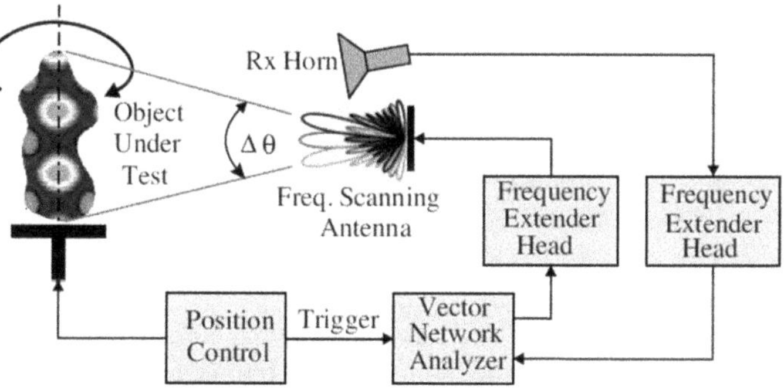

Fig. 8.2 Terahertz-frequency scanning antenna array used in imaging setup [22]

8.2 Related Work and Problem Formulation

A phased array antenna comprises of a number of radiating elements (antennas) each with a phase shifter. The beams are produced by shifting the phase of the signal emitted from each radiating element resulting in constructive or destructive interference in order to steer the beams in the desired direction. Therefore, the phenomenon of beam steering is about varying the direction of the main lobe of a radiation pattern created from the phased array antenna. However, in case of radio systems, the beam steering is achieved by either switching the antenna elements or by altering the relative phases of the RF signals which derive the antenna elements. A THz beam-steering technique has been exemplified from strip-line photoconductive phased array antenna using difference in frequency generation obtained from two spatially dispersed beams produced from an ultrafast laser [1]. Similarly, THz beam steering has been achieved by means of varying the angle of incidence of an ultrafast laser beam on to the photoconductive wide-gap area. However, such technique offers difficulty in quick steering of the THz beam over a wide angle [23]. By applying periodic bias voltages across the photoconductive antenna array, the THz beam generated from the antenna configuration also gets steered as reported in [24]. The only limitation to such technique is to control the period of the bias in addition to the period of the optical pulses in the train at each tuning frequency. Beam steering using micromechanical method [25] has also been reported. Using optical phase shifters [26], the microwave beam is steered through the control of phase of each element with high speed and wide angle.

With electronic 2-D steering capability (i.e., covering azimuth plane and elevation plane), a fixed frequency Fabry-Perot antenna configuration is demonstrated in [27]. The authors have presented the beam-steering capability of antenna by controlling the leaky-wave propagation constant across the antenna configuration. For THz standoff imaging, a programmable THz diffraction grating is achieved using array of metallic cantilevers for a wide-frequency band 0.15 THz to 0.9 THz [2]. The authors of [2] have used the switching technique between different grating patterns, so as to facilitate the tailoring of spatiotemporal profiles of the THz wave which results in steering the THz beam. The beam-steering and beam-forming phenomenon have been presented using array of ultra-wideband antenna elements [28], wherein a mode-locked laser is deployed to synchronize the jitter-free pulses radiated from each antenna element to create a radiated field pattern which gets steerable with the help of optical time delay technique. A scalable transmitter architecture based on distributed active radiation from the distributed active radiator (DAR) is designed at THz frequency with integrated beam scanning capabilities using silicon [29]. For the desired electromagnetic field profiles, the DAR with self-oscillating active electromagnetic structure contains two loops which carry out-of-phase currents at the fundamental frequency in addition to the in-phase currents at the second harmonic. The fundamental signal gets spatially filtered, at the same time as the second harmonic is radiated selectively. The simulation results presented in [29] show that at 0.28 THz, the beam steers in 2-D over 80° in both azimuth and elevation planes.

Similarly, the work has been initiated in the field of microstrip antennas for several THz applications [30]. In [31], a microstrip antenna is designed and fabricated at THz frequencies using simple printed circuit board (PCB) etching method and the antennas are tested up to 0.1THz. Moreover, with respect to beam forming [32, 33] at THz frequencies, reflect arrays offer unique potential. Reflect arrays combine the benefit of low profile of phased array and the high efficiency of parabolic antennas. However, there is a bandwidth limitation which results from the nonlinear phase response while using the reflect arrays. In [34], to enhance the bandwidth, the authors have used a single-layer stub-loaded resonator for constructing reflect arrays. Further, the multibeam functionality of the THz Luneburg lens antenna has been explored in [35], wherein a single THz system maintained a number of independent links concurrently. Such multibeam antennas lead to short-range THz radar devices that monitor several directions simultaneously. Earlier THz equipment had an obstacle to provide low cost and compact design for the widespread deployment of THz applications. However, with the surge of THz-integrated circuits design (THz-IC) in CMOS and SiGe HBT technologies over the last decade, the gap between the requirements and the reality of system cost and performance of THz components has become narrower. Moreover, features such as scalability, reconfigurability, and signal processing of silicon technology have enabled the research in design of complex THz ICs which further provides the expansion of functionality of THz systems [36, 37]. In [38], the authors have investigated the design of a wideband and high-efficiency 1 × 2 octagonal shorted annular ring on-chip antenna array at 320GHz and the results show improved gain values with reduced surface wave.

In this chapter, a novel and simple approach to realize terahertz (THz) dynamic two-dimensional (2D) beam-steering antenna configuration based on (2 × 2) photoconductive dipole phased array antenna with FSS is presented with its operation in the range 1.4–2.2THz for sensing and imaging the hidden explosives carried by some moving object.

8.3 Design of Photoconductive Dipole Phased Array Antenna with FSS

The resonant frequency of the photoconductive dipole antenna is determined using the relation $f_r = c/(2l_e[(1 + \varepsilon_d)/2]^{1/2})$, where f_r is the resonating frequency, and ε_d is the dielectric permittivity of GaAs substrate and is taken as 12.9. By putting the values of c, ε_d, and l_e, the resonant frequency of the designed photoconductive dipole antenna is 1.95 THz. Figure 8.3 shows the schematic of the four elements of photoconductive dipole antennas forming an array with FSS for improved directivity. Moreover, the dimensions of the superstrate, substrate, and ground are taken as (110 × 110)μm^2, with thickness 1 μm, 10 μm, and 0.35 μm, respectively. The spacing between photoconductive dipole antennas in an array arrangement is taken as λ/2 and, for(2 × 2) photoconductive dipole array, the spacing between two sets of transmission line is also considered as λ/2. In the designed photoconductive dipole

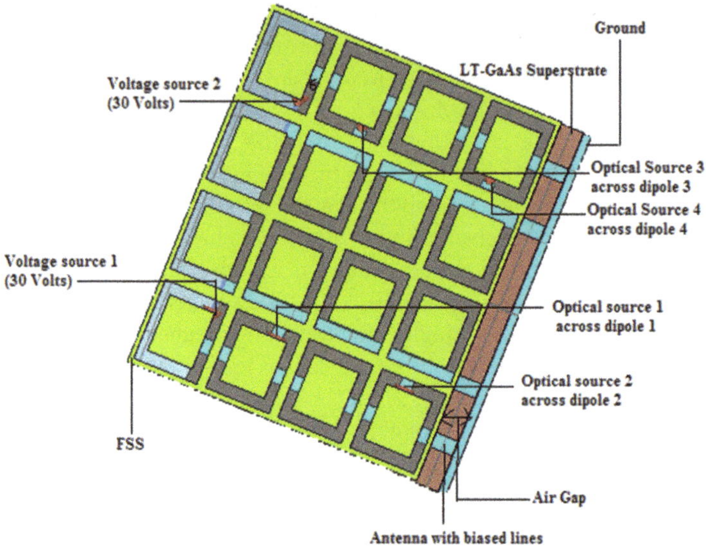

Fig. 8.3 Schematic of (2 × 2) photoconductive dipole array antenna used for phased array implementation for the beam-steering phenomenon operating at 1.95 THz [39]

antenna, the length of transmission lines is taken as 110 μm with width and thickness 10 μm and 0.35 μm, respectively. The FSS structure made of copper with thickness 0.35 μm is followed by the polytetrafluoroethylene substrate [39].

Initially, the progressive phase shift is performed in single direction (i.e., along x-axis only) and the beam steering occurs in E-plane, whereas only the change in magnitude is observed in H-plane. Similarly, on performing the progressive phase shift along y-axis, the beam steering occurs in H-plane along with slight change in magnitude of the main lobe of radiation characteristics in E-plane. However, when the progressive phase shift is applied in both the directions (x-axis and y-axis), then the beam-steering phenomenon occurs in both principal planes (E-plane and H-plane). The progressive phase shift is applied to the optical sources and the biased voltage sources are kept constant with 30 Volts.

The uniform current distribution is applied individual array element and the results are investigated for the scan angle ($0^\circ \leq \theta \leq 40^\circ$), where $\varphi = 0^\circ$. The enhancement in the directivity of small-gap photoconductive dipole antenna is achieved by using frequency selective surface (FSS) [40].

8.4 Numerical Analysis and Simulation Results

Initially, (2 × 2) small-gap photoconductive dipole antenna array with operating frequency 1.95THz is simulated using transient solver of CST Microwave Studio. The radiation patterns 3-D as well as 2-D view for directivity (dBi) are obtained in the

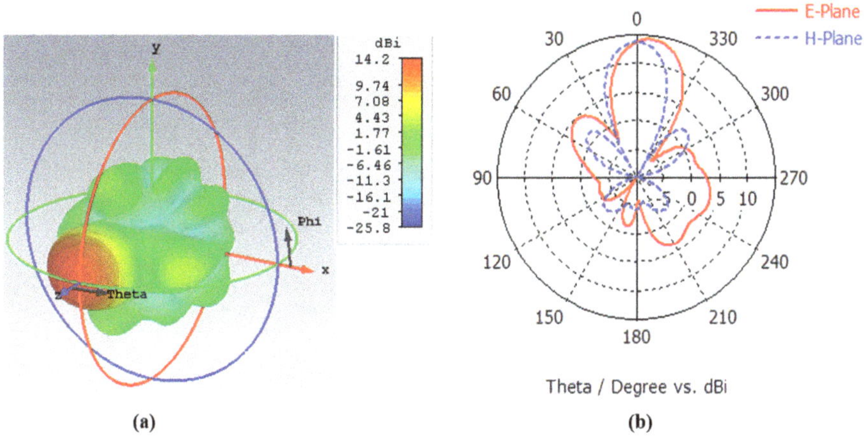

Fig. 8.4 A far-field radiation pattern (directivity) (**a**) 3-D view and (**b**) 2-D view, in both principal planes E and H

principal planes for the array configuration and are illustrated in Fig. 8.4. The simulated results in Fig. 8.4b show the value of directivity 14.2dBi in E-plane with main lobe direction at 354.0°. The angular width (3-dB) is 29.8° with the side lobe level – 9.6 dB. However, in H-plane, the directivity is 13.7 dBi with its main lobe direction 0°. The angular width (3-dB) is 26.6° having a side lobe level − 11.7 dB. Here, the amplitude and the phase of all the four optical sources are kept uniform with 1 W power input and 0° phase shift. However, for the design of phased array antenna and its simulation in CST Microwave Studio, the transient solver supports several excitation patterns, such as uniform distribution, cosine distribution, Taylor distribution, Binomial distribution, Chebyshev distribution, and Gaussian distribution.

To determine the beam-steering characteristics of the (2 × 2) small-gap photoconductive, dipole antenna array configuration for THz imaging application, we kept the input power from the optical source constant for the simplicity and ease of feasibility of the antenna configuration for the desired application. Moreover, a uniform distribution of amplitude excitation is provided to each unit of photoconductive dipole array antenna. First of all, the progressive phase shift of 10° is performed along x-axis of the optical sources across the respective antenna elements. Figure 8.5 shows the simulation results, wherein the beam steering occurs in E-plane and H-plane, and only the magnitude changes.

Moreover, it is observed that the proposed phased array antenna configuration successfully scans for ($0^{\circ} \leq \theta \leq 40^{\circ}$), where $\varphi = 0^{\circ}$. The values obtained from the progressive 10° phase shift along the x-axis for both the principal planes are listed in Table 8.1.

While scanning at every 10°, the main beam shifts by 3° in E-plane; however, only the magnitude of the main lobe reduces in the H-plane. However, the decrease in the magnitude of the main lobe of the radiation characteristics is contributed by the occurrence of the mismatch between the antenna and the feed network with an

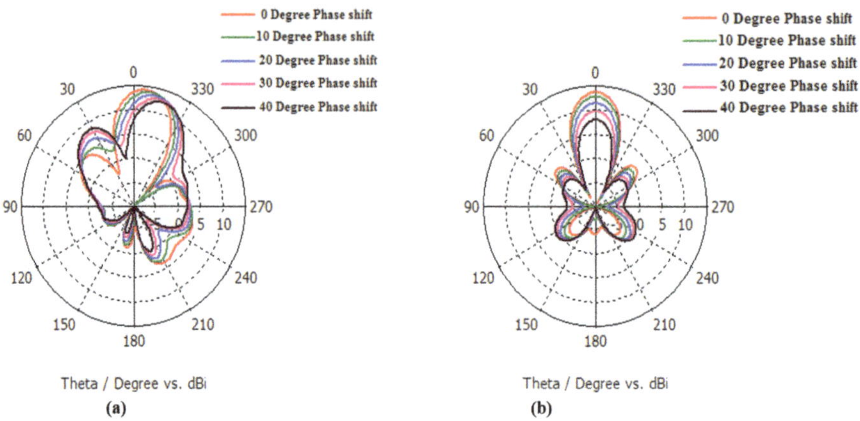

Fig. 8.5 The radiation characteristics (directivity, dBi) with beam steering at angle 0° (red), 10° (green), 20° (blue), 30° (pink), and 40° (brown) progressive phase shift along x-axis of (2 × 2) smcall-gap photoconductive dipole array antenna in (**a**) E-plane and (**b**) H-plane

Table 8.1 The beam steering of (2 × 2) small-gap photoconductive dipole array antenna for ($0^\circ \leq \theta \leq 40^\circ$) with 10° progressive phase shift along x-axis of the array antenna configuration

Progressive phase shift (degree)		Main lobe directivity (dBi)	Main lobe direction (degree)	Angular width (dB)	Side-lobe level (dB)
0	E-plane	14.2	354	29.8	−9.6
10		13.8	351	30.9	−8.7
20		13.3	348	31.1	−6.8
30		12.9	345	31.2	−5.1
40		12.4	342	31.4	−3.9
0	H-plane	13.7	0	26.6	−11.7
10		12.7	0	26.7	−11.7
20		11.4	0	26.8	−11.4
30		9.71	0	26.9	−9.4
40		7.93	0	27.0	−7.3

increase in the phase shift. To determine the beam-steering phenomenon in H-plane, the progressive phase shift is applied along the y-axis of the (2 × 2) small-gap photoconductive, dipole antenna array configuration.

For the scanning range ($0^\circ \leq \theta \leq 40^\circ$) with $\varphi = 0^\circ$ and 10° progressive phase shift along y-axis of the array configuration, it is observed that the beam steering occurs in H-plane with very small change in the magnitude of main lobe of the radiation characteristics in E-plane. The simulation results are shown in Fig. 8.6 and are listed in Table 8.2. On providing the 10° progressive phase shift along y-axis of the array antenna elements, the main beam shifts by 1° in H-plane; however, only a slight change in the magnitude of the main lobe is observed in the E-plane.

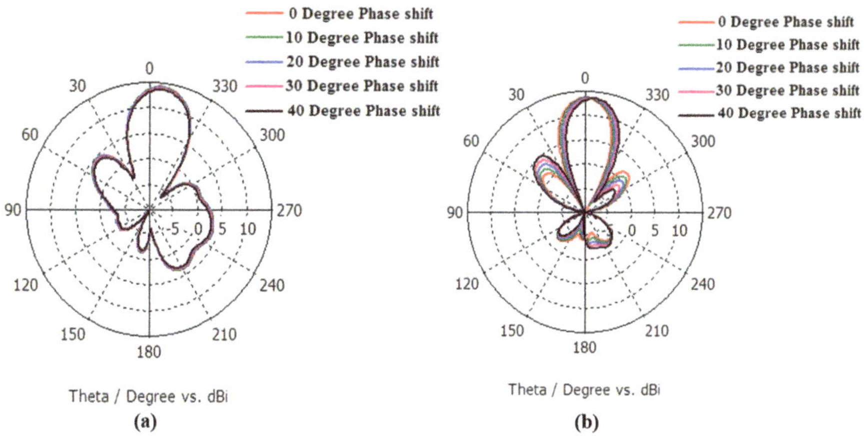

Fig. 8.6 The radiation characteristics (directivity, dBi) with beam steering at angle 0° (red), 10° (green), 20° (blue), 30° (pink), and 40° (brown) progressive phase shift along y-axis of (2 × 2) small-gap photoconductive dipole array antenna in (**a**) E-plane and (**b**) H-plane

Table 8.2 The beam steering of (2 × 2) small-gap photoconductive dipole array antenna for (0° ≤ θ ≤ 40°) with 10° progressive phase shift along x-axis of the array antenna configuration

Progressive phase shift (degree)		Main lobe directivity (dBi)	Main lobe direction (degree)	Angular width (dB)	Side lobe level (dB)
0	E-plane	14.2	354	29.8	−9.6
10		14.2	354	29.8	−9.6
20		14.1	354	29.9	−9.5
30		13.9	354	30.0	−9.4
40		13.7	354	30.1	−9.2
0	H-plane	13.7	0	26.6	−11.7
10		13.6	359	26.7	−11.3
20		13.5	358	26.8	−10.3
30		13.4	357	26.8	−9.4
40		13.3	356	26.9	−8.6

For the THz imaging application to detect the hidden explosives, there is a need of scanning the suspicious object in both principal planes. Therefore, it is essential to make the ability of the highly directive small-gap photoconductive dipole array antenna to scan or to steer the main lobe of the radiation characteristics in both the principal planes. To achieve this, the progressive phase shift is applied in both directions, that is, in the X–Y directions of the (2 × 2) small-gap photoconductive dipole antenna array antenna configuration.

Figure 8.7 shows the simulation results, wherein the beam steering occurs in both E-plane and H-plane. The values obtained from the progressive 10° phase shift along the x-axis and y-axis for both principal planes are listed in Table 8.3.

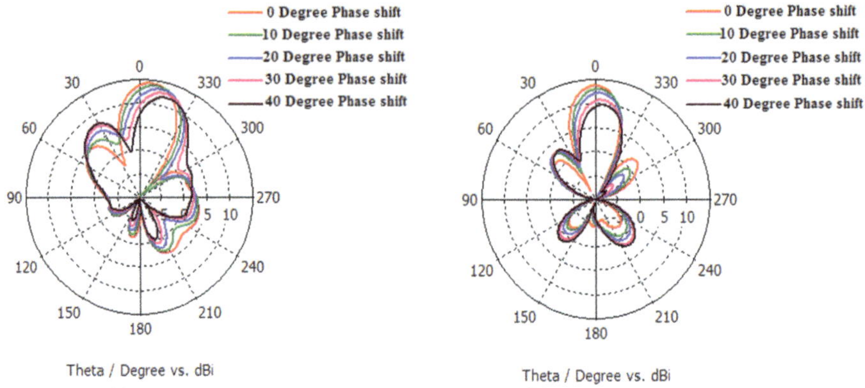

(a) (b)

Fig. 8.7 The radiation characteristics (directivity, dBi) with beam steering at angle 0° (red), 10° (green), 20° (blue), 30° (pink), and 40° (brown) progressive phase shift along x-axis and y-axis of (2 × 2) small-gap photoconductive dipole array antenna in (**a**) E-plane and (**b**) H-plane

Table 8.3 The beam steering of (2 × 2) small-gap photoconductive dipole array antenna for (0° ≤ θ ≤ 40°) with 10° progressive phase shift along x-axis and y-axis of the array antenna configuration

Progressive phase shift degree)		Main lobe directivity (dBi)	Main lobe direction (degree)	Angular width (dB)	Side lobe level (dB)
0	E-plane	14.2	354	29.8	−9.6
10		13.8	351	30.9	−8.7
20		13.2	348	31.1	−6.7
30		12.6	345	31.0	−5.1
40		11.9	342	31.5	−3.8
0	H-plane	13.7	0	26.6	−11.7
10		12.9	359	26.7	−10.1
20		12.2	358	26.9	−9.1
30		10.7	357	27.3	−7.6
40		9.7	356	27.7	−6.7

It is observed from the simulation results that the proposed phased array antenna configuration successfully scans for (0° ≤ θ ≤ 40°) with φ = 0°, and progressive phase shift of 10° in both principal planes provides the beam steering by 3° and 1° in E-plane and H-plane, respectively. Therefore, the obtained beam-steering capabilities of highly directive (2 × 2) small-gap photoconductive dipole array antenna using progressive phase shift at 1.95 THz operating frequency is useful to detect the hidden explosive, such as RDX and HMX, concealed in some baggage or carried by some object because such explosives show their absorption spectra at this frequency.

8.5 Summary

In this chapter, highly directive (2 × 2) small-gap photoconductive dipole phased array antenna at 1.95 THz is presented which can be used in the detection of hidden explosives in THz imaging application. The scanning of the beam is primarily performed in single direction (x-axis); hence, the beam gets steered in one plane, that is, in E-plane. With every 10° shift of the scan angle, the highly directive beam gets steered by 3°. However, when the progressive phase shift is applied in y-axis, then scanning of the beam occurs in H-plane only. In this case, with every 10° shift of the scan angle, the highly directive beam gets steered by 1° in H-plane with small reduction in the magnitude of beam in the same direction for E-plane. Moreover, for the full electronic beam steering, there is a requirement to shift the beam in both principal planes. Therefore, the progressive phase shift is applied along both the x-axis and y-axis. For the scanning range ($0^{\circ} \leq \theta \leq 40^{\circ}$) with $\varphi = 0^{\circ}$, for every 10° progressive phase shift, the main beam gets steered by 3° in E-plane and by 1° in H-plane, respectively.

References

1. K. Maki, C. Otani, Terahertz beam steering and frequency tuning by using the spatial dispersion of ultrafast laser pulses. Opt. Express **16**(14), 10158–10169 (2008)
2. Y. Monnai, K. Altmann, C. Jansen, H. Hillmer, M. Koch, H. Shinoda, Terahertz beam steering and variable focusing using programmable diffraction gratings. Opt. Express **21**(2), 2347–2354 (2013)
3. P.U. Jepsen, D.G. Cooke, M. Koch, Terahertz spectroscopy and imaging –modern techniques and applications. Laser and Photonics Reviews **5**(1), 124–166 (2011)
4. K. Kawase, Y. Ogawa, Y. Watanabe, H. Inoue, Non-destructive terahertz imaging of illicit drugs using spectral fingerprints. Opt. Express **11**(20), 2549–2554 (2003)
5. W.L. Chan, K. Charan, D. Takhar, K.F. Kelly, R.G. Baraniuk, D.M. Mittleman, A single-pixel terahertz imaging system based on compressed sensing. Appl. Phys. Lett. **93**(12), 121105/1–121105/3 (2008)
6. P.Y. Chen, M. Farhat, A.N. Askarpour, M. Tymchenko, A. Alu, Infrared beam-steering using acoustically modulated surface plasmons over a graphene monolayer. J. Optics **16**(9), 094008/1–094008/9 (2014)
7. I.H. Libon, S. Baumgärtner, M. Hempel, N.E. Hecker, J. Feldmann, M. Koch, P. Dawson, An optically controllable terahertz filter. Appl. Phys. Lett. **76**(20), 2821–2823 (2000)
8. R. Kersting, G. Strasser, K. Unterrainer, Terahertz phase modulator. Electron. Lett. **36**(13), 1156–1158 (2000)
9. S. Busch, B. Scherger, M. Scheller, M. Koch, Optically controlled terahertz beam steering and imaging. Opt. Lett. **37**(8), 1391–1393 (2012)
10. H.T. Chen, W.J. Padilla, M.J. Cich, A.K. Azad, R.D. Averitt, A.J. Taylor, A metamaterial solid-state terahertz phase modulator. Nat. Photonics **3**(3), 148–151 (2009)
11. H.T. Chen, J.F. O'Hara, A.K. Azad, A.J. Taylor, R.D. Averitt, D.B. Shrekenhamer, W.J. Padilla, Experimental demonstration of frequency-agile terahertz metamaterials. Nat. Photonics **2**(5), 295–298 (2008)
12. H.T. Chen, W.J. Padilla, J.M.O. Zide, A.C. Gossard, A.J. Taylor, R.D. Averitt, Active terahertz metamaterial devices. Nature **444**(7119), 597–600 (2006)

13. C.F. Hsieh, R.P. Pan, T.T. Tang, H.L. Chen, C.L. Pan, Voltage-controlled liquid-crystal terahertz phase shifter and quarter-wave plate. Opt. Lett. **31**(8), 1112–1114 (2006)
14. H.Y. Wu, C.F. Hsieh, T.T. Tang, R.P. Pan, C.L. Pan, Electrically tunable room-temperature 2π liquid crystal terahertz phase shifter. IEEE Photon. Technol. Lett. **18**(14), 1488–1490 (2006)
15. B. Scherger, M. Reuter, M. Scheller, K. Altmann, N. Vieweg, R. Dabrowski, J.A. Deibel, M. Koch, Discrete terahertz beam steering with an electrically controlled liquid crystal device. J. Infrared Millim Terahertz Waves **33**(11), 1117–1122 (2012)
16. N.M. Froberg, B.B. Hu, X.C. Zhang, D.H. Auston, Terahertz radiation from a photoconducting antenna array. IEEE J. Quantum Electron. **28**(10), 2291–2301 (1992)
17. K. Uematsu, K. Maki, C. Otani, Terahertz beam steering using interference of femtosecond optical pulses. Opt. Express **20**(20), 22914–22921 (2012)
18. D.M. Sheen, D.L. McMakin, T.E. Hall, Three-dimensional millimeter-wave imaging for concealed weapon detection. IEEE Trans. Microw. Theory Tech. **49**(9), 1581–1592 (2001)
19. M.T. Ghasr, D. Pommerenke, J.T. Case, A. McClanahan, A. Aflaki-Bbeni, M. Abou-Khousa, S. Kharkovsky, K. Guinn, F.D. Paulis, R. Zoughi, Rapid rotary scanner and portable coherent wideband Q-band transceiver for high-resolution millimeter-wave imaging applications. IEEE Trans. Instrum. Meas. **60**(1), 186–197 (2008)
20. H.P. Tran, F. Gumbmann, J. Weinzierl, L.P. Schmidt, A fast scanning w-band system for advanced millimetre-wave short range imaging applications, in *3rd European Radar Conference, EURAD*, vol. 2006, (Manchester, UK, 2006, 13–15 September), pp. 146–149
21. C.H. Tseng, T.H. Chu, An effective usage of vector network analyzer for microwave imaging. IEEE Trans. Microw. Theory Tech. **53**(9), 2884–2891 (2005)
22. R. Camblor Diaz, S. Ver Hoeye, C. V'azquez, G. Hotopan, M. Fern'andez, F.L. Heras, Submillimeter wave frequency scanning 8X1 antenna array. Prog. Electromagn. Res. **132**, 215–232 (2012)
23. B.B. Hu, J.T. Darrow, X.C. Zhang, D.H. Auston, Optically steerable photoconducting antennas. Appl. Phys. Lett. **56**(10), 886–888 (1990)
24. N.M. Froberg, B.B. Nu, X.-C. Zhang, D.H. Auston, Terahertz radiation from a photoconducting antenna array. IEEE J. Quantum Electron. **28**(10), 2291–2301 (1992)
25. T.D. Drysdale, E.D. Walsby, D.R.S. Cumming, Measured and simulated performance of a ceramic micromechanical beam steering device at 94 GHz. Appl. Opt. **47**(13), 2382–2385 (2008)
26. Y. Kamiya, Y. Murakami, W. Chojo, M. Fujise, An electro-optic BFN for array antenna beam forming. IEICE Trans. Electron. **78**(8), 1090–1094 (1995)
27. R. Guzmán-Quirós, A.R. Weily, J.L. Gómez-Tornero, Y.J. Guo, A fabry-perot antenna with two-dimensional electronic beam scanning. IEEE Trans. Antenn. Prop. **64**(4), 1536–1541 (2016)
28. E.E. Funk, C.H. Lee, Free-space power combining and beam steering of ultra-wideband radiation using an array of laser-triggered antennas. IEEE Trans. Microw. Theory Tech. **44**(11), 2039–2044 (1996)
29. K. Sengupta, A. Hajimiri, A 0.28 THz power-generation and beam-steering array in CMOS based on distributed active radiators. IEEE J. Solid State Circuits **47**(12), 3013–3031 (2012)
30. M.S. Rabbani, H. Ghafouri-Shiraz, Size improvement of rectangular microstrip patch antenna at MM-wave and terahertz frequencies. Microw. Opt. Technol. Lett. **57**(11), 2585–2589 (2015)
31. M.S. Rabbani, H. Ghafouri-Shiraz, Improvement of microstrip patch antenna gain and bandwidth at 60 GHz and X bands for wireless application. Microw. Antenn. Prop. **10**(11), 1167–1173 (2016)
32. D. Headland, Y. Monnai, D. Abbott, C. Fumeaux, W. Withayachumnankul, Tutorial: Terahertz beamforming, from concepts to realization. APL Photonics **3**, 051101 1–32 (2018)
33. M.S. Rabbani, H.G. Shiraz, Liquid crystalline polymer substrate-based THz microstrip antenna arrays for medical applications. IEEE Antenn. Wirel. Propag. Lett. **16**, 1533–1536 (2017)

34. X. You, R.T. Ako, W.S.L. Lee, M.X. Low, M. Bhaskaran, S. Sriram, C. Fumeaux, W. Withayachumnankul, Tterahertz reflectarray with enhanced bandwidth. Adv. Opt. Mater. **7**(20), 190079 1–190079 8 (2019)
35. D. Headland, W. Withaya Chumnankul, R. Yamada, M. Fujita, T. Nagatsuma, Terahertz multi-beam antenna using photonic crystal waveguide and Luneburg lens. APL Photonics **3**, 126105 1–19 (2018)
36. P. Hillger, J. Grzyb, R. Jain, U.R. Pfeiffer, Terahertz imaging and sensing applications with silicon-based technologies. IEEE Trans. Terahertz Sci. Tech. **9**(1), 1–19 (2019)
37. H. Chu, Y.X. Guo, T.-G. Lim, Y.M. Khw, X. Shi, 135-GHz micromachined on-chip antenna and antenna array. IEEE Trans. Antennas Propag. **60**(10), 4582–4588 (2012)
38. H. Zhu, X. Li, Z. Qi, J. Xiao, A 320 GHz octagonal shorted annular ring on-chip antenna array. IEEE Access **8**, 84282–84289 (2020)
39. I.Malhotra, K.R. Jha, G. Singh, Design of highly directive lens-less photoconductive dipole antenna array with frequency selective surface for terahertz imaging applications. Optik: International Journal for Light and Electron Optics **173**, 206–219 (2018)
40. I. Malhotra, K. R. Jha, G. Singh, Beam steering characteristics of highly directive photoconductive dipole phased array antenna for terahertz imaging application. Optical and Quantum Electronics **51**(27), 1–19 (2019)

Chapter 9
Terahertz Near-Field Imaging and Sensing

9.1 Introduction

Terahertz wave is an important branch of the electromagnetic spectrum as it facilitated both the fundamental research on spectroscopic studies as well as its applications in the field of wireless communications and imaging. The terahertz imaging systems are broadly classified as: (1) far-field imaging systems and (2) near-field imaging systems. However, both the systems have been explored and utilized by researchers in several imaging applications as per the salient features of both types of imaging systems. Further, the major limitation of terahertz far-field imaging system is the occurrence of problem of diffraction-limited spatial resolution which is key to THz wavelengths in comparison to that of the visible radiation [1, 2]. Using the Rayleigh criterion, the optical resolution (R) in the object plane is determined by: $R = \frac{0.61\lambda}{\text{NA}} \approx \lambda / 2\text{NA}$, where λ is the wavelength and NA is the numerical aperture [3–5]. Since the resolution is directly proportional to the wavelength and inversely proportional to the numerical aperture of the optical components, therefore, it could be easily understood that in the imaging system, minimal resolvable feature corresponds to the minimum size in the object plane of the smallest object which an optical system can resolve. Therefore, an imaging system operating at 1 THz provides a resolution of approximately 0.3–0.5 mm considering the loss of resolution caused by lenses used on the beam path of the system. Therefore, the effective and wide spreading of terahertz images in the biomedical research and diagnostics is challenging. Moreover, the wavelength-limited resolution is caused by diffraction of propagating waves, and to overcome this problem it is generally suggested to collect terahertz pulses in the near field. It means, the terahertz pulses are determined at a distance with the sample comparable to the wavelength in terahertz near-field imaging. Further, in the near-field imaging, the evanescent waves which propagate very near to the sample and get decayed rapidly can be detected by placing a terahertz detector in the proximity of the sample. Therefore, the terahertz near-field imaging can break the diffraction limit of $\lambda/2$ and can provide the

I. Malhotra, G. Singh, *Terahertz Antenna Technology for Imaging and Sensing Applications*, https://doi.org/10.1007/978-3-030-68960-5_9

resolution three magnitudes higher than that of terahertz far-field imaging. Consequently, using several designs of detectors and sources, several techniques have been explored to enable sub-wavelength near-field terahertz imaging [6, 7].

The first terahertz near-field imaging set-up was established by Hu and Nuss in 1995 [8]. The system was based on optoelectronic terahertz time-domain spectroscopy. The terahertz time-domain waveforms were down-converted from terahertz to kHz frequency range with frequency analysis of waveform for each pixel using a digital signal processor in real time so as to extract the compositional information at the point of interest. Figure 9.1 shows the schematic of terahertz imaging system wherein the object is raster-scanned using a two-axis motorized stage [8]. In the set-up, the terahertz transients were focused to a diffraction-limited spot on the sample (a digital IC) and the transmitted terahertz waveforms were acquired. Further, the processing of the acquired terahertz waveform was processed in real time at each point of the sample.

In Fig. 9.1, the sequence of occurrence of events in the set-up is numbered accordingly. Initially, a self-mode-locked Ti:sapphire laser acting as a femtosecond optical source provides pulses of 100 fs duration at 800 nm wavelength. These pulses of approximately 80nW then excite the photoconductive surface of terahertz transmitter which is a strip-line-based photoconductive antenna to generate terahertz radiations from its surface. In the set-up of the imaging system, two pairs of off-axis paraboloid reflectors are used to collimate the terahertz radiations before and after the object (an IC) under scanning. First, the collimated terahertz radiation emitted from the terahertz strip-line antenna is allowed to focus down to a diffraction-limited spot on an IC using paraboloids. Likewise, the second set of paraboloids collect and further focus the reflected terahertz radiation from an IC onto the optically gated terahertz dipole detector which is a photoconductive dipole antenna. Moreover, in order to improve the coupling of terahertz radiation to air, a high-resistivity silicon hyper-hemispherical substrate lenses are also proposed to the back of substrates of both terahertz transmitting and receiving antennas of the set-up. The function of scanning delay-line is to down-convert the

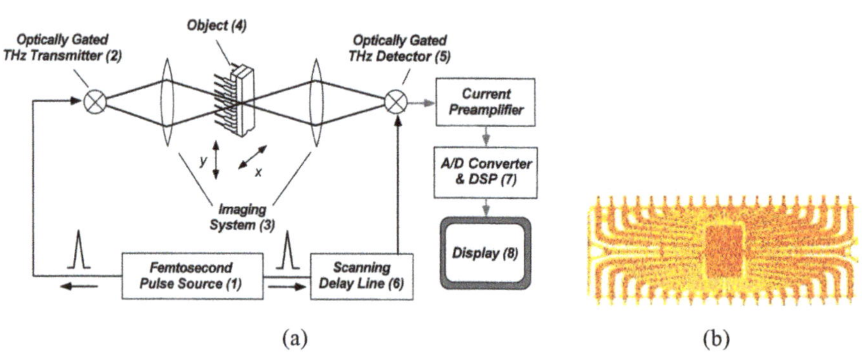

Fig. 9.1 To scan an integrated circuit (IC) chip (**a**) schematic of terahertz imaging system and (**b**) terahertz image of packed semiconductor IC chip [8]

terahertz waveform into an audio range thereby providing the down-conversion ratio twice the ratio of the speed of delay-line to speed of light. For 1 terahertz scanning beam, the delay-line maintaining the speed of 15 cm/s down-converts the signal to 1 kHz waveform in the set-up. The imaging system provides a spectral resolution of 20 GHz with the maximum amplitude of the scan to be 0.75 cm. The detected terahertz signal is then amplified with the help of current pre-amplifier which is further digitized and processed using digital signal processing (DSP) chip on the flight for each pixel. A single commercial signal processing board such as data translation DT3818 has been used in the set-up which has an analog-to-digital converter integrated on the same chip. Therefore, the spectral information at each pixel is obtained and processed by a single chip with minimal processing time. The authors in [8] have shown the terahertz image acquisition, analysis, and its display at the rate of 12 pixels/s and suggested the means to increase the rate. Based on the differences in the interface of the materials of an IC with terahertz radiation, the obtained image has the variation in the appearance of part of an IC. The plastic packaging material of the chip did little absorption in the terahertz spectral range while metal pins of an IC fully absorb the terahertz radiations and doped semiconductors partially absorb the radiations. Therefore, the obtained image from the set-up revealed the clear area of silicon chip as well as the metal leads connected to the package pins.

A new human body security inspection imaging technology is coming up using sparse linear array multiple-input multiple-output (MIMO) synthetic aperture radar (SAR) systems [9]. Such terahertz near-field human security inspection is based on an array of terahertz transmitters which illuminates the human carrying the concealed weapon and accordingly, records the back-scattered radiation coherently with the help of array of terahertz receivers. In [9], the authors have proposed the use of efficient computational 3D imaging algorithms such as back-propagation, fast-factorized back-projection, and range migration algorithm. The use of these imaging algorithms provides the feasibility of parallel processing on a graphics processing unit in order to accelerate the image generation. The terahertz near-field MIMO-SAR technology is a new area of exploration and there are some basic technical issues which need to be taken care of such as: (a) integration of radar chip with an array, (b) need of fast imaging algorithm, (c) requirement to implement a method to extract echo characteristics, and (d) technique to retrieve information of hidden objects from multiple view angles.

9.2 State-of-the-Art Terahertz Near-Field Imaging

The terahertz near-field imaging methodologies are broadly categorized based on probe design and known as the aperture type and aperture-less type. However, in both cases, the ultra-high resolution which breaks through the diffraction limit is predicated on the generation and exploitation of evanescent waves in the vicinity of probe apex [10]. These imaging modalities are in use by several applications such as micro/nanostructured imaging [11, 12], non-destructive access to the surface and sub-surface structural information of the sample [13], biomedical imaging [14, 15], and crystal characterization [16]. In [17], the authors have designed and demonstrated a near-field imaging probe having a concave aperture and its design is based on size-optimized tapered frustum. The authors have evaluated the performance of novel design probe using finite-difference time-domain electromagnetic simulation and near-field imaging experiments as well. The simulation results illustrate an increase in the electric field intensity which is transmitted through the aperture by a factor of 3 within the near-field domain outside the aperture [17].

Further, to overcome the diffraction limit the subwavelength resolution imaging requires a sharpened superfocusing. Conventionally, high numerical aperture superfocused lenses were used for superfocusing. However, it makes the imaging system bulky and expensive. Recently, relying on abrupt amplitude and phase discontinuities, an ultrathin platform based on metasurfaces has been explored to generate arbitrary wavefronts over subwavelength thickness. In [18], the authors have demonstrated a dual-field superfocusing of terahertz wavelengths with the help of near-field metasurface thereby simultaneously tailoring the amplitude and phase of the evanescent field within electrostatics and magnetostatics limits. The reported simulation results hold useful information for the practical implementation of metasurfaces for significant applications such as terahertz beam steering, laser-based microscopy, non-contact sensing, imaging, and lithography. Likewise, based on the concept of chirality, which is being used in optics to study some special phenomenon caused by the interaction of light and matter (also known as chiral optical responses of matter), the chiral metasurfaces have been proposed in [19] so as to achieve an improved optical responses in the near-field imaging. The authors of [19] have introduced a patterned and electrically doped graphene into the metamirror which consists of double-layer C-shaped split-ring resonators. In the presented work, the electrical doping of graphene makes changes in the absorption loss of metasurface cavity and therefore the structure gets enabled and shows the switching between a chiral metasurface and an ordinary metal mirror. Further, such "on" and "off" states result in an efficient modulation of the terahertz circular dichroism in a large dynamic range.

Another promising technology which has been developed for near-field imaging applications is to produce an image by introducing a scene with a series of spatially resolved subwavelength patterns while recording the correlated intensity on a single-element detector [20]. The terahertz single-pixel near-field imaging carries salient features such as: (a) does not require long recording time as is needed under a fragile probe of near-field raster scanning method [21], (b) reduce the cost of imaging by a terahertz array detector [22, 23], and c) able to break through the diffraction limit thereby making it possible to capture micrometer-scale details in a more simple and faster way. In [24], the authors have combined the use of spatial light modulator with digital micromirror device to shape the optical pump beam. Further, with reference to the spatial light modulators, the research has developed in the following three directions, which are as follows: (a) for better resolution, the smaller subwavelength details could be observed using thin spatial light modulators so as to reduce the diffraction effects, (b) to make spatial light modulators operational with fast switching speed which will reduce the imaging time thereby making the real-time imaging possible, and (c) use of external stimulants which can provide deep modulation depth. The reported simulation results reveal spatial light modulators performance with 60% terahertz peak amplitude modulation depth with broad bandwidth under relative low pump fluence.

Similarly, the development of extreme subwavelength metallic nanostructures based on scattering-type scanning near-field optical microscopy (s-SNOM) has enabled, the near-field nano-imaging in the 50 GHz–5 THz spectral region for accessing local conductivity and phase coexistence [11, 25–35]. Moreover, using s-SNOM near-field imaging via nanostructures has greatly impacted near-field material contrast with the use of lateral size of nanostructures while the spatial resolution is practically independent of it [36]. Further, the contrast is mainly affected by the connectivity of the metallic structures to a larger metallic ground plane. Therefore, such state-of-the-art terahertz near-field imaging technologies are progressively showing great interest by researchers to develop advanced products for efficient terahertz imaging and sensing applications.

9.3 Terahertz Near-Field Measurements

Most of the investigations related to enhanced terahertz transmission are conducted in the far-field. However, in order to determine how light emerges from terahertz antenna arrays, near-field measurements are vital [37–42], as is performed in recent near-infrared and visible-wavelength near-field array measurements [43–45]. The presence and interaction of evanescent fields, the surface waves as well as the spatial distribution of the surface charge density are the near-field information which are inevitably get lost in the far-field measurements. Moreover, measuring an individual component of the terahertz near-field with sub-wavelength spatial resolution when it emerges from the surface of structures also illustrates how the field interacts with the aperture. In [46], the authors have presented the measurements of complete terahertz electric near-field distribution (*Ex*, *Ey*, and *Ez*) in both the time-domain and frequency-domain for subwavelength apertures and subsections of subwavelength aperture arrays. To image and characterize subwavelength structures in the near-field, the terahertz time-domain spectroscopy (THz-TDS) with current transient terahertz generation and electro-optic detection is used [47]. The advantage of using THz-TDS is that the three-electric near-field components in the time domain can be measured easily and provides both the field amplitude as well as the relative phase. It is an important distinctive feature of THz-TDS in comparison to the measurements conducted at optical wavelengths wherein only typical optical power is measured. Further, when the boundary conditions for the perpendicular component of the electric field (*Ez*) are considered at the metal–dielectric (i.e., antenna–air) interface then measuring the near-electric field instead of optical power proves to be advantageous. However, such conditions state that Ez is discontinuous across the metal–dielectric interface by an amount determined by the surface charge density ($\sigma = Ez/\epsilon$).

The set-up of near-field imaging where it is presented how the samples (array of subwavelength microstructures with 90 μm periodicity) are imaged [46, 48] is shown in Fig. 9.2a. Initially, a terahertz light is focused onto subwavelength structures using an off-axis parabolic mirror. Using the near-infrared probe beam (with wavelength λ = 800 nm) through counter-propagating, the electric near-field is detected. The probe is kept tightly focused to a small spot immediately under the

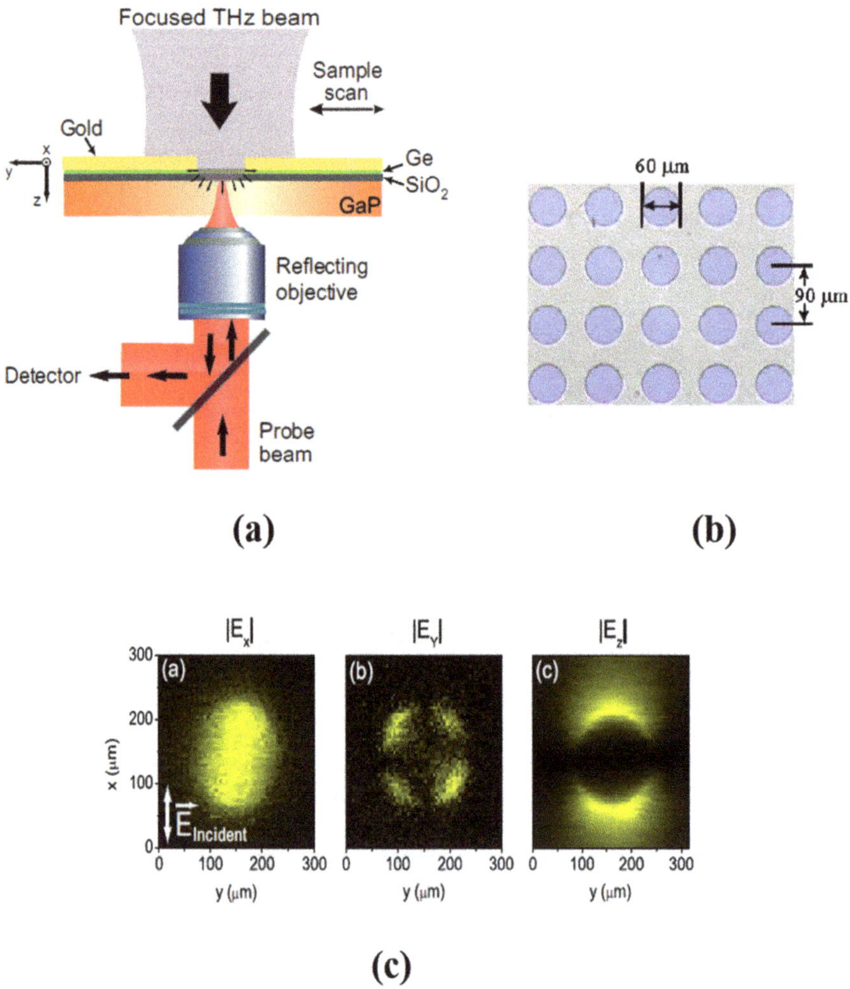

Fig. 9.2 (**a**) Terahertz near-field imaging set-up to image microstructures, (**b**) microscope image of an array subsection, and (**c**) terahertz electric near-field for a single, isolated, and circular aperture at 0.2THz [46]

metal structure in the set-up. Next is the function of standard electro-optic detection set-up where the reflected elliptically polarized probe beam by the terahertz near-field is received. To measure the long terahertz transients in minimal time, the delay between the probe-beam and terahertz beam is modulated at 50 Hz. In the set-up samples are raster scanned in the focal plane of the terahertz beam and accordingly, the terahertz near-field is measured pixel by pixel by keeping the incident beam and the probe fixed.

In [46], the x, y, and z components of terahertz electric near-field are measured by selecting a suitable probe beam polarization along with crystal orientation for the electro-optic detection crystal. A microscopic image of an array of circular micro-structured array having a periodicity of 90 μm is shown in Fig. 9.2b. The structure of apertures is arranged in a square lattice and is formed using a gold layer of 200 nm thickness which is deposited onto GaP detection crystal of thickness nearly 300 μm. The thickness of gold layer is thick enough to prevent significant direct transmission of terahertz radiation in the regions between apertures as the skin depth of gold is only 80 nm at 1 THz. Moreover, the integration of array directly on top of detection crystal helps to avoid complications which basically arise from Fabry–Perot effects between the sample and detector. The amplitude images of x, y, and z components of the measured terahertz electric near field for a single circular aperture of an array are shown in Fig. 9.2c. The bright regions represent high-field amplitude and dark areas represent low-field amplitude. Therefore, the x-component seems to be strongest in the middle of the aperture with non-spherical symmetric spatial distribution of the field. It follows from the incident field polarization which breaks the spherical symmetry. Moreover, it also represents that at the metal surface there can be no field component parallel to the metal. However, the y-component of the field consists of four lobes inside the aperture with adjacent lobes 180° out-of-phase. Similarly, the z-component of field, which in general is directly related to the surface charge density, is concentrated at the top as well as the bottom of aperture edges where each lobe is also 180° out-of-phase. As a concluding remark, this can be traced out that at THz range where metals act much like perfect conductor, the boundary conditions for the time-dependent electric field suggest that the field must be perpendicular to the metal surface. Accordingly, using Bouwkamp's model for the diffraction of light through subwavelength aperture, the near-field z-component of the electric field at the shadow side of an infinitely thin and perfectly conducting metal at $z = 0$ [42] is

$$E_Z = E_0 \left({}^{4i}\!/_{3} \right) ka \left[\left({}^{a}\!/_{\rho} \right) / \sqrt{\rho^2 / a^2 - 1} \right] \cos\varnothing, \rho > a$$

where a, ρ, Φ, E_0, and k are the radius of circular aperture, the radial coordinate, the azimuthal angle, the electric field amplitude, and wavevector of the incident radiation, respectively. Further, such terahertz near-field measurements help to make out strategies to increase the local field strength of specially designed holes for non-linear optical applications such as infrared near-field microscopy and edge enhanced Raman scattering as discussed in the next section.

9.4 Near-Fields of Various Subwavelength Holes

The near-field images of spectral amplitude and phase of the electric field show the formation, propagation, and attenuation of surface waves, and allow the investigators to distinguish between propagating and stationary modes. At microwave and far-infrared frequencies, the electromagnetic properties of metallic meshes and hole arrays were investigated and the work was performed in the 1960s and 1970s [49–52]. However, the experimental evidence for enhanced transmission at the terahertz range through periodic, subwavelength hole array is initially reported in [53] where the arrays are constructed in doped silicon wafers. As it is known nanoscopic structures of size smaller than the wavelength of light have the ability to display both electric and magnetic response and it is even true in a case where the constituent materials of all the structure have a near-unity magnetic permeability [54]. Interestingly, apart from metamaterials [55], the magnetic response of nanoscopic structures has been neglected in early times. However, recently researchers have started developing several techniques to map the optical magnetic field and even have achieved means to quantify the emission of light from magnetic dipoles [56–59]. Moreover, the investigations into Mie scattering of dielectric spheres are performed where it has been found that the interference between the magnetic and electric field responses results in directional emission [60–62].

Recently, the nanoplasmonic structure having extraordinary optical transmission and metamaterials has received considerable attention which has a dielectric-filled hole in the metallic film [63, 64]. Even for such a simple structure, few things have discussed about the presence of a magnetic optical response but still its magnitude is expected to be comparable to the electric response. In [65], the authors have quantified the optical response of subwavelength holes using a surface plasmon polariton (SPP) launcher and a single subwavelength hole which is perforated in 200 nm thick gold film having a support with glass substrate. Figure 9.3 shows the scanning electron micrograph of the sample in which SPP propagates from the slit along the positive x-direction toward the subwavelength hole. In the set-up, the slit is illuminated

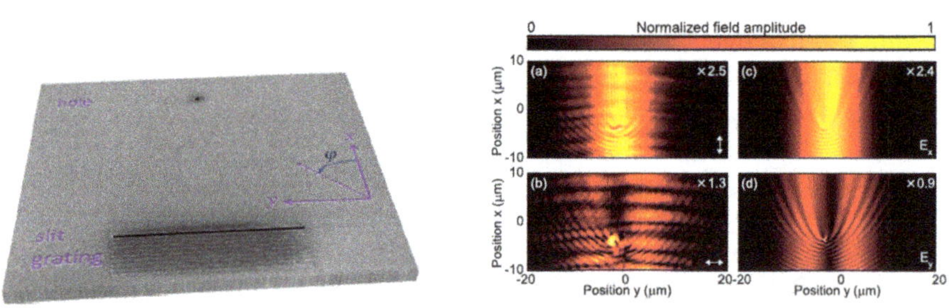

(a) (b)

Fig. 9.3 (**a**) Scanning electron micrograph of the sample, and (**b**) polarization resolved near-field maps of an SPP beam scattering from a hole for both x and y electric field orientations (the factors written at the top right corners represent the amplitude relative to the color scale) [65]

from the glass side using a continuous laser source thereby launching SPPs and with the help of grating the radiations are directed efficiently toward the hole [66]. In Fig. 9.3b, the near-field maps of an SPP beam scattering from a hole are shown wherein the parabolic fringes occur due to the interference between the incident SPP and the scattered wave.

According to the coordinate system shown in Fig. 9.3a, the SPPs propagate along the x-axis toward the hole and are transverse in the plane along the y-axis. The angle subtends between in-plane vector and x-axis is Φ. The scattering event which occurs when the SPP wave interacted with the hole gets imaged using the near-field microscope [67]. In order to quantify the interaction, an analytical model for both incident field ($E^{\text{in}}, H^{\text{in}}$) and scattered field ($E^S, H^S$) can be designed. Using the Fourier sum of plane waves, the three components of the incident Gaussian SPP beam above the film are considered as:

$$E_x^{\text{in}}(x,y,z) = -C\frac{w_{\text{sp}}}{k_0\kappa_{\text{sp}}}e^{iw_{\text{sp}}z}\sum_{\kappa_y,\kappa_x} k_x e^{-\left(\frac{\alpha^2}{2}\right)\kappa_y^2 e^{i(\kappa_x x+\kappa_y y)}} \tag{9.1a}$$

$$E_y^{\text{in}}(x,y,z) = -C\frac{w_{\text{sp}}}{k_0\kappa_{\text{sp}}}e^{iw_{\text{sp}}z}\sum_{\kappa_y,\kappa_x} k_y e^{-\left(\frac{\alpha^2}{2}\right)\kappa_y^2 e^{i(\kappa_x x+\kappa_y y)}} \tag{9.1b}$$

$$E_Z^{\text{in}}(x,y,z) = C\frac{\kappa_{\text{sp}}}{k_0}e^{iw_{\text{sp}}z}\sum_{\kappa_y,\kappa_x} e^{-\left(\frac{\alpha^2}{2}\right)\kappa_y^2 e^{i(\kappa_x x+\kappa_y y)}} \tag{9.1c}$$

In this the limits of simulation that reflect the explicit separation of the in-plane wave vector into its components are $k_y \in [-k_0, k_0]$ and $\kappa_{\text{sp}}^2 = \kappa_x^2 + \kappa_y^2$, respectively. Here, $k_0 = 2\pi/\lambda$ is the wave vector of the incident light of wavelength, λ and $\kappa_{sp} = k_0\sqrt{(\varepsilon+1)/\varepsilon}$ and $w_{\text{sp}} = -k_0/\sqrt{(\varepsilon+1)}$ are the in-plane and out-of-plane SPP wave vectors, respectively, and the complex dielectric function of gold at 1550 nm is $\varepsilon = -115 + 11i$. Here, C and α determine the amplitude and width of the incident SPP beam. The corresponding magnetic field H^{in} is computed from Eq. (9.1) using Ampere's law. Similarly, an analytical expression for the SPPs radiated by the hole dipoles can be determined. For plasmonic scattering, such radiation is dominated by an out-of-plane electric dipole p_z and an in-plane magnetic dipole m_y, and therefore the expression for E^S can be written as:

$$\begin{aligned} E^S &= -2\pi i\rho^0 e^{iw_{\text{sp}}z}\Big[\kappa_0^2\kappa_{\text{sp}} H_1^{(1)}(\kappa_{\text{sp}}r)\cos\phi\, m_y \\ &\quad + ik_0\kappa_{\text{sp}}^2 H_0^{(1)}(\kappa_{\text{sp}}r)\, p_z\Big]\left((\hat{r}) - \frac{\kappa_{\text{sp}}}{w_{\text{sp}}}(\hat{z})\right) \end{aligned} \tag{9.2}$$

here, $\rho^0 = \frac{\varepsilon}{(1+\varepsilon)^{1.5}(1-\varepsilon)}$ and $H_m^{(1)}$ are Hankel functions, $r = \sqrt{(x-x_0)^2 + (y-y_0)^2}$ is the displacement from the hole position at (x_0, y_0) and $\hat{r} = (\cos\varphi\,\hat{x}, \sin\varphi\,\hat{y})$. The total field of the scattering event, which includes contributions from both the incident and scattered SPPs can then be written as: (E^{in} + E^{s}). Moreover, one can expect to observe SPPs on the silver/air interface with the wavelength $\lambda_{SPP} = \lambda\left(\frac{\mathrm{Re}(\varepsilon_m)+1}{\mathrm{Re}(\varepsilon_m)}\right)^{1/2} \approx 508\,\mathrm{nm}$, taking into account the value of dielectric permittivity of Ag, $\varepsilon_{Ag} \approx -11.7 + 0.4i$ [68].

Similarly, in [40], the authors have investigated metallic microstructures carrying holes using time-resolved terahertz near-field imaging. In the set-up, the emission and detection of a single cycle terahertz pulses is observed by using photoconductive antennas which is optically gated by the output of a mode-locked Ti:sapphire laser system [69]. Figure 9.4a shows the metal sample (which are microstructures carrying holes having dimensions as diameter, d = 300 μm, the spacing from center-to-center of holes, s = 450 μm and thickness, l = 300 μm) and Fig. 9.4b shows terahertz near-field imaging set-up using photoconductive antennas. The terahertz wave propagates through the microstructures and the transmitted electric field is detected. The silicon-on-sapphire detector chip which is mounted with the silicon layer with H-shaped electrodes is kept at the backside of the sample facing toward it. Since the incident terahertz pulses are set to be x-polarized, therefore, the detector is also oriented to be sensitive to the x-component of the electric field. Then, the spatially resolved measurements are made by moving the detector chip along with the probe laser in all coordinates. However, the spatial resolution is limited by the finite dimensions of the probe beam focus as well as of photoconductive gap between the detector electrodes. Since in the set-up, the authors [40] during the investigation have kept the detector electrodes very near to the metal surface of the imaging sample so capacitive coupling cannot be ignored. Further, as the detector has to move over the entire image area during one measurement, therefore, the capacitive influence if gets localized on the area of the photoconductive gap becomes isotropically distributed over the image and so its influence over the observed field pattern gets minimal.

In the observed results, there is the emergence of spherical wave which radiates from the hole into the free space. Moreover, the Ex component of the field is observed to be confined to the hole area and two wave contributions which appear

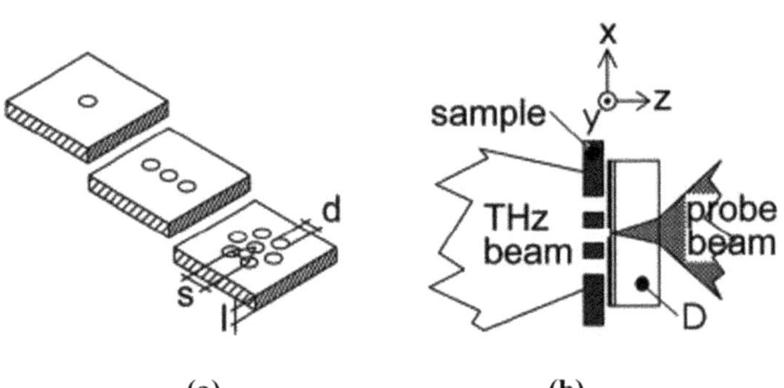

Fig. 9.4 (**a**) Three different types of metal samples investigated using the set-up, (**b**) THz near-field imaging set-up [40]

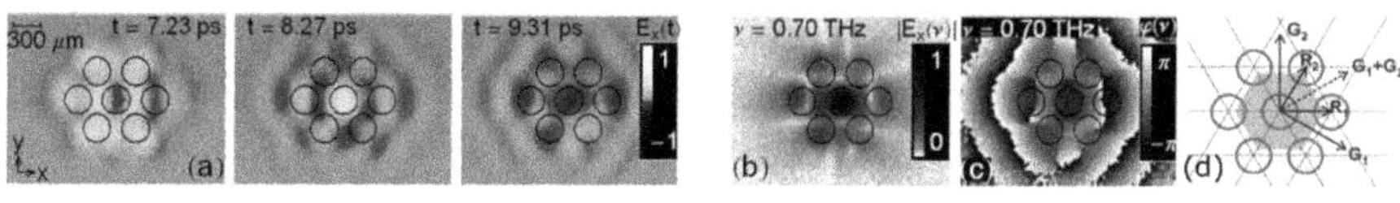

Fig. 9.5 Time sequence of the x-polarized THz electric field transmitted through the surface of a hexagonal hole array [40]

along the surface away from the hole appears at later times. This happens because of the residual tangential component of the spherical wave. Figure 9.5 shows the time sequence of the field transmitted through hexagonal hole array sample and accordingly Fig. 9.5b, c represents the amplitude and phase images at 0.7 THz. Figure 9.5d corresponds to the realization of primitive vectors with gray area representing the first Brillouin zone.

Under the normal incidence resonant coupling of an electromagnetic wave to a surface mode in a two-dimensional grating structure occurs when the wave vector of the surface wave matches the in-plane momentum provided by the hole array $\vec{G} = i\overrightarrow{G_1} + j\overrightarrow{G_2}$, where $\overrightarrow{G_1}$ and $\overrightarrow{G_2}$ are the primitive reciprocal lattice vectors, and i and j are integers. This condition restricts the continuous spectrum of surface waves that emerges from a single hole to a discrete spectrum and is represented by wave vectors of a hexagonal lattice as:

$$\left|\vec{k}_{\mathrm{sw}}\right| = \frac{2\pi}{s}\left(\sqrt{\frac{4}{3}\left(i^2 - ij + j^2\right)}\right),$$

Here, s is the separation between the holes. Moreover, at terahertz frequencies, the surface wave behaves almost like light and thus, $\left|\vec{k}_{\mathrm{sw}}\right| \approx \frac{2\pi\nu}{c}.\sqrt{\varepsilon_d}$, with ν being the optical frequency and ε_d the dielectric constant of the medium in contact with the metal, that is, air. Therefore, the discrete surface wave spectrum in a two-dimensional hole array is due to the interaction of continuous; however, the anisotropic spectrum of surface waves emerges from a single hole with the discrete spectrum of lattice momenta. Moreover, in-between the holes the phase also gets oscillated due to standing surface waves. Hence, the lattice represents a surface wave resonator with a discrete spectrum of eigenmodes. Therefore, the periodic arrays of holes perforated in metal films have been considered as an attractive interest in the aspect of enhanced transmission [70–72].

9.5 Kirchhoff Formalism for Near-Field Estimate

For the perfect transmission, besides the periodicity of hole arrays, several researchers also demonstrate that individual shapes of rectangles can even induce enhanced transmission [38, 73–77]. In [78], the authors have determined extraordinary far-field transmissions caused by strong near-field enhancements of sub-micrometer

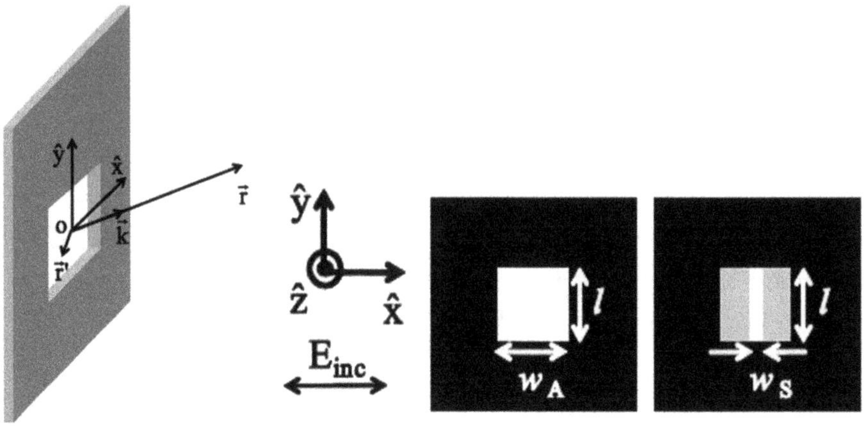

(a) (b)

Fig. 9.6 (**a**) Diffraction geometry for a plane screen with a square aperture at the center and (**b**) reference aperture and the aperture with slit sample assuming the incident beam with horizontal direction of polarization [84]

width apertures in the terahertz range. Therefore, to measure such near-field enhancement, electro-optic measurements of terahertz near-field are performed [42, 79]. However, it is technically difficult to do direct measurements of the terahertz near-field for extreme sub-wavelength holes which are below sub-micrometer size and this is because of the limited resolution of the near-field imaging as the spot size of the optical probe beam is typically in the order of few micrometers. Secondly, there is a need of sample to be fabricated on a specially oriented detection crystal which can enhance optical properties of metallic holes and the near-field distribution [80, 81]. Looking toward the difficulties in direct near-field measurements, a new approach has also been proposed based on the measurement of the near-field enhancement via the far-field transmission [82]. Further, the estimation of near-field enhancement via far-field transmission can be achieved without direct near-field measurements. In this technique, the transmission through a finite aperture is used as the reference signal and the near-field enhancement given by a sample can be read from the transmission ratio between signals with and without the sample.

The relation between the near-field and far-field is derived from the Kirchhoff integral formalism [83]. Considering Fig. 9.6, when an observation point is located far from the hole then the diffracted electric field can be approximated by the Kirchhoff integral as:

$$\vec{E}(\vec{r}) = \frac{ie^{ikr}}{2\pi r}\vec{k}\int \hat{n}\times\vec{E}(\vec{r'})\times e^{-i\vec{k}\vec{r'}}\,da'$$

where A represents the aperture opening, r is the distance from origin O to the observation point, $\hat{n}$ is the surface normal, and $\vec{k}$ is the wave vector in the direction of observation point.

For a terahertz beam polarized in the horizontal direction (x-axis) and impinging upon the aperture at the normal incidence, the transmitted horizontal (x-axis) electric fields can be defined only through the reference aperture as $E_{\mathrm{f}}^{\mathrm{A}}$ and through a slit sample attached on the reference aperture as $E_{\mathrm{f}}^{\mathrm{S}}$, respectively [84]. Then, the normalized amplitude $\alpha(w)$ is defined by the ratio between two amplitudes:

$$\alpha(w) \equiv \left| \frac{E_{\mathrm{f}}^{\mathrm{S}}(w)}{E_{\mathrm{f}}^{\mathrm{A}}(w)} \right|$$

According to the Kirchhoff integral formalism, the transmitted amplitude through the aperture, measured at the diffraction center $r = (0, 0, r)$, is proportional to the x-component of the near-field distribution in the aperture E_{n}^{A}.

$$E_{\mathrm{f}}^{\mathrm{A}} = \frac{e^{ikr}}{i\lambda r} \int_0^{w_{\mathrm{A}}} \int_0^{l} E_{\mathrm{n}}^{\mathrm{A}}(x',y')\,dx'\,dy' = \frac{e^{ikr}}{i\lambda r} \left\langle E_{\mathrm{n}}^{\mathrm{A}} \right\rangle w_{\mathrm{A}} l,$$

where the bracket $\langle \quad \rangle$ means the averaged near-field over the width w, that is, $\left\langle E_{\mathrm{n}} \right\rangle = \frac{1}{w} \int_0^{w} E_{\mathrm{n}}(x')\,dx'$. Unlike the case of the x-component, the averaged y- and z-components of the near-field are negligible due to the odd symmetric field distributions in the opening, thus can be neglected [84]. In the similar manner, the transmitted far-field through the slit $E_{\mathrm{f}}^{\mathrm{S}}$ is:

$$E_{\mathrm{f}}^{\mathrm{S}} = \frac{e^{ikr}}{i\lambda r} \int_0^{w_{\mathrm{S}}} \int_0^{l} E_{\mathrm{n}}^{\mathrm{S}}(x',y')\,dx'\,dy' = \frac{e^{ikr}}{i\lambda r} \left\langle E_{\mathrm{n}}^{\mathrm{S}} \right\rangle w_{\mathrm{S}} l$$

By dividing the two equations, a simple relation between the near- and far-field becomes

$$\frac{\left| \left\langle E_{\mathrm{n}}^{\mathrm{S}}(w) \right\rangle \right|}{\left| \left\langle E_{\mathrm{n}}^{\mathrm{A}}(w) \right\rangle \right|} = \frac{\alpha(w)}{\beta}$$

where $\beta = \frac{w_{\mathrm{S}}}{w_{\mathrm{A}}}$ denotes the ratio of widths between the reference aperture and the slit. By considering the field enhancement by the reference aperture with respect to the incident terahertz beam, $\left| \left\langle E_{\mathrm{n}}^{\mathrm{S}}(w) \right\rangle \right| = \gamma(w).\left| \left\langle E_{\mathrm{inc}}(w) \right\rangle \right|$, the near-field enhancement factor measured in the far-field, is finally determined as:

$$\frac{\left| \left\langle E_{\mathrm{n}}^{\mathrm{S}}(w) \right\rangle \right|}{\left| \left\langle E_{\mathrm{inc}}(w) \right\rangle \right|} = \frac{\gamma(w).\alpha(w)}{\beta}$$

Therefore, the near-field enhancement can be estimated from the far-field measurement using the hole-to-aperture ratio.

9.6 Summary

The terahertz wave exhibits many features such as non-ionizing, non-invasive, phase sensitive to polar substances, spectral fingerprinting, relatively good resolution, coherent detection properties, and penetration capabilities. The traditional imaging methods such a magnetic resonance imaging and computerized tomography cause radiation damage to bio-tissues. However, the terahertz imaging provides quick, non-destructive, and accurate imaging of bio-tissues. Accordingly, the imaging systems are based on either near-field or far-field set-up depending upon the application. The terahertz near-field imaging approach provides time and spatially resolved measurements of the electric field distribution in the vicinity of structured samples with subwavelength spatial resolution. The state-of-the-art measurement techniques of near-field imaging are in progress for several application areas. Using the Kirchhoff formalism, the near-field enhancement can be estimated from the far-field measurements and are analytically discussed. The practical implementations of metasurface or slits or holes hold promise for near-field terahertz beam steering, laser-based microscopy, non-contact sensing, imaging, and lithography applications. Moreover, the terahertz field enhancements would trigger various future researchers such as terahertz nonlinearity measurement, terahertz-controlled nano-optic devices, and single molecule detection using terahertz waves.

References

1. W.L. Chan, J. Deibel, D.M. Mittleman, Imaging with terahertz radiation. Rep. Prog. Phys. **70**, 1325–1379 (2007)
2. A.J.L. Adam, Review of near-field terahertz measurement methods and their applications. J. Infrared Millimeter Terahertz Waves **32**(8–9), 976–1019 (2011)
3. J.W. Goodman, *Introduction to Fourier Optics*, 2nd. edn. (McGraw-Hill, New York, 1996)
4. M. Born, E. Wolf, *Principles of Optics*, 7th edn. (Cambridge University Press, Cambridge, UK, Dec. 2019)
5. J.B. Pawley, *Handbook of Biological Confocal Microscopy* (Springer, Berlin/Heidelberg, 2006)
6. T. Yuan, Z. Xu, X.C. Zhang, Development of terahertz wave microscopes. Infrared Phys Technol **45**, 417–425 (2004)
7. F. Blanchard, A. Doi, T. Tanaka, K. Tanaka, Real-time, subwavelength terahertz imaging. Ann. Rev. Mater. Res **43**, 237–259 (2013)
8. B.B. Hu, M.C. Nuss, Imaging with terahertz waves. Opt. Lett. **20**(16), 1716–1718 (1995)
9. J. Liu, J. An, R. Zhou, Y. Yang, Terahertz near-field MIMO-SAR technology for human security inspection. J. Opto-Electron. Eng **47**(5), 190682/1–11 (2020)
10. D. Yan, H. Zhang, D. Xu, W. Shi, C. Yan, P. Liu, J. Shi, J. Yao, Numerical study of compact terahertz gas laser based on photonic crystal fiber cavity. J. Lightwave Technol. **34**(14), 3373–3378 (2016)
11. P. Klarskov, H. Kim, V.L. Colvin, D.M. Mittleman, Nanoscale laser terahertz emission microscopy. ACS Photonics **4**, 2676–2680 (2017)
12. S.C. Chen, L.H. Du, K. Meng, J. Li, Z.H. Zhai, Q.W. Hi, Z.R. Li, L.G. Zhu, Terahertz wave near-field compressive imaging with a spatial resolution of over $\lambda/100$. Opt. Lett. **44**, 21–24 (2019)

13. K. Moon, P. Hongkyu, J. Kim, Y. Do, S. Lee, G. Lee, H. Kang, H. Han, Subsurface nanoimaging by broadband terahertz pulse near-field microscopy. Nano Lett. **15**(1), 549–552 (2015)
14. H. Chen, S.H. Ma, X. Wu, W. Yang, T. Zhao, Diagnose human colonic tissues by terahertz near-field imaging. J. Biomed. Opt. **20**(3), 036017/1–036017/5 (2015)
15. L. Mavarani, P. Hillger, T. Bucher, J. Grzyb, U.R. Pfeiffer, Q. Cassar, A. Al-Ibadi, T. Zimmer, J.P. Guillet, P. Mounaix, G. MacGrogan, Near sense- advances towards a silicon-based terahertz near-field imaging sensor for Ex vivo breast tumour identification. Frequenz **72**(3–4), 1–7 (2018)
16. P. Alonso-Gonzalez, A.Y. Nikintin, Y. Gao, A. Woessner, M.B. Lundeberg, A. Principi, N. Forcellini, W. Yan, S. Velez, A.J. Huber, K. Watanabe, Acoustic terahertz graphene plasmons revealed by photocurrent nanoscopy. Nat. Nanotechnol. **12**(1), 31–35 (2017)
17. N. Wang, X. Zhang, J. Liang, T. Chang, H. L. Cui, Novel configuration of aperture-type terahertz near-field imaging probe. Journal of Physics D: Applied Physics **53**, 295102/1–8 (2020)
18. F.Y. Han, P.K. Liu, Terahertz near-field metasurfaces : Amplitude-phase combined steering and electromagnetostatic dual-field superfocusing. Adv. Opt. Mater **8**(6), 1901331/1–12 (2020)
19. J. Li, Y. Yang, J. Li, Y. Zhang, L. Wu, Z. Zhang, M. Yang, C. Zheng, J. Li, J. Huang, F. Li, T. Tang, H. Dai, J. Yao, Metal-graphene hybrid active chiral metasurfaces for dynamic terahertz wavefront modulation and near field imaging. Carbon **163**, 34–42 (2020)
20. Y. Fang, A. Armin, P. Meredith, J. Huang, Accurate characterization of next-generation thin-film photodetectors. Nat. Photonics **13**(1), 1–4 (2019)
21. N.C.J. Vandr Valk, P.C.M. Planken, Electro-optic detection of subwavelength terahertz spot sizes in the near field of a metal tip. Appl. Phys. Lett. **81**(9), 1558/1–1558/3 (2002)
22. H.T. Chen, W.J. Padilla, J.M. Zide, A.C. Gossard, A.J. Taylor, R.D. Averitt, Active terahertz metamaterial devices. Nature **444**(7119), 597–600 (2006)
23. J.F. Federici, B. Schulkin, F. Huang, D. Gary, R. Barat, F. Oliveira, D. Zimdars, THz imaging and sensing for security applications-explosives, weapons and drugs. Semicond. Sci Technol **20**(7), S266–S280 (2005)
24. X. Yang, Z. Tian, X. Chen, M. Hu, Z. Yi, C. Ouyang, J. Gu, J. Han, W. Zhang, Terahertz ingle-pixel near-field imaging based on active tuneable subwavelength metallic grating. Appl. Phys. Lett. **116**, 241106 1–4 (2020)
25. H.T. Chen, R. Kersting, G.C. Cho, Terahertz imaging with nanometer resolution. Appl. Phys. Lett. **83**(15), 3009–3011 (2013)
26. K. Wang, D.M. Mittleman, N.C.J. Van Der Valk, P.C.M. Planken, Antenna effects in terahertz apertureless near-field optical microscopy. Appl. Phys. Lett. **85**(14), 2715–2717 (2004)
27. G. Dai, G. Geng, X. Zhang, J. Wang, T. Chang, H.L. Cui, W-band near-field microscope. IEEE Access **7**(16), 48060–48067 (2019)
28. P. Alonso-Gonzalez, A.Y. Nikitin, Y. Gao, A. Woessner, M.B. Lundeberg, A. Principi, N. Forcellini, W. Yan, S. Velez, A.J. Huber, K. Watanabe, Acoustic terahertz graphene plasmons revealed by photocurrent nanoscopy. Nat. Nanotechnol. **12**(1), 31–35 (2017)
29. H.G. von Ribbeck, M. Brehm, D.W. van der Weide, S. Winnerl, O. Drachenko, M. Helm, F. Keilmann, Spectroscopic THz near-field microscope. Opt. Express **16**(5), 3430–3438 (2008)
30. A.J. Huber, F. Keilmann, J. Wittborn, J. Aizpurua, R. Hillenbrand, Terahertz near-field nanoscopy of mobile carriers in single semiconductor nanodevices. Nano Lett. **8**(11), 3766–3770 (2008)
31. F. Kuschewski, H.G. Von Ribbeck, J. Döring, S. Winnerl, L.M. Eng, S.C. Kehr, Narrow-band near-field nanoscopy in the spectral range from 1.3 to 8.5 THz. Appl. Phys. Lett **108**(11), 113102/1–113102/3 (2016)
32. H.T. Stinson, A. Sternbach, O. Najera, R. Jing, A.S. Mcleod, T.V. Slusar, A. Mueller, L. Anderegg, H.T. Kim, M. Rozenberg, D.N. Basov, Imaging the nanoscale phase separation in vanadium dioxide thin films at terahertz frequencies. Nat. Commun. **9**(1), 3604/1–3604/9 (2018)

33. J. Zhang, X. Chen, S. Mills, T. Ciavatti, Z. Yao, R. Mescall, H. Hu, V. Semenenko, Z. Fei, H. Li, V. Perebeino, Terahertz nanoimaging of graphene. ACS Photonics **5**(7), 2645–2651 (2018)
34. C. Liewald, S. Mastel, J. Hesler, A.J. Huber, R. Hillenbrand, F. Keilmann, All-electronic terahertz nanoscopy. Optica **5**(2), 159–163 (2018)
35. M.C. Giordano, S. Mastel, C. Liewald, L.L. Columbo, M. Brambilla, L. Viti, A. Politano, K. Zhang, L. Li, A.G. Davies, E.H. Linfield, Phase-resolved terahertz self-detection near-field microscopy. Opt. Express **26**(14), 18423–28435 (2018)
36. X. Chen, X. Liu, X. Guo, S. Chen, H. Hu, E. Nikulina, X. Ye, Z. Yao, H.A. Bechtel, M.C. Martin, G.L. Carr, Q. Dai, S. Zhuang, Q. Hu, Y. Zhu, R. Hillenbrand, M. Liu, G. You, THz near-field imaging of extreme subwavelength metal structures. ACS Photonics **7**, 687–694 (2020)
37. S. Ducourtieux, S. Gresillon, J.C. Rivoal, C. Vannier, C. Bainier, D. Courjon, H. Cory, Imaging subwavelength holes in chromium films in scanning near-field optical microscopy: Comparison between experiments and calculation. Eur. Phys. J. Appl. Phys. **26**(1), 35–43 (2004)
38. M.A. Seo, A.J.L. Adam, J.H. Kang, J.W. Lee, K.J. Ahn, Q.H. Park, P.C.M. Planken, D.S. Kim, Near field imaging of terahertz focusing onto rectangular apertures. Opt. Express **16**, 20484–20489 (2008)
39. O. Mitrofanov, M. Lee, J.W.P. Hsu, L.N. Pfeiffer, K.W. West, J.D. Wynn, J.F. Federici, Terahertz pulse propagation through small apertures. Appl. Phys. Lett. **79**, 907–909 (2001)
40. A. Bitzer, M. Walther, Terahertz near-field imaging of metallic subwavelength holes and hole arrays. Appl. Phys. Lett **92**, 231101/1–231101/3 (2008)
41. A. Bitzer, H. Merbold, A. Thoman, T. Feurer, H. Helm, M. Walther, Terahertz near-field imaging of electric and magnetic resonances of a planar metamaterial. Opt. Express **17**, 3826–3834 (2009)
42. A.J.L. Adam, J.M. Brok, M.A. Seo, K.J. Ahn, D.S. Kim, J.H. Kang, Q.H. Park, M. Nagel, P.C.M. Planken, Advanced terahertz electric near-field measurements at sub-wavelength diameter metallic apertures. Opt. Express **16**, 7407–7417 (2008)
43. M. Mrejen, A. Israel, H. Taha, M. Palchan, A. Lewis, Near-field characterization of extraordinary optical transmission in sub-wavelength aperture arrays. Opt. Express **15**, 9129–9138 (2007)
44. H. Gao, J. Henzie, T.W. Odom, Direct evidence for surface plasmon-mediated enhanced light transmission through metallic nanohole arrays. Nano Lett. **6**, 2104–2108 (2006)
45. S.C. Hohng, Y.C. Yoon, D.S. Kim, V. Malyarchuk, R. Muller, Ch. Lienau, J.W. Park, K.H. Yoo, J. Kim, H.Y. Ryu, Q.H. Park, Light emission from the shadows: Surface plasmon nano-optics at near and far fields. Appl. Phys. Lett. **81**, 3239–3241 (2002)
46. J. R. Knab, A. J. L.Adam, M. Nagel, E. Shaner, .A. Seo, D. S. Kim and P. C. M. Planken, "Terahertz near-field vectorial imaging of subwavelength apertures and aperture arrays", Opt. Express, vol. 17, no. 17, pp. 15072–15086, 2009
47. G. Zhao, R.N. Schouten, N.C.J. van der Valk, W.T. Wenckebach, P.C.M. Planken, Design and performance of a THz emission and detection setup based on a semi-insulating GaAs emitter. Rev. Sci. Instrum. **73**, 1715–1719 (2002)
48. R. Chakkittakandy, J.A. Corver, P.C.M. Planken, Quasi-near field terahertz generation and detection. Opt. Express **16**, 12794–12805 (2008)
49. C.-C. Chen, Transmission of microwave through perforated flat plates of finite thickness. IEEE Trans. Microwave Theory Technol **MTT-21**, 1–6 (1973)
50. R. Ulrich, M. Tacke, Submillimeter waveguiding on periodic metal structure. Appl. Phys. Lett. **22**, 251–253 (1973)
51. A. Mitsuishi, Y. Otsuka, S. Fujita, H. Yoshinaga, Metal mesh filters in the far infrared region. Jpn. J. Appl. Phys. **2**, 574–577 (1963)
52. C.-C. Chen, Transmission through a conducting screen perforated periodically with apertures. IEEE Trans. Microwave Theory Technol **MTT-18**, 627–632 (1970)
53. J. Gomez Rivas, C. Schotsch, P. Haring Bolivar, H. Kurz, Enhanced transmission of THz radiation through subwavelength holes. Phys. Rev. B **68**, 201306 (2003)
54. C.J. Bouwkamp, Diffraction theory. Rep. Prog. Phys. **17**, 35 (1954)

55. I. Sersic, M. Frimmer, E. Verhagen, A.F. Koenderink, Electric and magnetic dipole coupling in near-infrared split-ring metamaterial arrays. Phys. Rev. Lett. **103**(21), 213902/1–213902/3 (2009)
56. M. Burresi, D. van Oosten, T. Kampfrath, H. Schoenmaker, R. Heideman, A. Leinse, L. Kuipers, Probing the magnetic field of light at optical frequencies. Science **326**, 550–553 (2009)
57. R.L. Olmon, M. Rang, P.M. Krenz, B.A. Lail, L.V. Saraf, G.D. Boreman, M.B. Raschke, Determination of electric-field, magnetic-field and electric-current distributions of infrared optical antennas: A near-field optical vector network analyser. Phys. Rev. Lett. **105**(16), 167403/1–167403/3 (2010)
58. H.W. Kihm, S.M. Koo, Q.H. Kim, K. Bao, J.E. Kihm, W.S. Bak, S.H. Eah, C. Lienau, H. Kim, P. Nordlander, N.J. Halas, N.K. Park, D.-S. Kim, Bethe-hole polarization analyser for the magnetic vector of light. Nat. Commun. **2**(1), 1–6 (2011)
59. T.H. Taminiau, S. Karaveli, N.F. van Hulst, R. Zia, Quantifying the magnetic nature of light emission. Nat. Commun. **3**(1), 979/1–979/6 (2012)
60. I. Staude, A.E. Miroshnichenko, M. Decker, N.T. Fofang, S. Liu, E. Gonzales, J. Dominguez, T.S. Luk, D.N. Neshev, I. Brener, Y. Kivshar, Tailoring directional scattering through magnetic and electric resonances in subwavelength silicon nanodisks. ACS Nano **7**(9), 7824–7832 (2013)
61. S. Person, M. Jain, Z. Lapin, J.J. Saenz, G. Wicks, L. Novotny, Demonstration of zero optical backscattering from single nanoparticles. Nano Lett. **13**(4), 1806–1809 (2013)
62. Y.H. Fu, A.I. Kuznetsov, A.E. Miroshnichenko, Y.F. Yu, B. Lukyanchuk, Directional visible light scattering by silicon nanoparticles. Nat. Commun. **4**(1), 1527/1–1527/6 (2013)
63. N. Rotenberg, M. Spasenovic, T.L. Krijger, B. le Feber, F. J. Garc'ıa de Abajo, and L. Kuipers, "Directional visible light scattering by silicon nanoparticles". Phys. Rev. Lett. **108**, 127402/1–127402/3 (2012)
64. J.-M. Yi, A. Cuche, F. de Leon-Perez, A. Degiron, E. Laux, E. Devaux, C. Genet, J. Alegret, L. Martın-Moreno, T.W. Ebbesen, Diffraction regimes of single holes. Phys. Rev. Lett. **109**(2), 023901/1–023901/3 (2012)
65. N. Rotenberg, T.L. Krijger, B. le Feber, M. Spasenovic, F.J.G. de Abajo, L. Kuipers, Magnetic and electric response of single wavelength holes. Phys. Rev. B **88**, 241408/1–241408/5 (2013)
66. F. Lopez-Tejeira, S.G. Rodrigo, L. Mart'ın-Moreno, F.J. Garcıa-Vidal, E. Devaux, T.W. Ebbesen, J.R. Krenn, I.P. Radko, S.I. Bozhevolnyi, M.U. Gonzalez, J.C. Weeber, A. Dereux, Efficient unidirectional nanoslit couplers for surface plasmons. Nat. Phys. **3**(5), 324–328 (2007)
67. M. Sandtke, R.J.P. Engelen, H. Schoenmaker, I. Attema, H. Dekker, I. Cerjak, J.P. Korterik, F.B. Segerink, L. Kuipers, Novel instruments for surface plasmon polariton tracking in space and time. Rev. Sci. Instrum. **79**(1), 013704/1–11 (2008)
68. D.V. Permyakv, I.S. Mukhin, I.I. Shishkin, A.K. Samusev, P.A. Belov, Y.S. Kivshar, Mapping electromagnetic fields near subwavelength hole. JETP Lett. **99**(11), 722–726 (2014)
69. D.R. Grischkowsky, Optoelectronic characterization of transmission lines and waveguides by terahertz time-domain spectroscopy. J. Sel. Top. Quantum Electron **6**(6), 1122–1135 (2000)
70. C. Genet, T.W. Ebbesen, Light in tiny holes, in *Nanoscience and Technology: A Collection of Reviews from Nature Journal*, ed. by P. Rodgers, (World Scientic Publication, Singapore, 2009), pp. 205–212
71. E. Popov, M. Nevière, S. Enoch, R. Reinisch, Theory of light transmission through subwavelength periodic hole arrays. Phys. Rev. B **62**(23), 16100/1–16100/9 (2000)
72. W.L. Barnes, W.A. Murray, J. Dintinger, E. Devaux, T.W. Ebbesen, Surface plasmon plaritons and their role in the enhanced transmission of light through periodic arrays of subwavelength holes in a metal film. Phys. Rev. Lett. **92**, 107401/1–107401/3 (2004)
73. J.W. Lee, M.A. Seo, D.H. Kang, K.S. Khim, S.C. Jeoung, D.S. Kim, Terahertz electromagnetic wave transmission through random arrays of single rectangular holes and slits in thin metallic sheets. Phys. Rev. Lett. **99**(13), 137401/1–137401/3 (2007)

74. K.J.K. Koerkamp, S. Enoch, F.B. Segerink, N.F. van Hulst, L. Kuipers, Strong influence of hole shape on extraordinary transmission through periodic arrays of subwavelength hole. Phys. Rev. Lett. **92**, 183901/1–183901/4 (2004)
75. K.L. Van der Molen, K.J. Klein Koerkamp, S. Enoch, F.B. Segerink, N.F. Van Hulst, L. Kuipers, Role of shape and localized resonances in extraordinary transmission through periodic arrays of subwavelength holes: Experiment and theory. Phys. Rev. B **72**, 045421/1–045421/9 (2005)
76. H. Cao, A. Nahata, Influence of aperture shape on the transmission properties of a periodic array of subwavelength apertures. Opt. Express **12**(16), 3664/1–3664/9 (2004)
77. F.J. García-Vidal, E. Moreno, J.A. Porto, L. Martín-Moreno, Transmission of light through a single rectangular hole. Phys. Rev. Lett. **95**(10), 103901/1–103901/3 (2005)
78. M.A. Seo, H.R. Park, S.M. Koo, D.J. Park, J.H. Kang, O.K. Suwal, S.S. Choi, P.C.M. Planken, G.S. Park, N.K. Park, Q.H. Park, D.S. Kim, Terahertz field enhancement by a metallic nano slit operating beyond the skin-depth limit. Nat. Photonics **3**(3), 152–156 (2009)
79. A.J.L. Adam, J.M. Brok, P.C.M. Planken, M.A. Seo, D.S. Kim, THz near-field measurements of metal structures. C. R. Physique **9**(2), 161–168 (2008)
80. J.H. Kang, J.-H. Choe, D.S. Kim, Q.H. Park, Substrate effect on aperture resonances in a thin metal film. Opt. Express **17**(18), 15652–15658 (2009)
81. L. Guestin, A.J.L. Adam, J.R. Knab, M. Nagel, P.C.M. Planken, Influence of the dielectric substrate on the terahertz electric near-field of a hole in a metal. Opt. Express **17**(20), 17412–17425 (2009)
82. D.J. Park, S.B. Choi, Y.H. Ahn, F. Rotermund, I.B. Sohn, C. Kang, M.S. Jeong, D.S. Kim, Terahertz near-field enhancement in narrow rectangular apertures on metal film. Opt. Express **17**(15), 12493–12501 (2009)
83. M. Born, E. Wolf, *Principles of Optics*, 7th edn. (Cambridge University Press, Cambridge UK, 1999)
84. J.S. Kyoung, M.A. Seo, H.R. Park, K.J. Ahn, D.S. Kim, Far field detection of terahertz near field enhancement of sub-wavelength slits using Kirchhoff integral formalism. Opt. Commun. **283**, 4907–4910 (2010)

Chapter 10
Terahertz Technology for Biomedical Application

10.1 Introduction

Terahertz waves are the next frontier in science and technology, and they occur between the infrared and microwave regime of the electromagnetic spectrum. There is a dearth of research and development in advanced imaging and sensing technology in the terahertz region compared to the science and technology in the microwave and optical regime of the spectrum. Because the terahertz range is rich in spectroscopic features, terahertz-based science and technology will not only have an impact on material characterization and identification but also have potential applications in the fields of communications, imaging, medical diagnostics, health monitoring, environmental control, chemical and biological sensing, as well as security and quality control applications. Therefore, research involving the terahertz regime of the spectrum is one of the most promising areas of study for transformational advances in imaging and sensing as well as other interdisciplinary fields. This regime of the spectrum was unexplored/unutilized until the past three decades; the terahertz time-domain spectroscopy technique was invented by Grischkowsky and his colleagues [1–3]. However, the development of ultrafast lasers [4, 5] and imitative techniques such as photoconductive switching and sampling [6, 7] and electro-optic generation and detection [6] made it feasible to study. In other words, the ultrafast technology has helped to develop efficient terahertz sources and detectors, though the fabrication of compact terahertz systems remains a challenge. Furthermore, the development of various other free-space terahertz generation and detection techniques has been explored, including the photoconductive switching method. However, the most prominent ones among them are based on optical rectification for terahertz generation [8] and on the electro-optic effect for its detection [9]. These techniques facilitated the study of terahertz radiation over a much larger bandwidth than that of the photoconductive switching and allowed measurement flexibility with real-time imaging [10]. The free-space terahertz spectroscopy systems have been extensively applied in

I. Malhotra, G. Singh, *Terahertz Antenna Technology for Imaging and Sensing Applications*, https://doi.org/10.1007/978-3-030-68960-5_10

scientific investigations, especially in semiconductor physics [11, 12] and gas sensing [13]. One of the advantages of terahertz spectroscopy is that it allows the samples to be directly inserted in the terahertz beam path and displays an intuitive coherent response in the time domain, and a readily achievable high signal-to-noise ratio is another advantage.

Owing to recent advances in the development of terahertz sources and detectors, the terahertz imaging techniques have great potential in various fields, such as non-destructive testing and non-contact measurements, and large-scale market products. Initially, terahertz imaging was employed for several applications as aforementioned. The terahertz radiation strongly interacts with materials because the order of their wavelength corresponds to that of a molecular-level structure. The terahertz frequency range accommodates some semiconductor transport parameters, the rotational energy of gas molecules, and that of the collective vibrations' mode. The associated energies of hydrogen bonding modes of molecular structures also lie in the terahertz frequency range. It is an important frequency range in which the characteristics of the materials can be identified [14–16]. Terahertz radiation has unique characteristics that are not present in other regimes of the electromagnetic spectrum because the energy level of a terahertz wave does not pose an ionization hazard like X-ray radiation. It is also characterized by its high transmittance for non-metallic and non-polar materials and high reactivity for high-molecular components. Furthermore, we know that the use of molecular spectroscopy as an investigative tool for the study of molecular structures is increasingly confronted with the term 'terahertz spectroscopy'. Currently, the traditional terahertz 'frequency domain' spectroscopy has been augmented with the powerful terahertz 'time domain' spectroscopy, as shown in Fig. 10.1. In contrast to optical and infrared spectroscopy, which measure only the intensity of light at specific frequencies, terahertz time-domain spectroscopy is able to determine both the amplitude and the phase of the terahertz pulse, thus allowing the evaluation of both the absorption coefficient and the refractive index. Terahertz technology has novel imaging modality with potential applications in molecular spectroscopy. Various common materials and living tissues are transparent/semi-transparent to terahertz radiation and exhibit 'terahertz fingerprints', permitting them to be imaged and identified. This imaging technique provides a higher resolution image than that of microwave imaging because the resolution is defined by the beam diameter at the terahertz wavelength. Scanning near-field microscopy with a modulated terahertz field is likely to further improve the resolution to yield terahertz images at a resolution of tens of microns.

The key elements necessary to yield the molecular information are: (a) the use of high specificity imaging probes, (b) suitable amplification strategies and (c) sensitive systems capable of producing high-resolution images. Furthermore, the merit of terahertz radiation comes from its sensitivity to the motion of water molecules because the rotational and vibrational energies of water molecules are within the terahertz spectral region. The imaging with terahertz waves was first illustrated using a leaf with varying water content and a packaged integrated circuit [17]. Although the experiments were rather simplistic, they were appreciated by

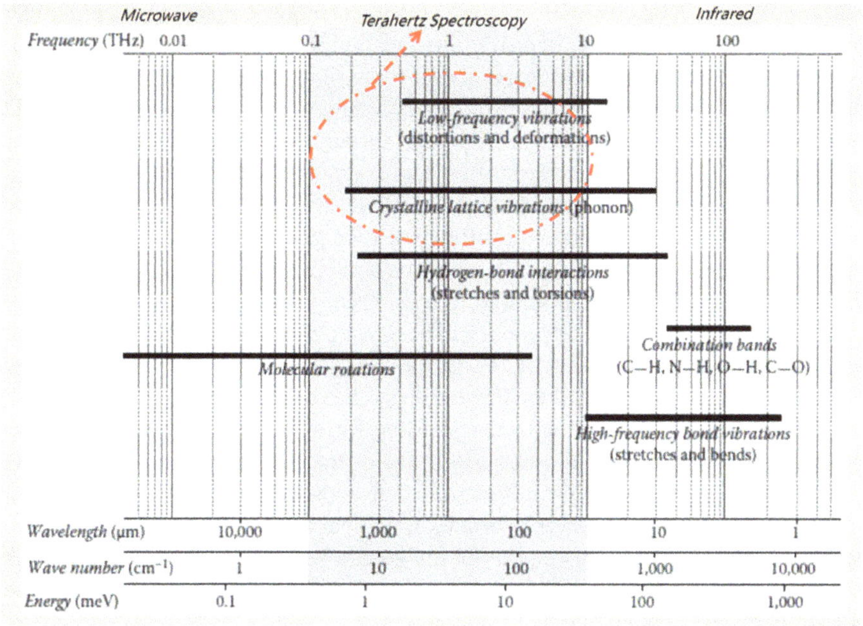

Fig. 10.1 Characteristic energies in the electromagnetic spectrum around the terahertz frequency region

researchers for the excellent spectroscopic information contained in the images. In this case, the information came from the fact that terahertz waves are sensitive to water molecules and are absorbed by them, while they are transparent to dielectrics such as plastics because of the characteristic energies of molecules' rotational and vibrational motions in the terahertz region, as shown in Fig. 10.1, where the chemicals and biological molecules can be identified by their characteristic resonant peaks. The resonant energies coming from the bonds and motions of molecules [18] and the ability to identify molecules is the most unique and important feature of spectroscopy using terahertz waves, which is a feature that electromagnetic waves in other spectral ranges rarely possess, as shown in Fig. 10.1. This advantage has been exploited for the detection of explosives [19] and illicit drugs [20] as well as for biological molecules. It also provides a reasonably good spatial resolution of less than a few hundred microns, and this diffraction-limited resolution is maintained even in biological cells and tissues because the dimensions of cells are comparable to or smaller than the wavelength of terahertz waves. However, visible- or infrared-light cannot have such a resolution due to Rayleigh scattering by the cells. These useful features of terahertz waves have been exploited in the imaging of human skin cancer [21].

Thereafter, many researchers realized the potential of terahertz radiation in medical applications and investigated it further to assess its success in one of the major medical imaging modalities using human breast and oral cancers [21, 22]. Furthermore, the developments in terahertz imaging for medical diagnosis led to the realization of a terahertz molecular imaging technique employing nanoparticle probes which dramatically improve the sensitivity of measurements [23, 24]. Furthermore, the tomography technique is a non-invasive measurement which is used to obtain information in three dimensions using sequences of transmitting 2D images. As the terahertz waves easily pass through non-metallic and non-polar materials, the tomography images of objects composed of these materials simply show the inner regions of such objects. There are two types of tomography techniques: (1) based on terahertz diffraction tomography (DT) and (2) based on terahertz computed tomography (CT). The scattered or transmitted terahertz images are measured by rotating the sample to a different projection angle and then the images are reconstructed as 3D images. The terahertz Fresnel lens tomography and terahertz holography are introduced in the text later as novel techniques. The variation of the focal length of the Fresnel lens relies on the frequency. It is possible to yield the images on the same image plane for a sample but from a different position after applying the Fresnel lens in the diffraction topography system. The terahertz holography technique differs from conventional holography because it can record the amplitude and phase information without splitting the source beam into reference and sample beams and a 3D image can be achieved with it by using a simple system. Furthermore, digital holography methods have to be used because the terahertz waves are in the invisible range.

This chapter briefly reviews the significance of terahertz waves for imaging and sensing applications. Furthermore, the background for biomedical imaging and sensing applications as well as how terahertz waves interact with biomolecules are explored. Additionally, various biomedical imaging techniques are compared, considering different potential parameters. Moreover, we discussed the millimetre and terahertz biosensors and their sensing techniques with potential scopes. Finally, we present a summary of the chapter.

10.2 Applications of Terahertz Imaging

In the field of non-destructive and contact-free terahertz imaging and sensing, terahertz spectroscopy has great potential [25, 26]. The continuing advances in the development of terahertz systems, as well as the appearance of the first related commercial products, indicate that the introduction of terahertz systems is rapidly approaching the large-scale market. Selected industrial applications for terahertz systems, comprising parallel monitoring of compounding processes, plastic weld joint inspection, birefringence analysis of fibre-reinforced components, water distribution monitoring in polymers and plants, as well as quality inspection of food products employing both continuous-wave (CW) and pulsed terahertz systems are well explored. Because of the ability to non-destructively pass through non-polar solid objects and to produce images of the interiors of objects, terahertz electromagnetic waves have attracted increased attention from different research groups [26]. Furthermore, their non ionizing features have been explored for use in security screening [27] and biomedical [28] applications. The terahertz signal is strongly absorbed by intermolecular bonds such as the hydrogen bonds present in water and the N–H bonds present in proteins, revealing that the imaging systems are sensitive to the slight changes involving increased water content—for example, blood flow—which can arise due to the presence of diseased tissue [39]. Therefore, using terahertz signals in medical imaging applications for both screening and diagnostic purposes has been explored, resulting in the development of suitable technologies and procedures. Recent advances in the technologies for generation, manipulation and detection of terahertz radiation have been revolutionizing the field and motivating the exploration of a large number of possible applications of terahertz imaging and sensing, communication and spectroscopy. Moreover, the applications of terahertz imaging techniques are very diverse, as shown in Fig. 10.2, and several potential fields for their imaging applications are as follows.

Biomedical Because terahertz radiation does not damage the material used for imaging and sensing, terahertz imaging and sensing is a useful tool in the area of biomedical applications such as diagnosis of breast cancer and skin cancer, monitoring of a tablet film coating [30], drug testing, and terahertz endoscope

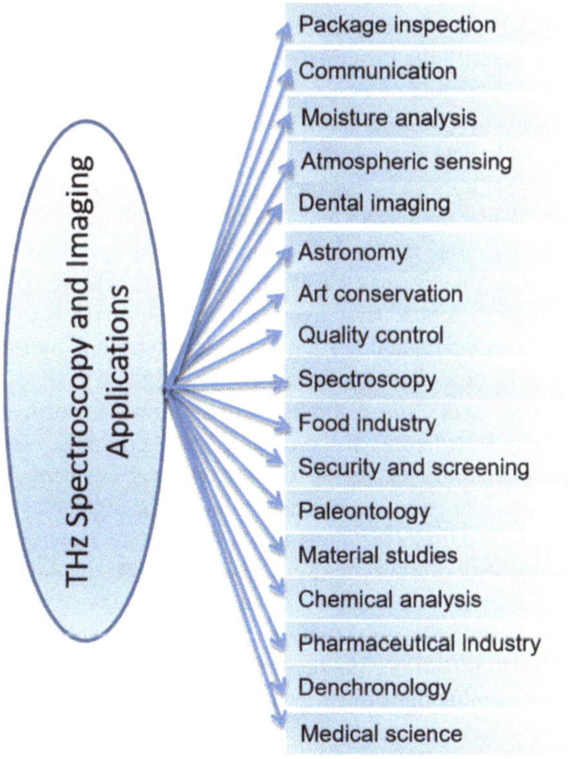

Fig. 10.2 Terahertz spectroscopy and imaging applications

medical diagnostics [31]. The resolution of a terahertz image is comparable to that obtained through magnetic resonance imaging (MRI); however, the equipment that is required is relatively simple. The analysis of change in terahertz optical constants, as well as change in cell structure, enables the detection of the disease-tissue. Frozen tissue from the human brain [32], porcine tissue [33] and rabbit liver tumour tissue [34] all presented different terahertz amplitudes for different cell types. The pharmaceutical applications of terahertz imaging have been developed to characterize the crystalline performance of pharmaceutical medications and solid drugs [35]. The evaluation of coating thickness and uniformity in coated pharmaceutical tablets is included to measure and image the terahertz chemical properties [32]. A mixture of multiple drugs has been verified by the chemical components of drug spectral characteristics in the terahertz range. Furthermore, a miniaturized terahertz device has been developed for imaging the organs within the human body, and this terahertz endoscope system is based on the optical fibre technology which measures terahertz absorption as well as the refractive index of the sidewall of the digestive organ [36].

Because terahertz phase imaging provides many exciting new imaging modalities, it is rapidly being applied for (a) biomedical use [37], (b) food quality control [38] and (c) metrology/morphology [39]. The water content and distribution in bio-tissues is an important factor in determining and monitoring biological function. It can not only be associated with the differences between healthy and diseased tissues but also be used to identify the locations of fat and muscle. Terahertz radiation is a viable tool for water content measurement due to its high sensitivity to water content. Wang et al. [40] theoretically analysed the errors in estimating water concentrations in biological tissues. In order to reduce the effect on the results of any error in the sample thickness, thicker samples are preferred. Wallace et al. [41] used image Basal cell carcinoma ex vivo and in vivo in a clinical setting. The system had a usable frequency range of 0.1–3 THz with an average power of approximately 100 nW. The time-domain terahertz waveforms have been achieved by sweeping the optical delay through the entire terahertz pulse at a rate of 15 Hz. The difference between basal cell carcinoma and normal tissue is determined by terahertz pulsed imaging which has the potential to delineate tumour edges. The diseased tissue had differentiable terahertz properties compared with those of the normal tissue, which manifested itself as a broadening of the reflected terahertz pulse. Breast cancer is one of the most commonly occurring diseases among women. As a result of technological advances in early breast cancer detection, segmental mastectomy is increasingly common and limits the total breast removal rate. However, the precision in accurately delineated tumour margins remains weak, leading to secondary surgery. In fact, following histopathologic examination of excised tissues, up to 20% of patients are shown to have tumours at the margins [42]. Terahertz phase imaging can be used to investigate the potential of breast tissue diagnosis. Terahertz phase imaging has been demonstrated to have the potential to aid in the accurate identification of the locations and the removal of breast cancer intra-operatively by studying the ex vivo breast cancer [43].

Security and Military: The terahertz imaging technique has been used to detect materials hidden within visible–opaque packing materials such as paper and plastic bags and clothes and to assist the people involved in the Food and Drug Administration (FDA) in detecting harmful drugs or dangerous weapons at different public places. Security applications at public places, such as airports, already use terahertz imaging and sensing techniques for security applications at 500 GHz by employing arrays of Schottky barrier diodes with silicon photonic bandgap crystal and heterodyne detection. The terahertz imaging with a tunable terahertz wave source has been used to develop non-destructive inspection techniques for detection and identification of illegal drugs hidden in mail envelopes, passengers and baggage [20]. Explosives and narcotics have their fingerprints in the terahertz regime of the spectrum; thus, the ability of terahertz imaging to identify these chemicals is valuable for applications in security-related fields [44, 45]. Furthermore, non-destructive 3D chemical mapping and identification of different chemicals buried in the 3D matrix can be performed using terahertz pulsed imaging techniques [46].

Food Inspection: For the monitoring and inspection of food quality, detection of pesticides and antibiotics, an appreciable amount of research has been conducted using terahertz imaging and sensing techniques. The water content in food is especially employed for detection because the terahertz absorption is significantly high for polar liquids [47]. The moisture content in dried food such as pecans [48], crushed wheat grain [49] and oil [50] has been explored. By employing different absorption fingerprints, pesticides [51] and antibiotics [52] have been detected, and metallic and non-metallic foreign bodies [53] have been detected in the food matrix. The strong analytic ability offered by terahertz imaging has been applied to detect materials hidden in visible–opaque materials such as carbohydrate foods and paper and plastic bags [54].

Semiconductor Industry: The non-contact electrical probe terahertz imaging systems can achieve fast scanning through real-time imaging and be used to perform the imaging of arrays devised on large areas in general, and approximately in the millimetre to centimetre range. By employing the transmission or reflection terahertz imaging configuration without the need for any contact or pre-processes, the electrical properties of semiconductor devices such as mobility, conductivity, carrier density and plasma oscillations can be precisely characterized. Terahertz imaging can be successfully used to measure the electron density without needing to contact the samples [17, 55]. Thus, this technology is expected to help improve the yield rates of semiconductor devices through their inspection, such as probing of resistivity or disconnected metal lines within the circuits. The terahertz imaging technique is also employed to enhance the efficiency or local overload of units or arrays of devices by mapping the heat distributions [56]. Additionally, the terahertz near-field microscopy on semiconductor transistors allows accessing the carrier concentration and mobility with a resolution of 40 nm [57].

Industrial Applications: Terahertz imaging has been explored for industrial applications with a focus on imaging considering some single point measurement scenarios. Some illuminating scenarios comprise examples from the polymer processing industry, the food industry and plant breeding. We only discuss a small excerpt of the emerging terahertz applications here, and for additional information, we refer the reader to the excellent reviews reported in [58]. The polymer processing industry is continuously investigating new process control and non-destructive testing (NDT) techniques to refine production techniques and open new markets for the plastics industry, as many challenging measurement tasks of the production lines remain unanswered. The polymers and terahertz waves form a fruitful symbiosis, as many plastic materials are transparent to terahertz radiation and serve as a base material for passive terahertz system components [59]. The terahertz imaging systems reveal valuable information regarding the composition of polymers and the inner structure of polymeric components. The potential applications of terahertz technology occur throughout the value creation chain, starting with offline measurements for the characterization of molecular properties, considering the following inline process monitoring with real-time feedback, and referring finally to quality inspection of the polymeric components. To fully understand the potential of terahertz sensors for polymer testing applications, a closer look at the underlying physics is mandatory. The most prominent interactions between terahertz waves and polymers are collective motions of large molecules, which appear as broad absorption bands. Some far-infrared absorption spectroscopy studies have already investigated skeletal vibrations, liquid lattice modes, hydrogen bonds and intermolecular vibrations in the crystalline phase [60]. However, the terahertz TDS systems have a clear advantage here, as the simultaneous extraction of the refractive index, the absorption coefficient and the sample thickness [61] offer information that complements and exceeds the far-infrared absorption data. Exemplarily, we will discuss the contribution that THz TDS can provide to the identification of the glass transition temperature. The glass transition temperature in amorphous or semi-crystalline polymers often limits their operating temperature range. The terahertz image based on the phase delay clearly reveals the presence of contamination. In the same manner, we have demonstrated the detection of plastic inclusions and small stones. THz TDS can even differentiate between intended ingredients, such as nuts, and dangerous contaminations so that it holds the potential to become an established method in the quality inspection of food products. However, before that vision can become a reality, the imaging speed has to be increased, so that close to real-time images can be obtained.

Other Applications: For imaging in astronomy, a sensitive heterodyne spectroscopic technique has been implemented in the terahertz regime of the spectrum to map remote objects. These spectroscopic techniques are becoming more important for the detection of thermal radiation in planetary/astronomical sciences, investigations of the dynamics of astronomical objects and of the planetary atmosphere [62], the distribution of gas molecules in the stratosphere, etc. Both passive imaging, in which thermal radiation is detected without additional illumination using a thermal detector, and the sensing of reflected/scattered radiation with illumination by a narrowband source are very dependent on the sensitivity of thermal detectors. Therefore, the development of low-noise thermal receivers that effectively operate in the terahertz region is a top priority in applications that use terahertz imaging to detect thermal radiation from targets. An important point in terahertz imaging is that the resolution of the image is fundamentally limited by the diffraction of the wavelength, similar to what occurs in other conventional imaging systems. The resolution remains challenging for medical applications because, in the conventional systems, the wavelength of the components corresponds to the range of a few hundred microns. Furthermore, a near-field imaging technique has been explored to overcome the diffraction limits, which improves the resolution limit up to a value of $\lambda/4$. Recently, a near-field microscopy technique was used to achieve a resolution of $\lambda/3000$ at 2.54 THz [57] for a single frequency and of the order of $\lambda/600$ or a broadband system [63]. In the near future, these advanced techniques will be used in various fields, ranging from theoretical studies to biomedical applications.

A plethora of applications for terahertz systems has been introduced in the literature [64], ranging from biological/medical imaging, non-destructive testing (NDT), security scanning and process control to next-generation wireless communication systems [65]. The enormous inherent potential of terahertz technology led to a rapid development of terahertz systems, and, today, the first commercial products are available. While measurement speeds still require improvement and system costs are relatively high, terahertz systems are on the brink of large-scale market introduction. Among the many realizations of terahertz systems, two basic principles can be differentiated: either electromagnetic picosecond pulses, which have spectral components ranging from approximately 100 GHz to several terahertz, are generated, or CW radiation is employed [66]. While the former method offers the advantage of broad spectral information from a single scan, CW techniques are the first choice when sharp spectral features are of interest and a frequency resolution down to a few MHz is desired [64]. The pulsed terahertz systems have recently overcome a major drawback, namely high system costs. Several applications in NDT require imaging techniques. Methods such as X-ray or ultrasonic testing employing detector arrays generate images in or close to real-time. Conventional terahertz time-domain spectroscopy (THz-TDS) systems, as described above, are designed to generate highly precise spectroscopic information at a single point of a sample. Thus, to generate full images at high speeds, several aspects must be taken into consideration. Among the concepts for terahertz imaging, several noncoherent techniques, including microbolometer arrays have been presented in [67]. While increasing the measurement speeds, these approaches give only limited information due to the lack of phase information. The most basic coherent imaging can be achieved by raster scanning a sample through the THz focus and generating full spectroscopic information at each pixel. Considering a small image of 60 by 60 pixels, 3600 single measurements are required. Depending on the delay line concept, the desired lock-in time constants and other factors, recording a high-quality THz pulse can take longer than 30 s and can result in a total measurement time of 30 h for a full image. This example shows the necessity of parallel or accelerated measurements. A straightforward approach to parallelization is a multiantenna setup enabling a linear downscaling of the measurement time with the number of employed emitter and detector pairs. However, arrays have to be developed to counter the drastically increasing costs, for example, as demonstrated in [26, 68].

10.3 Background for Medical Imaging Applications

Recent technological progress in the development of terahertz instrumentation means that the commercial terahertz systems being produced have improved performance, making them easier to operate and more reliable. Terahertz phase imaging has recently received significant attention as an advanced imaging technology which combines the benefits of terahertz and phase imaging techniques. Wan et al. [69] reviewed the methodologies employed to capture and process terahertz phase images with current imaging systems using both pulsed- and continuous-wave imaging systems. Pulsed terahertz phase imaging is a coherent measurement that includes terahertz pulsed imaging (TPI) based on a femtosecond laser and holographic imaging in the time domain to allow the phase and amplitude information of the electric field to be recorded. The CW terahertz phase imaging is mainly based on digital holography, interferometry and ptychography. These systems can obtain the complex amplitude by capturing diffraction patterns and applying numerical reconstruction techniques. The biomedical applications of such terahertz systems are well highlighted in [69].

Extensive research has been conducted to investigate terahertz phase imaging systems and to study their biomedical applications. We concentrate on continuous-wave terahertz phase imaging, including different holographic interferometry systems, for example, the Mach-Zehnder, Michelson and Fabry-Perot interferometers. Several methods are explored to enhance CW terahertz image resolution and reconstruction. The longer wavelength terahertz radiation exhibits strong penetration, but the sources only provide lower output power. However, the optical terahertz sources (shorter wavelengths) provide high output power, but the radiation is more weakly penetrating. As examples, such sources and their drawbacks are QCL, which operate at low temperatures, and optical pump terahertz lasers, whose output power is unstable. Furthermore, challenges also exist regarding the development of terahertz array detectors with high sensitivity, large dynamic range and a large number of pixels. The lack of suitable beam shaping components to control the profile of the object illumination beam affects the image quality. The lack of terahertz imaging lenses with large numerical aperture makes it difficult to collect high-frequency components and thus reduces resolution. Therefore, it is necessary to further develop high-performance terahertz sources, detectors and imaging devices, which are the foundation and prerequisite for future work. Mait et al. [70] and Barbastathis et al. [71] explored two interrelated directions of future development which will involve the modern super resolution techniques of computational imaging and artificial intelligence in terahertz imaging. Further work will allow the establishment of the relationship between phase information and object morphology or internal structure. The work reported here represents only a small selection of recent terahertz phase imaging studies and advances. There are many more terahertz imaging approaches such as near field [72, 73], computed tomography [74, 75] and optical coherence tomography [76, 77] are proposed by researchers, respectively. Krid et al. [78] designed a reconfigurable bandwidth based on a concentric ring slot antenna using graphene which provides a tunable result in terms of bandwidth which varies between 255 GHz and 406 GHz. The use of graphene in terahertz has illustrated better performance than that of the metal, and the variation in chemical potential of graphene provides excellent performance properties such as a reconfiguration of the bandwidth with a significant change in the performance of the antenna. These results are promising for terahertz applications and particularly for the application of medical imaging. Poorgholam-Khanjari et al. [79] presented a compact and improved bandwidth planar Quasi Yagi-Uda antenna in the terahertz regime of the spectrum using the periodic ground layer and parasitic metamaterial elements. The periodic ground plane (grating structure) is used to dispense the current of the antenna to attain higher impedance matching, and two concentric rectangular parasitic loops are used to improve it. The ground plane used also led to an adjusted bandwidth and size reduction. In this structure, a dipole driver is exploited on both sides of a transparent glass with a refractive index of 1.95 and thickness of 1 μm as the substrate, and a microstrip-line is utilized for feeding the dipole element. The structure is optimized by parametric studies, and parameters for the best performance are presented for achieving higher bandwidth. We consolidated this antenna size more than 42.2% in comparison to the normal antenna, and the bandwidth reaches more than 27% at 1.84–2.42 THz. Thus, this antenna can be used for various applications in the terahertz spectrum, such as for medical imaging and future generation communication as well as systems for early skin cancer detection. Terahertz imaging is a truly interdisciplinary area of research and is still in its early stages of development. The common use of any system will depend on its price and robustness, particularly in the terahertz imaging and sensing equipment. To reduce the costs and expand the range of applications of terahertz technology will require the close cooperation and collaboration of physicists, biologist, engineers and medics. The schematics of terahertz imaging for biomedical applications are presented in Fig. 10.3.

10.4 Influence of Radiation on Biomolecules

Owing to the sensitivity of terahertz radiation absorption to the water content in tissue, as reported by Hu and Nuss [17], extensive spread time-domain spectroscopy and pulsed terahertz imaging systems for biomedical applications have emerged. The degree of hydration of tissue could be used as a measure of disease state [17]. A biomedical diagnosis is based on both structural and functional data, as well as

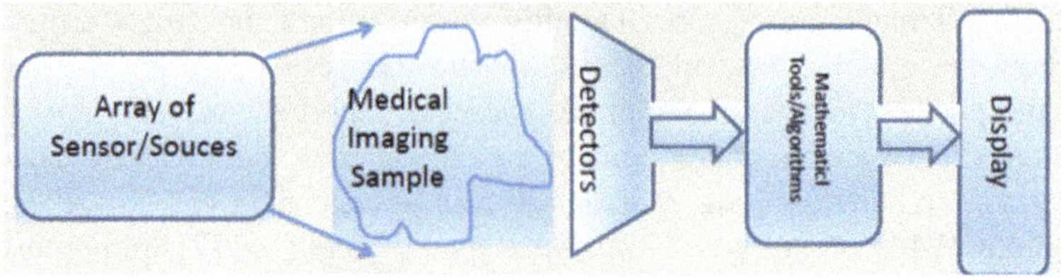

(a)

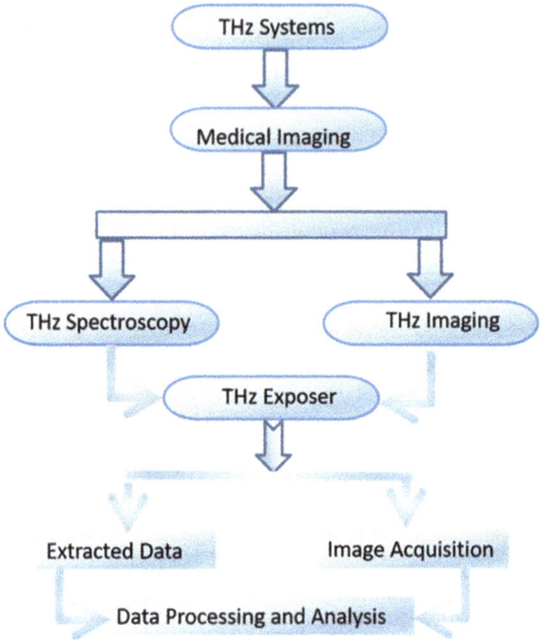

(b)

Fig. 10.3 Schematic of terahertz systems for medical imaging applications (**a**) general and (**b**) advanced

patient history and subjective symptoms. The structural imaging methods include X-ray, ultrasound imaging, MRI (magnetic resonance imaging), and classical and electron microscopy. Medical diagnostic techniques using infrared waves are functional, passive and non-invasive methods for analysing physiological functions related to body thermal homeostasis or organ temperature; therefore, their results cannot be directly compared to the results obtained by structural imaging methods. It is used to detect and locate temperature distribution characterized by a non-physiological increase or decrease in temperature at the body surface. Currently, the improved hardware detection systems with additional advanced software solutions make it possible to incorporate atomical and physiological information by image fusion, which helps to generate information on affected areas. Terahertz radiation has a variety of features which make it a viable spectroscopic imaging technique.

- Terahertz radiation has low photon energy and matches with the rotational- and vibrational-molecule modes as well as intermolecular vibrations energy levels. Thus, these low-frequency motions allow the identification of biomolecules, characterizing their spectral features in the terahertz region [29, 80].
- The low photon energy is insufficient to cause atoms ionization. Therefore, it is suitable as well as attractive for non-invasive biomedical imaging and has a great potentiality which could be applied to in vivo real-time studies without causing biological ionizing damage [81, 82].
- Similar to the extremely low-frequency electromagnetic fields and their possible human health effects, the effects of terahertz radiation are of interest because of the expanding terahertz technologies in biomedical applications if the thermal effects are not considered, it does not directly damage the DNA, but it could act as a co-inductor for non-thermal effects [83].
- Recent studies provide evidence of potential thermal and non-thermal effects induced by exposure to terahertz radiation [84, 85].
- The biological response to this stimulus is due to different parameters: (1) the physical settings of the terahertz radiation-like frequency, mean and peak power, radiation intensity, continuous or pulsed radiation, the degree of coherence, etc., (2) the physical and biological properties of the irradiated biological object (refractive index, absorption features, type of cells, tissues, organs) and (3) the design of the exposure [85].
- Terahertz radiation is very sensitive to polar molecules, and strong absorption will take place, for example, with water [82], limiting terahertz penetrability into the fresh tissues from tens to hundreds of microns [35] and the capability of diagnostic only for superficial layers in vivo.
- The high sensitivity of water content can be used as an endogenous marker for the differentiation between healthy and pathological tissues [81, 82].
- In particular, time-domain spectroscopy (TDS) and imaging are insensitive to the thermal background and show a high signal-to-noise ratio (SNR) [86].
- Terahertz pulse imaging coherent detection ensures the record of the temporal profile of the terahertz electric field; thus, both the terahertz amplitude and phase can be estimated and based on this information, and the broadband absorption coefficient and refractive index of a medium are determined without using Kramers–Kronig relations [86]. In general, the absorption is due to the chemical constituents of the material, and its spectral signatures can be measured, thus providing useful information. As a consequence, TPI has the potential to realize images with both morphological and functional information in the terahertz frequency range. For all these reasons, terahertz radiation is considered very important and deserves special attention.

Although various technological improvements have been made to facilitate biomedical applications of this powerful analytical technique, it has not yet properly come into the scientific background of many potential end users. Therefore, to achieve these fundamental objectives, an interdisciplinary approach is needed with scientists, spectroscopists, biologists and clinicians who must effectively communicate and understand each other's requirements and challenges [87]. The terahertz wave is ineffective to ionize atoms and molecules as well as drive cancer and genetic mutations due to low photon energy. However, thermal effects that arise owing to temperature variation during conditions under the powerful terahertz radiation cause other health effects, as reported in [88, 89]. Therefore, research regarding terahertz radiation effects on living organisms to determine the ultimate dosage and exposure time of terahertz/millimetre waves is important for the evaluation of the transient effects on health and for the elaboration of safety standards.

10.5 Comparison of Different Medical Imaging Techniques

The medical imaging techniques are considered as a tool for learning/diagnosis of significant information about neurobiology and human behaviours. The fundamental concept of a medical imaging system is shown in Fig. 10.3a, and it consists of an array of sensors/sources of energy which pass through/penetrate the human body/living things and are absorbed or attenuated at differing levels, according to the density and atomic number of the different tissues, creating signals. These signals are detected by special detectors compatible with the energy source, and then mathematically manipulated to create an image. The obtained images indicate the energy in the human tissue, leading to a classification based on the energy applied to the body. Kasban et al. [90] reviewed different medical imaging techniques on the basis of their concepts, advantages/disadvantages and applications. Various medical imaging techniques have been compared considering the imaging features such as: (1) the image quality which is represented by the spatial resolution with better contrast, (2) the system available that can be represented by the system cost and the availability of real-time information and (3) the safety that can be represented by the effect of ionizing radiation on the patient, and the effect of heating on the body. The comparison of different medical imaging techniques is summarized in Table 10.1. According to the energy sources, many different techniques can be used to yield a look inside the patient. These techniques are based on a signal passed through the patient and its interaction with the tissues of the patient. By detecting the signal coming out of the body/sample by transmission or reflection, an image of the inside/sample of the patient can be formed. The interesting techniques for imaging are X-ray radiography, X-ray computed tomography (CT), magnetic resonance imaging (MRI), ultrasonography, elastography, optical imaging, radionuclide imaging (scintigraphy, positron emission tomography (PET), single photon emission computed tomography (SPECT)), thermography and terahertz imaging. The global diagnostic imaging market is projected to reach USD 33.5 billion by 2024 from USD 25.7 billion in 2019, at a CAGR of 5.5% from 2019 to 2024. The growth of this market is primarily driven by the increasing demand for early disease diagnosis and widening scope of clinical applications, rapidly growing geriatric population and subsequent increase in the prevalence of associated diseases, technological advancements in the diagnostic imaging industry and increasing investments, funds and grants by public–private organizations [91].

Terahertz radiation frequencies have unique imaging capability due to their high absorption in water; however, reflective terahertz imaging has distinct advantages over that of the transmission-based systems. In terahertz imaging, the probe beam interacts with the sample, and the detection signal is obtained by combining the probe laser with the terahertz radiation. The image of the sample can be built up due to the selective absorption of the terahertz radiation. The detector receives the signals, and by scanning the sample, an image can be formed [92, 93]. Terahertz imaging for medical applications are employed to detect cancer cells [94], imaging the tooth crown, in medical imaging experiments on different in vitro and some in vivo biological tissues and has illustrated promising results in differentiating certain features in tissues of concern [95]. Furthermore, it detects certain polar molecules due to these molecules' fingerprints in the terahertz range of frequencies of tissues. The potential features of terahertz medical imaging are: (1) safe and non-ionizing radiation, (2) uniquely sensitive to the vibration modes of water and (3) the terahertz radiation can penetrate many materials due to the long wavelength of the terahertz photons. In addition to this potential feature, terahertz medical imaging techniques have the following challenges. (1) Terahertz detectors are characterized by poor signal-to-noise ratio and slow processing, (2) the emitters produce only incoherent and low-brightness terahertz radiation [92], (3) terahertz sources require

Table 10.1 A comparison between the different biomedical imaging techniques

	Image quality		System availability		Safety	
Imaging technique	Spatial resolution	Good contrast	Cost	Real time information	Ionizing radiation effect	Heating effect
Radiography (X-ray)	1 mm	Soft tissue and fluid	Medium	No	Yes (active)	Low
CT (X-ray)	0.5 mm	Hard and soft tissues	High	No	Yes (active)	Low
MRI (magnetic and radio frequency fields)	0.5 mm	Hard and soft tissues	High	No	No (active)	Medium
Ultrasonography	1 mm	Soft tissues	Low	Yes	No (active)	Negligible
Elastography	200 μm	Soft tissues	Medium	Yes	No (active)	Low
Optical	100 nm	Soft tissues	Low	No	No (active)	Medium
Radionuclide	3 mm	Soft tissues	High	No	Yes (active)	Medium
IR (Thermography)	1.5 mm	Soft tissues	Low	No	No (passive)	No (passive)
THz	~50–1000 μm (dependent on wavelength)[2]	Soft tissues	High/ Medium (TDS/ CW)	No	No (active)	Dependent on radiation power

cryogenic operating temperatures [96], (4) low contrast between healthy and pathological tissues and (5) poor source performance [97]. The US Federal Communications Commission established maximum permissible exposure limits of 1 mW/cm^2 for 6 min in the 30–300 GHz frequency range [98].

The terahertz sources have been classified into two major classes: (1) continuous-wave sources and (2) pulsed wave sources. The CW sources typically operate at a single frequency (narrowband) with variable or tunable terahertz source wavelength. In general, the CW sources can generate higher output power (>10 mW) than that of the pulsed sources (~μW), for example, the quantum cascade lasers (QCL) [99], backward-wave oscillators (BWO) [100] and optically pumped terahertz lasers (OPTL) [101]. On the basis of terahertz sources, the terahertz imaging systems can be classified either as a pulsed terahertz imaging system or as a continuous-wave (CW) terahertz imaging system. In the pulsed terahertz imaging systems, the signal is generated and detected using the femtosecond laser-based technique and is characterized by a broadband signal in the frequency domain. The pulsed terahertz imaging has been developed for time-of-flight imaging, which can provide the tomographic information and one-shot imaging for video-rate real-time imaging. Unlike the pulsed terahertz imaging, the CW terahertz imaging is characterized by a fixed or limited tunable frequency range and does not provide information regarding the phase. Despite these shortcomings, the CW terahertz imaging has the potential for application because it provides fast detection, simple analysis, a compact size and a relatively low cost.

Several schematics of terahertz continuous-wave instrumentation also exist in addition to the pulsed systems. A single fixed radiation frequency in narrowband sources can be used in the typical CW system. Furthermore, the broadband radiation can be received with photo mixers in which two lasers emit in two adjacent spectral ranges in visible or near-infrared regions. The photo mixer translates the laser beams of two lasers into terahertz radiation, and for this kind of CW systems, the terahertz radiation power does not exceed 2 W independent of the radiation frequency or semiconductor material used as a photo mixer. The CW imaging allows for more compact and simple systems; however, the pulsed imaging measurements yield a broader range of information. In addition to the potential capabilities at the moment, the terahertz imaging is limited to a great degree as it could distinguish the tumours and normal tissues, serious inflamed and normal livers, but could not distinguish the grade of tumours and inflammation [102]. However, the terahertz waves do not penetrate far into the body, which is one of the main drawbacks, and it would be possible to project terahertz endoscopically such as in optical coherence tomography. For example, Nakajima et al. [103] have reported a terahertz radiation technique for the diagnosis of cancerous tissues, as they exhibit different hydration levels from normal tissues. The terahertz spectra of tissues with a high-water content follow a similar trend as that for pure water, but the absorption is not as high [104]. The water absorption limits the depth of tissue that may be imaged using terahertz to a few hundred micrometres in the skin and up to several millimetres in tissue with a fatter content such as the breast. Owing to the strong absorption by water in different applications such as biomedical and food, the terahertz imaging and spectroscopic systems can be applied for mapping tumour margins or the surface of food products but not with a great depth [43, 105]. Moreover, the frequency dependency of the absorption coefficient on salt, protein and DNA content is also observed. Furthermore, terahertz water absorption depends on protein structural changes, such as ligand binding or denaturing [104]. Among the terahertz imaging and spectroscopy systems for these purposes, systems such as the terahertz pulse imaging, terahertz time-domain spectroscopy and continuous-wave photo mixer systems are mainly used. Presently, terahertz spectroscopy and imaging used in 0.1–4 THz require femtosecond lasers and photoconductive broadband antenna sources and detectors for generation and detection of terahertz radiation, respectively. However, they are the most expensive components of the systems, and this is a reason for higher healthcare cost compared to other methods. Unlike the pulsed terahertz imaging, the CW imaging presented only yields intensity data and does not provide any depth, frequency-domain or time-domain information about the subject when a fixed-frequency source and a single detector or arrays of detectors are used. However, the CW imaging systems are sufficient for many imaging applications affording a compact, simple, fast and relatively low-cost system. Because it does not require a pump-probe system and the optics of CW systems are much simpler, their cost can be considerably reduced compared to that of the pulsed terahertz systems, and since it does not require a time delay scan, image formation can take place more quickly. Compared to infrared and microwave systems, the terahertz imaging, spectroscopy and communication systems and their components remain less developed. For this reason, at the moment, the terahertz wave instrument capabilities remain lacking in comparison with infrared or microwave system feasibilities.

The terahertz time-domain spectroscopy systems are based on the generation and detection of pulses which are of a few picoseconds in duration. Traditionally, the time-domain systems used femtosecond lasers to pump a photoconductive antenna or electro-optical crystals to generate the output terahertz pulses [106]. The ability to accurately detect terahertz radiation is at the heart of terahertz technology. For the pulsed terahertz systems, there are two commonly employed detection methods. In the case of photoconductive sampling, the systems based on an above-band-gap excitation in a semiconductor [3] are used. Typically, the fast photoconductors, such as radiation-damaged silicon or low-temperature-grown GaAs are used to provide the necessary sub-picosecond sampling resolution. Additionally, some other detectors are based on free-space electro-optic sampling, in which the terahertz field induces birefringence in an electro-optic medium [107]. This birefringence is sampled using an optical probe beam as a function of the delay between the probe pulse and terahertz pulse. The resulting sampling resolution is determined largely by the duration of the optical probe pulse.

The CW terahertz imaging system is the basic imaging system and additional information can be extracted from a CW system with a more complicated arrangement. The CW terahertz imaging technique provided an apparently more useful result compared to terahertz pulsed imaging systems in the case of specific application as well as data information. This is mainly a result of the relative resolutions of the systems presented. Obviously, if the beam widths were comparable, a pulse energy plot of the time-domain data would closely match the intensity data from the CW system. All four voids are visible in the pulsed terahertz image, however, which means that in this case, the two techniques were equally effective. It is possible to focus the terahertz pulses to a smaller spot, characterized by sub-picosecond pulses, as in [63]. Additionally, when a defect is detected on the pulsed image, the waveform from that section can be inspected to verify it. A wider beam was used in this case for the ease of alignment, but higher resolutions can readily be achieved [108], limited by the central wavelength of the pulse via the Rayleigh criterion. Furthermore, the depth resolution of a pulsed system can be increased beyond the spatial resolution possible with a CW far-field system at a comparable wavelength [109]. These data demonstrate that simplicity is the key advantage of the CW imaging technique. Because the frequency spectrum of the source is narrow, a large F-number Fresnel lens can be used and thus easily increase the resolution by focussing to a small spot. The image data are simple to process, and the resulting images can be interpreted intuitively. The price of this simplicity is that it is not as replete with information as the pulsed terahertz data. While delaminations were visible in the CW data in this test, it is essential to note that the true delamination areas, the centres of the patches, are not distinguishable from the surrounding areas. While the time-domain system failed to identify those defects, with an improvement of its resolution and waveform processing algorithms, the possibility remains to have a delamination detection mechanism, whereas the CW system would require a complete change in the data measured in order to have a chance of detecting such defects. As mentioned earlier, an advantage of the pulsed terahertz imaging method is that the depth information can be recovered. This is encoded in the waveform mainly in changes in pulse timing. A change in the temporal location of the reflected peak indicates a change in the optical path length to the substrate, while the appearance of smaller pulses that are offset from the main peak indicates the existence of scattering centres or interfaces within the sample that are partially reflecting the pulse. The time delay between these secondary pulses and the main pulse can easily be converted to depth [109]. CW radiation is also scattered and reflected by these features, and it is possible through the use of additional detectors or by performing additional scans of the sample to recover similar information. By changing the angle of the detector relative to the beam, it is possible to measure scattered light instead of the specular reflection from the substrate. Different angles correspond to scattering and reflection at different locations within the substrate, and thus the scattered intensity can display the features of defects within each measured stratum of the sample. For the diffuse reflection images, the beam was reflected normal to the substrate by the planar mirror. It can be seen that the large, natural void at the bottom centre of the panel scatters strongly near the middle layer of the sample, but weakly at other depth locations. It can thereby be roughly located within the z-dimension of the sample. Such a technique obviously decreases the simplicity, which is the main advantage of the CW modality, and the depth resolution is limited by the number of detectors or the number of scans performed. In a pulsed system, the depth resolution can exceed the imaging plane resolution by more than an order of magnitude, but in the case of CW imaging in this mode, it can, at best, only approach the planar resolution. Additionally, the intensity of the scattered radiation measured at large angles, and thus the signal-to-noise ratio, is quite low compared to when the specularly reflected beam is measured. Other aspects of the pulsed data may be recovered in a CW system, such as phase information using interferometry and frequency-domain information using multiple or tunable sources if the attainable frequency range is sufficient [110], but narrowband sources are generally less suitable for broadband spectroscopic imaging applications because of time constraints, although for some spectroscopic applications, such as looking for a specific absorption feature, a narrow-linewidth CW source may provide a higher spectral resolution. In general, when one attempts to acquire information other than transmitted intensity and phase from the CW system, the information is more limited or more complicated to access than that available through pulsed time-domain measurements. Another limitation of the CW modality is the availability of high output power, compact sources at terahertz frequencies with room temperature operation. Current Gunn diode technology allows for frequency-multiplied sources up to 1 THz, with the output power decreasing rapidly as a function of increasing frequency. Future devices may break this barrier, allowing for higher spatial resolutions [111]. Moreover, the weight of pulsed imaging systems can be reduced significantly by using a more compact laser, such as a fibre laser. The authors of [111] have introduced the working of TPI and discuss its basic principles and performances, and the state-of-the-art applications for biomedical systems.

10.6 Terahertz Biosensor

A detection device which converts a biological response into a measurable signal is known as a biosensor, and recent biosensor developments have reaped huge benefits in the healthcare industry. Biosensors offer the ability to detect several types of cancers, vital signs such as heart rate, biomolecules such as DNA, proteins indicative of genetic disorders and glucose levels to monitor diabetes. Biomatter is considered a dielectric substance and exhibits conductivity on the application of an electric field. This dielectric property is measured in terms of dielectric permittivity [112], which is utilized for sensing, and this methodology is employed in electromagnetic wave biosensors starting at the radio frequency up to the optical range biosensors. The crux of electromagnetic wave sensing is observing the unique dielectric permittivity of the cells and tissues to identify the abnormalities or measuring the changes in dielectric permittivity to quantify the concentration of biomolecules whose presence in high quantities may imply an ailment or disorder. The millimetre and terahertz biosensors employ capacitive sensing, which means measuring dielectric permittivity in terms of capacitance or scattering parameters using interdigitated capacitors, resonators and microstrip structures. In the current scenario, the electromagnetic wave biosensors are attracting a lot of attention due to potential benefits such as being minimally invasive, label-free and cost-effective. The variation of dielectric permittivity as a function of frequency is known as dispersion. A sharp drop in dielectric permittivity occurs at relaxation frequency marked by a loss in electromagnetic energy, which occurs when a particular polarization fails to keep up with the fast-changing electric field. Furthermore, this dielectric permittivity spectrum provides a unique dielectric signature and can be utilized for sensing purposes. Biological targets can be biomolecules such as proteins, DNA, biomarkers, pathogenic organisms, hormones or other medically relevant analytes such as glucose and medical parameters such as pulse, heartbeat, etc. The biosensing applications include identification of these biological targets such as cancer cells and detection of biomarkers, vital signs (respiration and heart rate) and biomolecules. For instance, in the glucose biosensors, the change in concentration of glucose on the detection surface leads to the change in dielectric permittivity and shift in relaxation frequency which is detected. In the case of biomarker detection, the biomarkers are captured at the detection surface, leading to a change in dielectric permittivity, which is sensed. In general, the dielectric permittivity is different for different biological tissues and provides unique dielectric signatures. Dead and tumour [113] cells differ in their cytoplasmic contents from living and healthy cells due to the presence of pores and blisters [114], providing a dielectric permittivity contrast. By directly measuring the unique dielectric permittivity curves of cells and tissues or by indirectly measuring the changes in dielectric permittivity, we can perform the biosensing applications. Biosensors are becoming widely popular in the market due to their ability to perform a quick diagnostic test near the patient without the need for laboratory analysis. Furthermore, the millimetre/THz wave sensors are categorized into labelled and label-free, direct and indirect and in vivo and ex vivo sensors. We refer to 'direct' sensors as those sensors which provide intrinsic biomatter information by directly subjecting biomatter to the waves and observe the unique dielectric permittivity signatures of biomatter. On the other hand, we refer to 'indirect' sensors as those which measure the change in dielectric permittivity or optical properties and are based on immobilization techniques which require bioreceptors that are fixed on the sensor's surface and subsequently capture target analytes. This biorecognition event, due to affinity binding between the target and receptor, leads to a change in dielectric permittivity, which can be sensed. Moreover, the millimetre/THz wave biosensors can also be categorized into in vivo and ex vivo biosensors. If the biosensors are able to sense without needing a sample to flow through them, that is, they are able to detect abnormalities or vital signs in a living organism by simply keeping the sensor near the patient, they are known as in vivo biosensors. Alternatively, in the ex vivo sensors, the measurements relate to tissue/blood from an organism in an external environment with minimal alteration from environmental conditions. Several common applications of millimetre/THz wave sensors include detection of cancer cells [115], glucose [116], DNA [113], fibroblast cells [117], other biomolecules [116] and vital signs [118] and are discussed in this section in detail.

To understand the concept of affinity binding, first we discuss some common terminology. (a) *Analyte:* is a substance which is being measured or identified. (b) *Bioreceptor:* is a biomolecule that recognizes the target analyte in a sample solution. (c) *Immobilization:* is the process of binding bioreceptors to the sensor's surface. (d) *Biomarkers:* are the measurable chemicals in an organism whose presence indicates a disease, infection or environmental exposure. The sensors extensively employ this bio-recognition process, where the bioreceptors are immobilized on the sensor surface, and the analyte to be detected is allowed to flow across the detection region to selectively interact with the already immobilized biomolecules causing a change in the dielectric permittivity. The sensors are designed to be extremely sensitive to changes in the dielectric permittivity on their surface, which is measured. Apart from immobilization, there is an alternative of using microfluidic channels, which are small channels allowing passage of analyte over the detection area of the sensor, sensitive to permittivity change. In these sensors, generally, capacitance sensing is employed. Thus, due to the change in capacitance, the resonant frequency changes, which is sensed.

Currently, the terahertz biosensors are gaining a lot of popularity and deserve a special mention. The terahertz frequency range [71] is of relevance as it coincides with the vibrational frequency of some important biomolecules; thus, changes in the concentration of biomolecules can be viewed as a shift in resonant frequency in the dielectric spectrum. The advantages of these terahertz biosensors over other electromagnetic wave sensors are their improved sensitivity, detection limit, the volume of sample and cost. The terahertz frequency band is in accord with the vibrational and rotational energy levels of important biomolecules, for instance, cancer biomarkers, and thus its spectrum contains abundant information about their configuration and conformation. Studying protein–protein interactions is vital to gain insight into several natural processes occurring inside every living being. When the different molecules interact, their molecular conformation changes, changing the location as well as the intensity of the characteristic resonant peaks in the spectrum, thus enabling it to be probed [72]. This technique is promising as it offers label-free and non-contact inspection of biomolecules. The terahertz and millimetre wave biosensors, as shown in Table 10.2, employ similar techniques as that used in radiofrequency and microwave biosensors. Bolívar et al. [120], distinguished the hybridized and denatured DNA by comparing the difference in a resonant frequency shift in ssDNA. Further, Gen et al. [119] have proposed the use of metamaterial-based terahertz sensors due to their high sensitivity for detection of biomolecules. The split ring resonators integrated with microfluidics have been designed in [71] from 0.1 to 1 THz for detection of liver cancer biomarkers in early-stage, namely alpha-fetoprotein (AFP) and glutamine transferase isozymes II (GGT-II). Nagel et al. [122] demonstrated the terahertz spectroscopic analysis of genetic sequences, and a distinct refractive index is observed for each binding state of the DNA sequences.

10.7 Summary

The various applications discussed could benefit from the valuable insights the terahertz imaging systems offer. The CW and pulsed terahertz imaging techniques have unique advantages which need to be considered if they are to be optimally applied because of their relative complexity. In order to compete, the pulsed terahertz time-domain measurement systems must provide useful information which is not accessible in a CW system. If the information is imprecise or too difficult to interpret, the advantages offered by a time-domain system quickly disappear. However, there are applications where the information provided by a CW system is simply inadequate, in which case one has no choice but to use a time-domain system. The simplicity of a CW system evaporates as one attempts to extract more information from it. Thus, a key advantage of the pulsed terahertz modality is flexibility. All of this information is recorded with one measurement from one detector and can be reprocessed to precisely find the information required for each application. The contrast between the two modalities is between simplicity and adaptability. For many applications, the additional information is unnecessary, and a pulsed system has no advantage. In others, where such information is or may become useful, or in applications where imaging is inextricably tied to time-domain information, the pulsed terahertz modality cannot be forgone. The infrared and terahertz technology applications today are broad in domains such as astronomy, military and surveillance, telecommunications, security, energy and food control, but it seems one of the most topical is biomedical applications, for example, for breast and colon cancer diagnostics, burn imaging, etc., especially in the cases where there is a need to accurately locate tumour margins when conservation of normal tissue is required. It can be expected that the potentiality of these technologies will continue to progress in diverse directions in biological and medical fields. Further work is needed for the scientific challenge to provide information not available from other techniques. One of the main barriers in providing healthcare conclusions by using terahertz

Table 10.2 Millimetre and terahertz wave biosensors

Terahertz biosensors							
Geng et al. [119]	0.4–1 T	ex vivo	Label-free	Indirect	Resonator using microfluidic channels	S-parameters/resonant frequency shift	Liver cancer biomarker detection
Haring Bolívar et al. [120]	0.3–0.8 T	ex vivo	Label-free	Indirect	Planar resonator	S-parameter/resonant frequency shift	Detection of DNA hybridization
Millimetre wave biosensors							
Wu et al. [115]	40 G	ex vivo	Label-free	Direct	CPW transmission line	S-parameter/resonant frequency shift	Identifying cancer cells
Shaforost et al., [121]	35 G	ex vivo	Label-free	Direct	WGM resonator	Inverse quality factor and resonant frequency shift	Nanoliter liquid characterization
Salim and Lim [116]	G–THz	ex vivo	Label-free	Indirect	Resonator using affinity binding	S-parameter/resonant frequency shift	Biomolecular detection, glucose detection and hyperthermia treatment

technologies is the cost of terahertz pulsed imaging and spectroscopy instrumentation that is mostly related to the high cost of short-pulse lasers needed for their applications. The nature of the interaction of terahertz waves with biomatter provides tremendous opportunities in the field of biosensing. Biomatter behaves as a dielectric, and hence any changes are traced in terms of dielectric properties: permittivity and conductivity. The dielectric signatures (permittivity vs frequency) can be obtained in a broadband fashion (over a wide frequency range) or in cases when that broad spectrum of information is of little use, permittivity changes only over a narrowband of frequencies can be obtained. The choice of the frequency band is based on whether intracellular information is needed; if so, frequencies in the microwave region are suitable because most of the dipolar relaxations experienced by water in tissues occur in this region. When an alternating electric field is applied to a cell in a suspension medium, several dispersions are observed. In this chapter, we have summarized the optical, millimetre-wave and terahertz biosensor modalities. Sensing is realized in two ways: either by directly observing the intrinsic EM wave properties or indirectly by measuring the change in EM wave properties. In direct optical sensors, the backscattered light from the arteries can be used to obtain information about vital signs such as heart rate, whereas, in indirect optical sensors, change in concentration related to biochemical reaction is detected by observing changes in optical properties such as fluorescence, reflection intensity, photoluminescence, interference pattern or refractive index. A more popular technique is that of proximity coupling, alternatively known as near field sensing, where the periodic movement is sensed by the near field of the resonator as a variable capacitance. In indirect sensing, permittivity change is sensed. Here, mostly capacitive sensing is employed, whose capacitance is a function of permittivity. This variable capacitance is placed in an oscillator circuit and changes in resonant frequency and scattering parameters are detected. A large amount of research is in progress for the development of terahertz imagers and sensors for the healthcare industry. Commercializing terahertz imagers and sensors will be greatly beneficial because they will pave the way for non-invasive sensing and facilitate continuous monitoring applications.

References

1. C. Fattinger, D. Grischkowsky, Point source terahertz optics. Appl. Phys. Lett. **53**, 1480–1482 (1988)
2. C. Fattinger, D. Grischkowsky, Terahertz beams. Appl. Phys. Lett. **54**, 490–492 (1989)
3. M. van Exter, C. Fattinger, D. Grischkowsky, Terahertz time-domain spectroscopy of water vapor. Opt. Lett. **14**, 1128–1130 (1989)
4. J.-H. Son, J.V. Rudd, J. Whitaker, Noise characterization of a self-mode-locked Ti:sapphire laser. Opt. Lett. **17**, 733–735 (1992)
5. D.E. Spence, P.N. Kaean, W. Sibbett, 60-fsec pulse generation from a self-mode-locked Ti:sapphire laser. Opt. Lett. **16**, 42–44 (1991)

6. D.H. Auston, K.P. Cheung, J.A. Valdmanis, D.A. Kleinman, Cherenkov radiation from femtosecond optical pulses in electro-optic media. Phys. Rev. Lett. **53**, 1555/1–1555/3 (1984)
7. M. Ketchen, D. Grischkowsky, T. Chen, et al., Generation of subpicosecond electrical pulses on coplanar transmission lines. Appl. Phys. Lett. **48**, 751–753 (1986)
8. X.C. Zhang, X. Ma, Y. Jin, et al., Terahertz optical rectification from a nonlinear organic crystal. Appl. Phys. Lett. **61**, 3080–3082 (1992)
9. Q. Wu, X.C. Zhang, Free-space electro-optic sampling of terahertz beams. Appl. Phys. Lett. **67**, 3523–3525 (1995)
10. S. Mickan, D. Abbott, J. Munch, X.-C. Zhang, T. Van Doorn, Analysis of system trade-offs for terahertz imaging. Microelectron. J. **31**, 503–514 (2000)
11. D. Grischkowsky, S. Keiding, M. van Exter, C. Fattinger, Far-infrared time-domain spectroscopy with terahertz beams of dielectrics and semiconductors. J. Opt. Soc. Am. B **7**, 2006–2015 (1990)
12. J.-H. Son, T.B. Norris, J.F. Whitaker, Terahertz electromagnetic pulses as probes for transient velocity overshoot in GaAs and Si. J. Opt. Soc. Am. B **11**, 2519–2527 (1994)
13. H. Harde, N. Katzenellenbogen, D. Grischkowsky, Terahertz coherent transients from methyl chloride vapor. J. Opt. Soc. Am. B **11**, 1018–1030 (1994)
14. D. Abbott, X.-C. Zhang, Special issue on T-ray imaging, sensing, and retection. Proc. IEEE **95**(8), 1509–1513 (2007)
15. L. Ho, R. Müller, K.C. Gordon, P. Kleinebudde, M. Pepper, T. Rades, Y. Shen, P.F. Taday, J.A. Zeitler, Applications of terahertz pulsed imaging to sustained-release tablet film coating quality assessment and dissolution performance. J. Control. Release **127**(1), 79–87 (2008)
16. H.H. Mantsch, D. Naumann, Terahertz spectroscopy: The renaissance of far infrared spectroscopy. J. Mol. Struct. **964**(1–3), 1–4 (2010)
17. B. Hu, M. Nuss, Imaging with terahertz waves. Opt. Lett. **20**, 1716–1718 (1995)
18. B.M. Fischer, M. Walther, P.U. Jepsen, Far-infrared vibrational modes of DNA components studied by terahertz time-domain spectroscopy. Phys. Med. Biol. **47**, 3807–3814 (2002)
19. Y. Chen, H. Liu, Y. Deng, et al., Spectroscopic characterization of explosives in the far-infrared region, in *Terahertz for Military and Security Applications II*, vol. 5411, (International Society for Optics and Photonics, 2004), pp. 1–8
20. K. Kawase, Y. Ogawa, Y. Watanabe, H. Inoue, Non-destructive terahertz imaging of illicit drugs using spectral fingerprints. Opt. Express **11**, 2549–2554 (2003)
21. R.M. Woodward, B.E. Cole, V.P. Wallace, et al., Terahertz pulse imaging in reflection geometry of human skin cancer and skin tissue. Phys. Med. Biol. **47**, 3853–3863 (2002)
22. Y.C. Sim, J.Y. Park, K.-M. Ahn, C. Park, J.-H. Son, Terahertz imaging of excised oral cancer at frozen temperature. Biomed. Opt. Express **4**, 1413–1421 (2013)
23. S.J. Oh, J. Choi, I. Maeng, et al., Molecular imaging with terahertz waves. Opt. Express **19**, 4009–4016 (2011)
24. J.-H. Son, Principle and applications of terahertz molecular imaging. Nanotechnology **24**, 214001/1-10 (2013)
25. P. Jepsen, D. Cooke, M. Koch, Terahertz spectroscopy and imaging: Modern techniques and applications. Laser Photonics Rev. **5**, 124–166 (2011)
26. C. Jansen, S. Wietzke, O. Peters, M. Scheller, N. Vieweg, M. Salhi, N. Krumbholz, C. Jördens, T. Hochrein, M. Koch, Terahertz imaging: Applications and perspectives. Appl. Opt. **49**(19), E48–E57 (2010)
27. J.F. Federici, B. Schulkin, F. Huang, D. Gary, R. Barat, F. Oliveira, D. Zimdars, THz imaging and sensing for security applications—Explosives, weapons and drugs. Semicond. Sci. Technol. **20**, S266–S280 (2005)
28. R.M. Woodward, V.P. Wallace, R.J. Pye, B.E. Cole, D.D. Arnone, E.H. Linfield, M. Pepper, Terahertz pulse imaging of ex vivo basal cell carcinoma. J. Investig. Dermatol. **120**(1), 72–78 (2003)
29. Y.S. Lee, *Principles of Terahertz Science and Technology* (Springer Science & Business Media, New York, 2009)

30. L. Ho et al., Terahertz pulsed imaging as an analytical tool for sustained-release tablet film coating. Eur. J. Pharm. Biopharm. **71**(1), 117–123 (2009)
31. Y.B. Ji, E.S. Lee, S.H. Kim, J.H. Son, T.I. Jeon, A miniaturized fiber-coupled terahertz endoscope system. Opt. Express **17**, 17082–17087 (2009)
32. G.M. Png, R. Flook, B.W.H. Ng, D. Abbott, Terahertz spectroscopy of snap-frozen human brain tissue: An initial study. Electron. Lett. **45**, 343–345 (2009)
33. H. Hoshina, A. Hayashi, N. Miyoshi, F. Miyamaru, C. Otani, Terahertz pulsed imaging of frozen biological tissues. Appl. Phys. Lett. **94**, 123901/1–123901/3 (2009)
34. J.Y. Park, H.J. Choi, K.S. Cho, K.R. Kim, J.H. Son, Terahertz spectroscopic imaging of a rabbit VX2 hepatoma model. J. Appl. Phys. **109**, 064704/1–064704/4 (2011)
35. J.A. Zeitler, P.F. Taday, D.A. Newnham, M. Pepper, K.C. Gordon, T. Rades, Terahertz pulsed spectroscopy and imaging in the pharmaceutical setting-a review. J. Pharm. Pharmacol. **59**(2), 209–223 (2007)
36. Y.B. Ji, E.S. Lee, S.H. Kim, J.H. Song, T.I. Jeon, A miniaturized fiber-coupled terahertz endoscope system. Opt. Express **17**(19), 17082–17087 (2009)
37. Q. Sun, Y. He, K. Liu, S. Fan, E.P. Parrott, E. Pickwell-MacPherson, Recent advances in terahertz technology for biomedical applications. Quant. Imaging Med. Surg. **7**(3), 345–355 (2017)
38. G. Ok, K. Park, H.J. Kim, H.S. Chun, S.W. Choi, High-speed terahertz imaging toward food quality inspection. Appl. Opt. **53**, 1406–1412 (2014)
39. M. Locatelli, M. Ravaro, S. Bartalini, L. Consolino, M.S. Vitiello, R. Cicchi, F. Pavone, P. De Natale, Real-time terahertz digital holography with a quantum cascade laser. Sci. Rep. **5**(1), 13566 (2015)
40. Y. Wang, H. Minamide, M. Tang, T. Notake, H. Ito, Study of water concentration measurement in thin tissues with terahertz-wave parametric source. Opt. Express **18**, 15504–15512 (2010)
41. V.P. Wallace, P.F. Taday, A.J. Fitzgerald, R.M. Woodward, J. Cluffff, R.J. Pye, D.D. Arnone, Terahertz pulsed imaging and spectroscopy for biomedical and pharmaceutical applications. Faraday Discuss. **126**, 255–263 (2004)
42. J. Landercasper, E. Whitacre, A.C. Degnim, M. Al-Hamadani, Reasons for re-excision after lumpectomy for breast cancer: Insight from the American Society of Breast Surgeons Mastery (SM) database. Ann. Surg. Oncol. **21**(10), 3185–3191 (2014)
43. A.J. Fitzgerald, V.P. Wallace, M. Jimenez-Linan, L. Bobrow, R.J. Pye, A.D. Purushotham, D.D. Arnone, Terahertz pulsed imaging of human breast tumors. Radiology **239**(2), 533–540 (2006)
44. M.K. Choi, A. Bettermann, D.W. Van Der Weide, Potential for detection of explosive and biological hazards with electronic terahertz systems. Philos. Trans. R. Soc. London, Ser. A **362**, 337–349 (2004)
45. Y.C. Shen, T. Lo, P.F. Taday, et al., Detection and identification of explosives using terahertz pulsed spectroscopic imaging. Appl. Phys. Lett. **86**, 241116/1–241116/3 (2005)
46. Y.C. Shen, P.F. Taday, D.A. Newnham, M.C. Kemp, M. Pepper, 3D chemical mapping using terahertz pulsed imaging, in *Terahertz and Gigahertz Electronics and Photonics IV*, vol. 5727, (International Society for Optics and Photonics, 2005), pp. 24–31
47. J.F. Federici, Review of moisture and liquid detection and mapping using terahertz imaging. J. Infrared Millim. Terahertz Waves **33**, 1–30 (2012)
48. B. Li, W. Cao, S. Mathanker, W. Zhang, N. Wang, Preliminary study on quality evaluation of pecans with terahertz time-domain spectroscopy. Proc. SPIE **7854**, 78543V (2010)
49. H.S. Chua, J. Obradovic, A.D. Haigh, P.C. Upadhya, O. Hirsch, D. Crawley, A.A.P. Gibson, L.F. Gladden, E.H. Linfield, Terahertz time-domain spectroscopy of crushed wheat grain, in *Proc. IEEE MTT-S International Microwave Symposium Digest*, (IEEE, 2005), pp. 1–4
50. S. Gorenflo, U. Tauer, I. Hinkov, A. Lambrecht, R. Buchner, H. Helm, Dielectric properties of oil–water complexes using terahertz transmission spectroscopy. Chem. Phys. Lett. **421**(4–6), 494–498 (2006)
51. Y. Hua, H. Zhang, Qualitative and quantitative detection of pesticides with terahertz time-domain spectroscopy. IEEE Trans. Microwave Theory Tech. **58I**(7), 2064–2070 (2010)

52. A. Redo-Sanchez, G. Salvatella, R. Galceran, E. Roldós, J.-A. García-Reguero, M. Castellari, J. Tejada, Assessment of terahertz spectroscopy to detect antibiotic residues in food and feed matrices. Analyst **136**(8), 1733–1738 (2011)
53. Y.K. Lee, S.W. Choi, S.T. Han, D.H. Woo, H.S. Chun, Detection of foreign bodies in foods using continuous-wave terahertz imaging. J. Food Prot. **75**, 179–183 (2012)
54. A.A. Gowen, C.P. O'Donnell, C. Esquerre, G. Downey, Influence of polymer packaging films on hyperspectral imaging data in the visible near-infrared (450-950 nm) wavelength range. Appl. Spectrosc. **64**, 304–312 (2010)
55. D. Mittleman, J. Cunningham, M. Nuss, M. Geva, Noncontact semiconductor wafer characterization with the terahertz Hall effect. Appl. Phys. Lett. **71**, 16–18 (1997)
56. T. Kiwa, M. Tonouchi, M. Yamashita, K. Kawase, Laser terahertz-emission microscope for inspecting electrical faults in integrated circuits. Opt. Lett. **28**, 2058–2060 (2003)
57. A.J. Huber, F. Keilmann, J. Wittenborn, J. Aizpurua, R. Hillenbrand, Terahertz near-field nanoscopy of mobile carriers in single semiconductor nanodevices. Nano Lett. **8**, 3766–3770 (2008)
58. C. Jansen, T. Hochrein, R. Wilk, S. Wietzke, M. Scheller, N. Krumbholz, C. Jördens, K. Baaske, M. Koch, Applications for THz systems. Optik Photonik **4**, 26–30 (2008)
59. A. Podzorov, G. Gallot, Low-loss polymers for terahertz applications. Appl. Opt. **47**, 3254–3257 (2008)
60. V.A. Bershtein, V.A. Ryzhov, Far infrared spectroscopy of polymers. Adv. Polym. Sci. **114**, 42–121 (1994)
61. M. Scheller, C. Jansen, M. Koch, Analyzing sub-100 μm samples with transmission terahertz time domain spectroscopy. Opt. Commun. **282**, 1304–1306 (2009)
62. R.T. Boreiko, A.L. Betz, Heterodyne spectroscopy of the 63 μm OI line in M42. Astrophys. J. Lett. **464**(1), L83–L86 (1996)
63. M. Wachter, M. Nagel, H. Kurz, Tapered photoconductive terahertz field probe tip with subwavelength spatial resolution. Appl. Phys. Lett. **95**(4), 041112/1-4 (2009)
64. A. Redo-Sanchez, X.-C. Zhang, Terahertz science and technology trends. IEEE J. Sel. Top. Quantum Electron. **14**, 260–269 (2008)
65. R. Piesiewicz, T. Kleine-Ostmann, N. Krumbholz, D. Mittleman, M. Koch, J. Schöbel, T. Kürner, Short-range ultrabroadband terahertz communications: Concepts and perspectives. IEEE Antennas Propag. Mag. **49**, 24–39 (2007)
66. N. Karpowicz, H. Zhong, J. Xu, K.-I. Lin, J.-S. Hwang, X.-C. Zhang, Comparison between pulsed terahertz time domain imaging and continuous wave terahertz imaging. Semicond. Sci. Technol. **20**, S293–S299 (2005)
67. A.W. Lee, Q. Hu, Real-time, continuous-wave terahertz imaging by use of a microbolometer focal-plane array. Opt. Lett. **30**, 2563–2565 (2005)
68. B. Pradarutti, R. Müller, G. Matthäus, C. Brückner, S. Riehemann, G. Notni, S. Nolte, A. Tünnermann, Multichannel balanced electro-optic detection for terahertz imaging. Opt. Express **15**, 17652–17660 (2007)
69. M. Wan, J.J. Healy, J.T. Sheridan, Terahertz phase imaging and biomedical applications. Opt. Laser Technol. **122**, 105859/1-12 (2020)
70. J.N. Mait, G.W. Euliss, R.A. Athale, Computational imaging. Adv. Opt. Photon. **10**, 409–483 (2018)
71. G. Barbastathis, A. Ozcan, G. Situ, On the use of deep learning for computational imaging. Optica **6**, 921–943 (2019)
72. R.I. Stantchev, B. Sun, S.M. Hornett, P.A. Hobson, G.M. Gibson, M.J. Padgett, E. Hendry, Noninvasive, near-field terahertz imaging of hidden objects using a single-pixel detector. Sci. Adv. **2**(6), e1600190/1–e1600190/7 (2016)
73. S. Chen, L. Du, K. Meng, J. Li, Z. Zhai, Q. Shi, Z. Li, L. Zhu, Terahertz wave near-field compressive imaging with a spatial resolution of over λ/100. Opt. Lett. **44**, 21–24 (2019)
74. D. Wang, B. Li, L. Rong, Z. Xu, Y. Zhao, J. Zhao, Y. Wang, C. Zhai, Extended depth of field in continuous-wave terahertz computed tomography based on Bessel beam. Opt. Commun. **432**(1), 20–26 (2019)

75. B. Li, D. Wang, L. Rong, C. Zhai, Y. Wang, J. Zhao, Application of continuous-wave terahertz computed tomography for the analysis of chicken bone structure. Opt. Eng. **57**(2), 023105 (2018)
76. T. Nagatsuma, H. Nishii, T. Ikeo, Terahertz imaging based on optical coherence tomography. Photon. Res. **2**, B64–B69 (2014)
77. C.L. Koch Dandolo, M. Lopez, K. Fukunaga, Y. Ueno, R. Pillay, D. Giovannacci, Y. Le Du, X. Bai, M. Menu, V. Detalle, Toward a multimodal fusion of layered cultural object images: Complementarity of optical coherence tomography and terahertz time-domain imaging in the heritage field. Appl. Opt. **58**(5), 1281–1290 (2019)
78. H.B. Krid, Z. Houaneb, H. Zairi, Reconfigurable graphene annular ring antenna for medical and imaging applications. Prog. Electromagn. Res. **89**, 53–62 (2020)
79. S. Poorgholam-Khanjari, F.B. Zarrabi, S. Jarchi, Compact and wide-band Quasi Yagi-Uda antenna based on periodic grating ground and coupling method in terahertz regime. Optik **203**, 163990, (2020)
80. X.C. Zhang, J. Xu, *Introduction to THz Wave Photonics* (Springer, Berlin/Heidelberg, 2010)
81. I. Zaytsev, I.N. Dolganova, N.V. Chernomyrdin, G.M. Katyba, A.A. Gavdush, O.P. Cherkasova, G.A. Komandin, M.A. Shchedrina, A.N. Khodan, D.S. Ponomarev, et al., The progress and perspectives of terahertz technology for diagnosis of neoplasms: A review. J. Opt. **22**, 013001 (2020)
82. X. Yang, X. Zhao, K. Yang, Y. Liu, Y. Liu, W. Fu, Y. Luo, Biomedical applications of terahertz spectroscopy and imaging. Trends Biotechnol. **34**, 810–824 (2016)
83. L. Manti, A. D'Arco, Cooperative biological effects between ionizing radiation and other physical and chemical agents. Mutat. Res. **704**, 115–122 (2010)
84. L.V. Titova, R. Rodriguez-Juarez, R. Woycicki, F.A. Hegmann, O. Kovalchuk, Intense THz pulses down-regulate genes associated with skin cancer and psoriasis: A new therapeutic avenue? Sci. Rep. **3**, 2363 (2013)
85. V.I. Fëdorov, D.S. Serdyukov, O.P. Cherkasova, S.S. Popova, E.F. Nemova, The influence of terahertz radiation on the cell's genetic apparatus. J. Opt. Technol. **84**, 509–514 (2017)
86. M. Naftaly (ed.), *Terahertz Metrology* (Artech House, London, 2015)
87. G. Bellisola, C. Sorio, Infrared spectroscopy and microscopy in cancer research and diagnosis. Am. J. Cancer Res. **2**(1), 1–21 (2012)
88. A.G. Pakhomov, Y. Akyel, O.N. Pakhomova, B.E. Stuck, M.R. Murphy, Current state and implications of research on biological effects of millimeter waves: A review of literature. Bioelectromagnetics **19**(7), 393–413 (1998)
89. S. Romanenko, P.H. Siegel, D.A. Wagenaar, V. Pikov, Effects of millimeter wave irradiation and equivalent thermal heating on the activity of individual neurons in the leech ganglion. J. Neurophysiol. **112**(10), 2423–2431 (2014)
90. H. Kasban, M.A.M. El-Bendary, D.H. Salama, A comparative study of medical imaging techniques. Int. J. Inf. Sci. Intell. Syst. **4**(2), 37–58 (2015)
91. https://www.marketsandmarkets.com/Market-Reports/diagnostic-imaging-market-411.html
92. E. Berry, G.C. Walker, A.J. Fitzgerald, N.N. Zinov'ev, M. Chamberlain, S.W. Smye, R.E. Miles, M.A. Smith, Do in vivo terahertz imaging systems comply with safety guidelines. J. Laser Appl. **15**(3), 192–198 (2003)
93. E.R. Mueller, Terahertz radiation: Applications and sources. Ind. Phys. **9**(4), 27–29 (2003)
94. Z.D. Taylor, R.S. Singh, D.B. Bennett, P. Tewari, C.P. Kealey, N. Bajwa, M.O. Culjat, A. Stojadinovic, J. Hubschman, E.R. Brown, W.S. Grundfest, THz medical imaging: In vivo hydration sensing. IEEE Trans. Terahertz Sci. Technol. **1**(1), 201–219 (2011)
95. B. Davoudi, Comparative Study and Analysis of THz Pulse and Continuous-wave Imaging towards Medical Application, M.Sc. thesis, University of Waterloo, 2009
96. C. Balocco, S.R. Kasjoo, X. Lu, Z. Linqing, Y. Alimi, S. Winnerl, P. Bao, Y. Luo, L. Kin, A.M. Song, Novel terahertz nanodevices and circuits, in *Proc. IEEE International Conference on Solid-State and Integrated Circuit Technology*, (IEEE, 2010), pp. 1176–1179

97. A. Stylianou, M.A. Talias, Nanotechnology supported THz medical imaging. F1000 Res. **2**, 100 (2013)
98. C. K. Chou, J. Andrea (eds.), *IEEE Standard for Safety Levels with Respect to Human Exposure to Radio Frequency Electromagnetic Fields, 3 kHz to 300 GHz, C95.1-2005* (IEEE, 2006)
99. B.S. Williams, Terahertz quantum-cascade lasers. Nat. Photon. **1**, 517–525 (2017)
100. M.Y. Wong, G.D. Sims, I.M. Stephenson, Operation of the backward-wave oscillator. Nature **188**, 803–804 (1960)
101. S.G. Pavlov, H.-W. Hübers, H. Riemann, R.K. Zhukavin, E.E. Orlova, V.N. Shastin, Terahertz optically pumped Si: Sb laser. J. Appl. Phys. **92**(10), 5632–5634 (2002)
102. C.H. Zhang, G.F. Zhao, B.B. Jin, Y.Y. Hou, H.H. Jia, J. Chen, P.H. Wu, Terahertz imaging on subcutaneous tissues and liver inflamed by liver cancer cells. Terahertz Sci. Technol. **5**(3), 114–123 (2012)
103. S. Nakajima, H. Hoshina, M. Yamashita, C. Otani, N. Miyoshi, Terahertz imaging diagnostics of cancer tissues with a chemometrics technique. Appl. Phys. Lett. **90**, 041102/1–041102/3 (2007)
104. S.S. Dhillon et al., The 2017 terahertz science and technology roadmap. J. Phys. D. Appl. Phys. **50**, 043001 (2017)
105. Y.V. Grachev, I.A. Kuklin, I.V. Gerasimov, O.A. Smolyanskay, S.A. Kozlov, V.G. Bespalov, Study of how the radiation of the frequency range 0.05–2 THz affects biological tissues of various thickness in medical diagnosis. J. Opt. Technol. **77**(11), 731–733 (2010)
106. J.L. Coutaz, F. Garet, V.P. Wallace, *Principles of Terahertz Time-Domain Spectroscopy: An Introductory Textbook* (CRC Press, 2018)
107. A. Nahata, A.S. Weling, T.F. Heinz, A wideband coherent terahertz spectroscopy system using optical rectification and electro-optic sampling. Appl. Phys. Lett. **69**(16), 2321–2323 (1996)
108. J. O'Hara, D. Grischkowsky, Quasi-optic THz imaging. Opt. Lett. **26**, 1918–1920 (2001)
109. J.L. Johnson, T.D. Dorney, D.M. Mittleman, Interferometric imaging with terahertz pulses. IEEE J. Sel. Top. Quantum Electron. **7**(4), 592–599 (2001)
110. S. Verghese, K. McIntosh, S. Calawa, W. Dinatale, E. Duerr, K. Molvar, Generation and detection of coherent terahertz waves using two photomixers. Appl. Phys. Lett. **73**, 3824 (1998)
111. J.T. Lu, J.C. Cao, Terahertz generation and chaotic dynamics in GaN NDR diode. Semicond. Sci. Technol. **19**, 451–456 (2004)
112. I. Malhotra, K.R. Jha, G. Singh, Terahertz antenna technology for imaging applications ¾ A technical review. International Journal of Microwave and Wireless Technologies **10**(3), 271–290 (2013)
113. L. Wang, Microwave sensors for breast cancer detection. Sensors **18**(2), 655/1–17 (2018)
114. S. Guha, F.I. Jamal, C. Wenger, A review on passive and integrated near-field microwave biosensors. Biosensors **7**(4), 42/1-18 (2017)
115. H.-W. Wu, Label-free and antibody-free wideband microwave biosensor for identifying the cancer cells. IEEE Trans. Microw. Theory Tech. **64**, 982–990 (2016)
116. A. Salim, S. Lim, Recent advances in the metamaterial-inspired biosensors. Biosens. Bioelectron. **117**(15), 398–402 (Oct. 2018)
117. A. Salim, S.-H. Kim, J.Y. Park, S. Lim, Microfluidic biosensor based on microwave substrate-integrated waveguide cavity resonator. J. Sens. **2018**, 1324145/1–13 (2018)
118. C. Li, X. Yu, C.-M. Lee, D. Li, L. Ran, J. Lin, High-sensitivity software-configurable 5.8-GHz radar sensor receiver chip in 0.13 μm CMOS for noncontact vital sign detection. IEEE Trans. Microw. Theory Tech **58**, 1410–1419 (2010)
119. Z. Geng, X. Zhang, Z. Fan, X. Lv, H. Chen, A route to terahertz metamaterial biosensor integrated with microfluidics for liver cancer biomarker testing in early stage. Sci. Rep. **7**, 16378 (2017)

120. P. Haring Bolívar, M. Nagel, F. Richter, M. Brucherseifer, H. Kurz, A. Bosserhoff, R. Büttner, Label–free THz sensing of genetic sequences: Towards THz biochips. Philos. Trans. R. Soc. London, Ser. A **362**, 323–335 (2004)
121. E.N. Shaforost, N. Klein, S.A. Vitusevich, A. Offenhäusser, A.A. Barannik, Nanoliter liquid characterization by open whispering-gallery mode dielectric resonators at millimeter wave frequencies. J. Appl. Phys. **104**, 074111 (2008)
122. M. Nagel, M. Forst, H. Kurz, THz biosensing devices: fundamentals and technology. Journal of Physics: Condensed Matter **18**, S601–S618 (2006)

Chapter 11
Terahertz Integrated Circuit Design

11.1 Introduction

Terahertz is the portion of the spectrum which is sandwiched in between two mature and technologically well-developed spectral bands—the microwaves and optical frequency regime (photonics) of the spectrum. The photonic techniques used for visible and infrared light fail at such long wavelengths, while the electronic techniques used for microwave and radiofrequency technologies are unsuitable for the high frequencies of the terahertz band. Despite the plethora of potential uses, because the terahertz sources tend to have lower power and terahertz detectors have lower sensitivity than devices in other regimes of the electromagnetic spectrum, there is limited uptake of technologies operating in the terahertz range; thus, this phenomenon is referred to as the terahertz gap [1]. There are the following two potential approaches which have been employed to cater to this gap: (1) to up-scale the technology developed in microwaves and (2) to down-scale the developed optics technology. Further, with the recent advancement of nanotechnology and nanophotonics, especially with the combination and melding of these two technologies, several technology advances in the terahertz regime of the spectrum to overcome the terahertz gap have been explored [2]. Recently, the terahertz technology has gained immense interest in various applications such as sensing, imaging, and communication owing to its unique characteristics. The terahertz radiation has been widely used in spectroscopy by using the vibration of molecules for a given terahertz radiation frequency, which can produce a unique fingerprint depending on the type of dielectric materials. To elaborate on the use of the electromagnetic spectrum in the terahertz range as such, the design of highly efficient, compact, and reliable sources and detectors is an indispensable requirement, which has several potential challenges. Enormous research efforts have been directed to bridge this terahertz gap with different approaches from high-frequency electronics and photonics. Until recently, the fully integrated terahertz transceiver in silicon technology has not been considered a promising solution due to the limited device performance and the large

I. Malhotra, G. Singh, *Terahertz Antenna Technology for Imaging and Sensing Applications*, https://doi.org/10.1007/978-3-030-68960-5_11

propagation losses coupled through a resistive silicon substrate in the terahertz range. However, it has witnessed revolutionary achievements in RF and millimeter-wave integrated circuit (IC) technology with the advancement of nanoscale silicon technology.

Considering the impact of radiofrequency integrated circuits (RFICs) with their potential features including compactness, low-cost, and mass production, the terahertz integrated circuit (THz-IC) has opened a new research direction in imaging, sensing, spectroscopy, and ultrafast wireless communication. The terahertz imaging and sensing capabilities are promising for nondestructive quality control, 3D imaging, radar, and gesture recognition in autonomous systems, vehicles, robotics, and industrial automation, particularly in the lower part of the spectrum. The need for compact, efficient, high-performance sensing devices which are operable at a low power and deployable at a large scale must play a significant role in integrated circuit technology. The integrated circuit technology of choices such as electronics, photonics, or electronics–photonics hybrids will depend on the application which can be particularly diverse, spanning from sensing, spectroscopy, and imaging to wireless communication. The essential requirement of efficient and widely reconfigurable chip-scale systems is the prime challenge to enable real-world applications in terahertz sensing, spectroscopy, and imaging. Naturally, it has illustrated promising capabilities, but its deployment has been limited due to the complexity of the technology at hand. The two most prominent examples involve the relatively narrow absorption lines of rotational or vibrational excitations in gases and the somewhat broader lines that result from lattice vibrations (phonons) in crystalline solids, especially in molecular crystals [3]. As a result, one can envision using terahertz spectroscopy for identification of the chemical composition of powders [4] or plumes [5], for example. The absorption strength of many polar molecules' peaks in the THz range and numerous techniques have been developed recently to enhance the sensitivity and specificity of gas sensing [6]. As with mid-infrared fingerprint spectroscopy, the ability of a sensor to offer both highly specific and sensitive detection relies crucially on the tuning range of the system, among other parameters.

Recently, the wireless infrastructure has relied on the incredible complexity and reconfigurability of current radiofrequency integrated circuits, which would not have been possible without the level of integration supported by the silicon integrated circuit technology. The field of terahertz technology is beginning to experience a similar revolution. Indeed, it can be argued that terahertz electronics is at an inflexion point and the field is moving from bridging the terahertz gap to the technology and application. Owing to its unique position in the spectrum, researchers across electronics and photonics have converged to address the technological challenges in this field. One result of this effort is the significant research progress in various technologies of the terahertz generation and detection over the past decades. Although, the power is by no means the only important criterion, it is a useful metric for illustrating recent progress. In case of integrated terahertz systems and technology, the focus has shifted from a device-centric approach to enabling complex systems that might have moderate efficiency, which is versatile and programmable. This includes properties such as the ability to synthesize or sense the terahertz signal in a reconfigurable manner across the spectrum and spatial distribution. The system which can be manufactured at low cost and operate at room temperature can be an extremely powerful tool to significantly advance the application space, particularly in imaging and sensing where a single modality is often not sufficient. The progress of terahertz sources and detectors has demonstrated the evolvement of integrated terahertz technology. There are other considerations also involved behind the evaluation of integrated optimal terahertz technologies, which include efficiency, ability to reconfigure the spectrum and field distribution, cost, and scalability. A qualitative illustration of technologies in this multidimensional space is illustrated in Fig. 11.1.

The vastly underutilized spectrum in the terahertz frequency range enables disruptive applications including 10-gigabit chip-to-chip wireless communications and imaging/spectroscopy. From the imaging applications point of view, it is considered to be one of the emerging and disruptive technologies [1, 2]. The terahertz imaging systems are key enabling components in applications such as security surveillance, non-destructive testing, biology, radio astronomy, and medical imaging. From the sensing and communications perspective, the terahertz frequency range makes it possible to explore new unprecedented ideas on super-precise sensing at the micrometer level and multi-10-gigabit instant wireless access at the centimeter level spacing between the transmitter and the receiver.

This chapter mainly explores the integrated terahertz transceivers for sensing and imaging applications. Since the antenna size shrinks quadratically as radiation frequency increases for a given gain, the on-chip antennas have great potential in the terahertz range by eliminating packaging issues for cost-effective, compact terahertz transceivers. To achieve a significantly high radiation efficiency, the investigation of the loss mechanisms of several on-chip antennas implemented in conventional

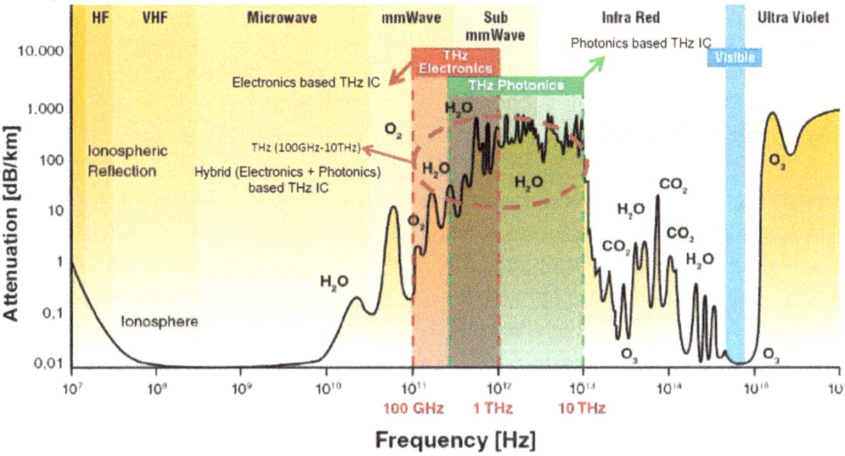

Fig. 11.1 The region of a terahertz electromagnetic spectrum with its involvement in different disciplines for THz-IC design

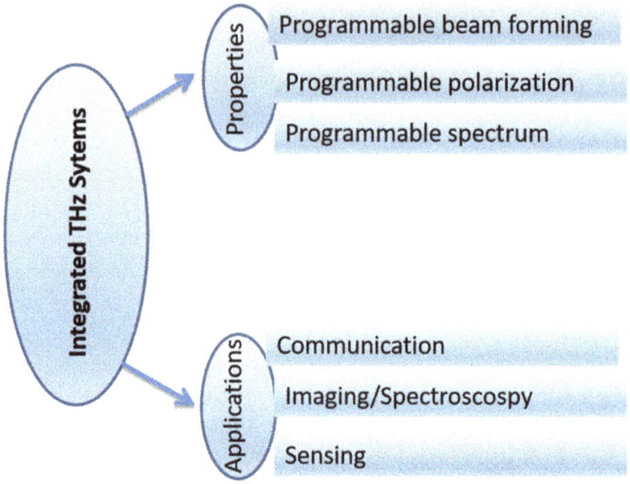

Fig. 11.2 Integrated terahertz systems with their potential applications and performance features

(Bi) CMOS technologies is a challenging task. This makes a lot of sense in the terahertz region considering the relatively efficient on-chip THz antennas and the reduced size of arrays with a half-wavelength pitch. This chapter begins with the derivation of conditions for maximizing power gain of active devices. Discussions of circuit topologies for terahertz sources, detectors, and transceivers with emphasis on their efficacy and scalability ensue, and this chapter concludes with a brief survey of interface options for channeling terahertz energy out of the chip (Fig. 11.2).

11.2 Silicon Technology for Terahertz Integrated Circuit

With the recent advancement of silicon technology, it becomes a potential candidate to replace existing systems for terahertz generation and detection devices and enabling easy integration with on-chip readout and control electronics. Therefore, the terahertz silicon integrated circuit (THz-SIC) has opened a new era in imaging, sensing, spectroscopy, and ultrafast wireless communication. The main motivation of scientists and technologists consisted of bringing together the optical and electrical waves to observe the continuity of the electromagnetic spectrum. To fulfil this purpose, two experimental approaches were possible, either the down-converting optical sources or multiplying electronic frequencies. Beyond 1THz and extending up to 10 THz, photonics-based systems comprising quantum-cascade lasers, silicon photonics, and photo mixer devices are expected to play a dominant role, particularly in systems where tunability of more than 1THz may be needed. In such systems, an efficient optical-to-terahertz and terahertz-to-optical converters are key future challenges. The terahertz systems will enjoy the ancillary benefits as the integrated photonics platforms continue to improve with the incorporation of passive devices and active devices such as high-speed modulators. To enable this heterogeneous future, research pathways across electronics and photonics, which have mostly developed independently, will need to continue to converge. In this direction, the scientific achievements have continuously improved during the twentieth century [2], but what aroused the keen interest in terahertz waves were the applications, starting with modern astronomy [7] and later developing in medical imaging [8], security [9–11], safety [12], communication [13], biological [14], and pharmaceutical industries [15].

Until now, the mass deployment of terahertz platforms suffers from distinct drawbacks including system cost, achievable integration level, and reliability. Currently, employing microelectronic packaging techniques for the electronic terahertz integrated systems cannot easily support the tolerance levels required for terahertz applications and are not scalable to large volumes. Therefore, the low-cost terahertz applications demand single-chip solutions which obviate the need for complex microelectronic assembly. The terahertz integrated circuits (THz-ICs) based on modern nanoscale CMOS and SiGe HBT/BiCMOS process technologies promise to fulfil this role, at least in the lower end of the terahertz spectrum. They concurrently offer economies of scale, small form factors, and unprecedented integration capability at the highest industry standards. The THz-ICs are, therefore, an interesting option for applications that require extremely compact terahertz microsystems for sensing [17–19] or active imaging components at a reasonable price/performance ratio [20]. In this regard, it is also important to note the ongoing research activities on the heterogeneous integration of III–V devices with foundry-level silicon technology [21], which show enormous potential for future medium-sized THz markets. However, the benefits of economic viability and system size are not the only drivers for research in THz-ICs. The design space that opens up by device plurality, reconfigurability, and terahertz-mixed signal cointegration promises to expand the functional scope of terahertz imaging systems. The potential applications, such as terahertz light field imaging based on CMOS detector focal-plane arrays (FPAs) [22], real-time sub-wavelength imaging with integrated near-field arrays [23], and broadband passive spectroscopy [24], have all been demonstrated based on THz-ICs that approach integration levels comparable to the ones exhibited in conventional consumer electronics. At present, the low speed of silicon transistors and the issues related to efficient terahertz signal escape [25] constitute an ongoing research challenge for both technology development and integrated circuit design.

However, the scaling of silicon technology poses considerable challenges for THz-ICs integration. However, the main purpose is to evaluate its current capabilities and performance for imaging and sensing applications such as: (1) advances in frequency performance which are inherently linked to a reduction in device breakdown voltages and (2) technological stack of foundry-level silicon technology presents an increasingly unfavorable environment for antenna and passive implementation in the terahertz frequency regime of the spectrum. The design of on-chip antennas in silicon technology is, in general, a challenging task because of potential multimode propagation issues [26, 27] as well as difficulties to yield appropriate high

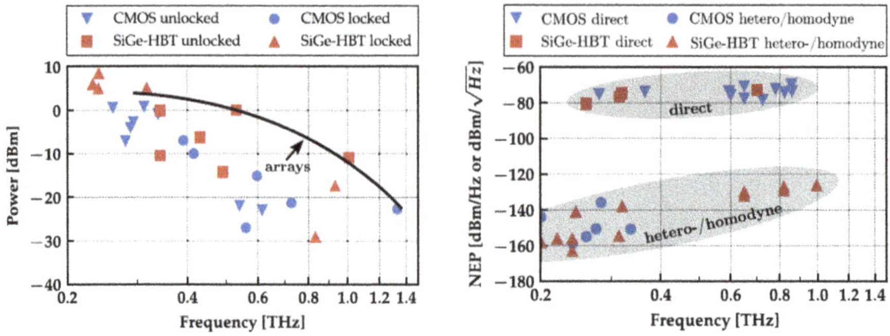

(a) (b)

Fig. 11.3 Comparison of state of the art on the basis of NEP (**a**) THz sources in CMOS and SiGe technologies and (**b**) direct power detectors and heterodyne/homodyne receivers in CMOS and SiGe technologies in the frequency range of 0.2–1 THz [16]

directivity of the antenna system which provide sufficient link-budget for practical applications that intended the wide use of backside radiating on-chip antennas with auxiliary silicon lenses [28]. Figure 11.3a summarizes the current state of the art for CMOS and SiGe HBT radiation sources above 250 GHz. The sources are grouped into unlocked oscillator-based sources and into oscillator- or multiplier-chain-based sources that can be locked to an external phase-stable reference signal. It is interesting to mention that no single best architecture has emerged so far; therefore, the researchers are interested to explore a large design space offered by silicon technologies to surmount the barriers related with limited device speed with terahertz wave escape from the chip-level design. The nonlinear frequency translation process of highly driven high-speed devices uses silicon terahertz sources above 300 GHz. The nonlinear terahertz front-end circuit is either directly enforced as a harmonic N-push oscillator or as a frequency multiplier circuit and both the approaches entail a conversion loss that increases rapidly with the harmonic order and drives the frequency. Thus, contrary to the typical 20 dB/decade roll-off for power amplifiers below the maximum frequency [29], the data-points indicate around 40 dB/decade power roll-off for single-radiator sources in the terahertz band. It is interesting to note that the small size of on-chip terahertz passives and the abundant availability of transistors has intended the research in coherent single-chip multi-element radiators. Moreover, as reported by several authors at different frequencies, the baseband signal processing functionalities of silicon technology have been further employed for reconfiguration of the radiation pattern of multi-element sources with the assistance of power-on control of asynchronous sources to diffuse area illumination at 530 GHz [30] or phased array functionality at 338 GHz and 400 GHz [31].

The availability of on-chip baseband signal processing circuits enables the integration of detectors into a chip-scale focal plane array with on-chip read-out circuitry which is the potential feature of silicon technology. In general, the receivers of silicon THz-IC have been focused on advancing the sensitivity and bandwidth of direct power detectors. Further, the silicon detectors offer significantly higher integration capabilities as compared to that of the recent room-temperature terahertz detector technologies. Due to the lack of low-noise preamplification in the terahertz band, the silicon power detectors are implemented as antenna-coupled direct detectors. Therefore, the predominantly exploited methods for terahertz direct detection are either non-quasi-static self-mixing in cold MOSFET channels [32–34] or rectification in the base-emitter junction of a high-speed HBT [27, 35]. At terahertz frequencies, both devices operate close to or above their cut-off frequencies defined by the carrier transit time, and their response and operation bandwidth are severely influenced by the efficiency of coupling the terahertz radiation into the intrinsic device. Since the broadband operation is typically desired for practical implementations to allow versatile application scenarios at a low cost, the classic narrowband RF matching techniques made way for innovative antenna-device codesign approaches. In general, the antenna systems based on backside-radiating on-chip with external hyper-hemispherical silicon lenses have been demonstrated to simultaneously provide $pW/\sqrt{}$Hz-level noise-equivalent power (NEP) across several hundreds of GHz and antenna directivity in the range of 20 dBi [33, 36]. The MOSFET direct power detectors are operated without drain-source bias and in the moderate inversion for the highest responsivity. Although, the device scaling can result in lower thermal noise of the MOSFET channel, so far reported detector performances such as responsivity, NEP, and operation bandwidth have been not considerably improved by migration to nano-scale CMOS technology nodes. The very high (kΩ-range) and frequency-dependent impedance of MOSFET operating in moderate inversion is one of the main reasons which makes efficient and broadband coupling to on-chip antenna very challenging. Figure 11.3b shows the sensitivity comparison between the state-of-the-art direct detectors and coherent hetero–/homodyne receivers. The heterodyne/homodyne receivers illustrate a better sensitivity order of 80 dB at 300 GHz as compared to the direct detectors. Further, to yield higher operation frequencies employing higher harmonic order results in a steep noise figure increase. However, a similar trend is not present for direct power detectors.

11.3 Terahertz Sources and Detectors Based on Silicon Technologies

The significance of terahertz electronics is self-evident and the general consensus among silicon terahertz circuit designers is that the silicon technology will be dominant for the lower end of the terahertz spectrum (up to 1 THz) in light of recent breakthroughs of silicon circuits in terms of effective isotropic radiated power (EIRP), phase noise, and receiver sensitivity. For different applications, the silicon circuits are similar to or even superior to III/V compound technologies and optical-based techniques in the terahertz frequency range. The silicon-based terahertz source and detectors are very promising alternatives to the existing terahertz technologies. The sources can be based on multiplier chains or oscillators and the detectors may use the non-linearity of different devices to operate as square-law power detectors. The silicon technology is used in the THz-IC design due to the use of intelligent signal processing on a single chip as well as dense array of detectors and radiators on a single wafer.

11.3.1 Terahertz Silicon-Based Sources

When we talk about the silicon terahertz sources, a plethora of options are available, which vary in functionality, complexity, and performance. The most important metrics of the terahertz sources are the output power and efficiency for the incoherent imaging applications, whereas for spectroscopy, the bandwidth is the most important specification. However, output power, power efficiency, tuning range, phase noise, harmonics, and spurious suppression are all important parameters for the most demanding application that is the terahertz communications. The first demonstration of a CMOS terahertz source by Seok et al. [37] at 410 GHz which radiated −42 dBm power has made tremendous progress toward the provision of suitable power levels for imaging applications with THz-ICs. Further, advances have been enabled by both new design methodologies for the proper synthesis of passive embedding networks to optimize the harmonic generation [38, 39] as well as the novel scalable system architectures. The sources are grouped into unlocked oscillator-based sources [40] and into the oscillator- or multiplier-chain-based sources that can be locked to an external phase-stable reference signal [41]. It is worth to mention that no single best architecture has emerged yet and researchers continue to explore a large design space offered by silicon technologies to overcome the barriers associated with the limited device speed and with terahertz wave escape from the chip level [25]. The silicon terahertz sources employ the nonlinear frequency translation process of strongly driven high-speed devices at above 300 GHz. The nonlinear terahertz front-end circuit is either directly implemented as a harmonic N-push oscillator or as a frequency multiplier circuit. Both the approaches entail a conversion loss that increases rapidly with the harmonic order and drive frequency. Therefore, contrary to the typical 20 dB/decade roll-off for power amplifiers below the maximum frequency $f_{\max}$, the data-points indicate around 40 dB/decade power roll-off for single-radiator sources in the terahertz band. Despite the fact that the multiplier chain sources may be used for coherent detection, they have two disadvantages which are the DC power consumption and the large layout area on silicon. The harmonic oscillators are however very compact and present an advantage for large array integration. There are two types of harmonic oscillator designs: the feedback design and the negative-resistance design.

A transistor in a specified technology can be characterized by the transistor short circuit current gain at the unit-gain frequency f_T and the maximum unit power gain frequency f_{max}. These two limits are defined by the technology process and are improved by device engineering. Beyond f_T/f_{max}, the transistor does not have gain. Below f_T/f_{max}, it is possible to build active circuits and overcome the technology limitations; therefore, harmonic generators are built. These generators consist of a combination of fundamental oscillators operating below the f_T/f_{max} of technologies. The harmonic component of the designed combination can be extracted at high frequencies. To achieve high power sources, a large number of such oscillators can be integrated on the same chip but increasing the number of elements would increase the power consumption by the same factor. In such a system, the total power consumption is a key element that limits the array size. Therefore, to achieve a large pixel count which favors diffused illumination, a control system is needed. In such oscillators, the stability condition criterion applies to maintain oscillations [42]. Now, modifying the stability conditions would bring the oscillator to operate in a different mode or to stop the oscillations. The control system would impact a specific oscillator in the array by modifying its stability. In a ring type of oscillator, this can be made by modifying or breaking the feedback loop. In a negative resistance-based oscillator, the control circuit would modify the stability criterion by inverting the resistance sign of the active element for instance. In both cases, the frequency of operation can be changed, or the oscillator can be turned off to minimize power consumption. For the terahertz oscillator sources, the oscillation conditions are sensitive to the designed passive elements, and active components such as switches can be difficult to implement. For this reason, it is more favorable to implement the control circuitry to impact the biasing conditions of each oscillator.

The terahertz source can be characterized by the following: (1) the total radiated power, (2) the frequency of operation, and (3) the DC power consumption. At terahertz frequencies, the lack of on-wafer measurement equipment such as probe stations makes the free space method the only possibility to characterize high-frequency sources. Therefore, an electronic terahertz source is usually equipped with an antenna which is characterized by its directivity at a specified frequency. In a free space system, the radiated power, frequency of operation, and directivity are verified optically. However, these measures are very critical for the characterization of a terahertz system. The DC power consumption is often referred to as the efficiency of the terahertz source and defined as the ratio radiated power to the DC power consumption.

11.3.2 Terahertz Silicon Detectors

In terahertz detectors, the nonlinearity features of active devices have been utilized to directly rectify a terahertz signal to a DC signal. The terahertz reception with detectors is incoherent, due to which only the amplitude information is recovered at the receiving side, which limits the terahertz detectors almost exclusively to incoherent terahertz imaging applications. The strength of terahertz detectors lies within their simplicity such as it does not need a local oscillator signal to perform terahertz down-conversion. The received terahertz signals self-mix themselves to DC through even-order nonlinearity of the device which makes scaling extremely easy, as only low-frequency routing is needed, whereas the local oscillator-driven mixer needs a cumbersome and power-hungry local oscillator tree which quickly becomes unmanageable when the array gets large. However, this flexibility comes with a cost—the gain and noise performance of detectors is quite limited, and the specification "responsivity" and "noise equivalent power (NEP)" are used in place of conversion gain and noise figures. The responsivity is defined as the voltage output divided by received power, and NEP is defined as the output noise voltage density divided by responsivity. The bandwidth of imaging applications is usually below 1 MHz; thus, the technologies with lower 1/f noise corner frequencies are preferred. The imaging principle of terahertz with detectors largely follow their optical counterpart—they use terahertz lenses to do the focusing. The problem is that terahertz wavelength is 2–3 orders of magnitude larger than that of the visible lights; thus, large and bulky terahertz optics are required for reasonable imaging resolutions. Therefore, a lot of effort is required to set up the imaging setup with the invisible terahertz waves which is the innate deficiency with incoherent imaging. However, the coherent imaging with terahertz transceivers could get rid of those optics.

The research in silicon THz-IC receivers is largely focused on advancing the sensitivity and bandwidth of direct power detectors. The silicon technology yields this feature and the availability of on-chip baseband processing circuits, enabling the integration of detectors into chip-scale focal plan array (FPA) with on-chip readout circuitry. Therefore, the silicon detectors offer significantly higher integration capabilities as compared with the current room-temperature terahertz detector technologies, such as Schottky barrier diodes, InP HEMT low-noise amplifiers, Golay-cells, microbolometers, and pyroelectric detectors [43]. Due to the lack of low-noise pre-amplification in the terahertz regime of the spectrum, the silicon power detectors are implemented as antenna-coupled direct detectors. Therefore, the predominantly exploited methods for terahertz direct detection are either non-quasistatic self-mixing in cold MOSFET channels [32] or having rectification in the base emitter junction of a high-speed HBT [35]. At terahertz frequencies, both device classes operate close to or above their cut-off frequencies defined by the carrier transit time, and their response and operation bandwidth are severely influenced by the efficiency of coupling the terahertz radiation into the intrinsic device. Since the broadband operation is typically desired for practical implementations to allow versatile application scenarios at low cost, classic narrowband RF matching techniques developed a way for innovative antenna-device code-sign approaches. In particular, the antenna systems based on backside-radiating on-chip primary antennas and external hyper-hemispherical silicon lenses have been demonstrated to simultaneously provide $pW/\sqrt{\text{Hz}}$-level noise-equivalent power (NEP) across several hundreds of GHz and antenna directivity values in the range of 20 dBi [36].

HBT-based terahertz detectors: In bipolar technology, a single diode can be built out of a transistor by connecting the collector and the base terminals. However, with the transistor third terminal, three basic single-stage topologies can be explored. For the same isolation reasons explained for the diode topology, RF filters and DC blocks are needed to avoid leakage of the RF power into the output load and vice versa. To analyze these detector stages, the same quasi-static analysis developed for the diode detector can be done and would lead to the same result $I\alpha\left(A_{RF}^{2}/4\right)$. However, it would only provide a DC analysis of the detected signal and would not give any information about the frequency response of the detector. Therefore, a Volterra series can be derived to incorporate the frequency-dependent effects of the detector.

CMOS-based terahertz detectors: Recent advances in terahertz detection with the use of CMOS technology have shown that this option has the potential to be a leading method of producing low-cost terahertz sensors with integrated readout systems. The article gives examples of some hints regarding radiation coupling and readout systems. Marczewski et al. [44] have illustrated that silicon CMOS technology is well adapted to the production of inexpensive imaging systems for sub-terahertz frequencies. The terahertz detectors embedded in this system were manufactured using the CMOS process.

The field effect transistor responsivity to terahertz radiation has made these devices useful for terahertz detection at room temperature. The early experiments have been focused on the use of terahertz for imaging and short-range communication applications. However, for the latter purpose, the Schottky diodes and HEMT transistors are predestined due to their higher operation speed. The imaging matrices based on detection on Schottky diodes integrated with the system of pixel amplifiers manufactured both in CMOS technology have been reported in [45]. Furthermore, after BiCMOS technology exploiting silicon–germanium (SiGe) heterojunction bipolar transistors (HBTs) [46] had been mastered, the devices and circuits produced in this process became competitive in both imaging and communication applications. The advanced systems for terahertz heterodyne detection developed with use of the BiCMOS process were recently reported in [47, 25, 48]. However, the standard MOS transistors still remain a valuable option for imaging and spectroscopy, where the detector operation speed is not critical. The complementary metal-oxide semiconductor (CMOS) technology has seen widespread use over the past dozen years for the fabrication of broadband terahertz detectors [49, 33]. An example proving the broadband capabilities of direct detection using FET–CMOS detectors may be the solution implemented in the standard 150 nm CMOS technology [50]. For the realization of research projects based on the CMOS process, there are several production tools such as the MPW service (multi-project wafer – a service integrating a number of different circuit designs from various teams within one substrate and offering a certain amount of chips for a reasonable price). It should be pointed out that CMOS technology allows for easy integration of terahertz detectors with read-out circuitry. A major disadvantage, however, is the large area required by the monolithically integrated antennas of each single detector. This area dominates and cannot be reduced because it is related to the THz radiation wavelength (sub-mm range).

The responsivity and the noise equivalent power (NEP) are a measure of the performance of terahertz direct detectors. The responsivity is a measure of the electrical output strength per input power. The output response of a direct detector can either be measured in current mode or voltage mode. The resulting unit of the responsivity can be expressed in [A/W] or [V/W] depending on the measurement mode. The NEP corresponds to the measured detector spot noise at a specific frequency for a 1-Hz bandwidth divided by the responsivity. The noise can be measured in current mode or in voltage mode, which leads to an NEP unit of $\left[W/\sqrt{\text{Hz}}\right]$. From a characterization point of view, three physical parameters have to be measured in order to determine the performance of a direct detector: (1) the output voltage or current, (2) the output noise, and (3) the input powers. Depending on the readout mode, the output DC voltage or current can be estimated by the analysis of the device behavior at terahertz frequencies and has been reviewed in the previous subsections of this chapter. However, the only measure of responsivity will not give any information about the SNR in the presence of a terahertz signal. In fact, if an amplifier is used, the signal is multiplied by the gain and also the noise. Therefore, the noise measurement will be used to calculate the NEP of the detector and give information about the SNR behavior of a detector for a specified amount of power. In an on-wafer setup, it is convenient to measure the electrical input power injected to the detector circuit by the probes. However, at terahertz frequencies, this is not possible due to the lack of suitable terahertz on-wafer equipment and optical measurement is needed. As a result, optical-based devices are the only possibility to build terahertz detectors. The sources are equipped with antennas as well as detectors. For an antenna coupled setup, the available optical input power to the detector can be calculated by the Friis transmission equation. An accurate measurement of the optical responsivity and NEP requires an accurate measurement of the electrical response, the noise, and the available power. This will be discussed in the next chapter where measurement results will be shown.

As it has been described in the previous section, a terahertz detector is possible in silicon technologies by using standard devices. However, the different theories around the detection mechanisms have led to the conclusion that the expected responsivities are very low. Therefore, an efficient read-out circuit is needed in order to extract with a small noise contribution the detector response to an incident terahertz power. This is more valid for large terahertz arrays where multi-pixel implementation can be achieved. The vastly under-utilized spectrum in the terahertz band enables disruptive applications such as 10 Gigabits chip-to-chip wireless communication and imaging applications. Aggressive scaling in feature size and $f_T/F_{\max}$ nano-scale silicon enable integration of a sophisticated system at a sub-terahertz system such as an array of9 detectors and phase array and MIMO transceivers with a massive number of elements. High-frequency requirements impose several challenges on the architecture and circuit design at these frequencies. Several techniques have been presented to address these challenges. The integrated circuit design challenges in silicon technology are as follows. (1) Lack of amplification near the $f_{\max}$ frequency range, (2) Weak output power and poor efficiency, (3) High-free space power loss, (4) Inefficient on-chip antenna gain and radiation efficiency, and (5) High phase noise and very small tuning range.

11.3.3 Terahertz Transceivers

The transceivers (transmitter and receiver) are usually integrated on a single chip and share the common terahertz frequency port through switches or duplexers. Park et al. [51] have exposed a fully integrated 260 GHz transceiver based on a shared leaky-wave antenna which is a lossy transmission line; thus, the transmitter and receiver ports could be placed on either end of the antenna. In this system, when the transmitter is working, then the receiver is turned off and vice versa; however, the problem with the leaky-wave antenna is that it is relatively long in this design. For FMCW radars, the leakage from the transmitter to the receiver results in strong interferences around DC [52], which raises the noise floor in the spectrum range. With the area permitting, the transmitter and receiver antenna should be separated further apart for better isolation. The measured crosstalk between the two antennas with a separation of about 1.8 mm in a 160 GHz FMCW radar transceiver is below 31 dB [53]. This isolation might be adequate for FMCW radar applications, but it is still wanting for communications. The transceiver-based terahertz imaging makes coherent imaging possible, as both the magnitude and phase information of the signal from targets is retained. With both information available, it is possible to get rid of bulky terahertz optics by sampling the terahertz field directly and do the focusing digitally. The terahertz field is usually sampled on a 2-D plane with different terahertz frequencies; this is fulfilled by raster scanning an FMCW transceiver. For a given point in space, the round-trip phase delay from the transceiver to that point is a function of its position and sampling frequency. By raster scanning the transceiver or the sample under different frequencies, its phase delay variation is orthogonal to every other point in the sampling space. This forms the basis for the 3-D imaging through the back-projection algorithm. 3-D imaging based on SiGe FMCW transceivers is reported by several groups [54], showcasing the great potential for low-cost THz imaging applications. For communication applications, the modulation scheme plays a major role in deciding the transceiver architecture. Low-complexity modulation schemes such as on–off keying (OOK) and binary phase-shift keying (BPSK) lead to robust and power-efficient design, but the spectrum efficiencies are relatively low. Modulation schemes such as 32 QAM and 128 QAM lead to a much higher spectrum efficiency, but they are quite demanding on linearity and phase noise performance, and they require image-rejection architectures as the spectra of QAM are asymmetric around the carrier. The upper sideband (USB) and lower sideband (LSB) of the spectra become each other's own image when converted to a baseband, and image-rejection is needed to avoid signal corruption. Image-rejection modulation/demodulation is difficult in the THz range as I/Q mixers are required. It is very difficult to guarantee phase and amplitude matching for the I/Q LO signal for adequate image-rejection at THz frequency.

A 210 GHz fundamental transceiver chipset with OOK modulation is demonstrated in a 32-nm SOI CMOS process [55]. Ideally speaking, power amplifier (PA)-based fundamental operation is more power-efficient than frequency multipliers. This helps to boost the efficiency of the whole system as PAs are usually the most power-hungry circuits in transceivers. Perhaps the most difficult part of this design is controlling the oscillator pulling effect. Since the PA works at the same frequency as the on-chip VCO, the significant coupling could occur between PA and VCO. The injection-locking effect would impact the phase noise performance heavily. The on-chip antenna used in this design only makes things more difficult. To improve the VCO performance, a stacked cross-coupled VCO topology is used to boost oscillation amplitude, improving its robustness in response to interferences.

A 240-GHz direction-conversion transceiver in SiGe BiCMOS technology is demonstrated with BPSK capability. BPSK is a constant-envelope modulation, which means the PA could be driven to saturation for better power efficiency. The spectra of BPSK modulation are symmetric around its carrier (symmetrically modulated), making direct conversion easier to implement as no image-rejection is needed. A 30 GHz LO signal is supplied to this transceiver, and on-chip 8 × 8 multipliers are used for the 240 GHz LO generation. This helps to alleviate the detrimental effect of LO spurs caused by multipliers since they are separated by twice the baseband bandwidth (15 GHz). An on-chip antenna with 1-dB bandwidth of 33 GHz is achieved partly due to the local back etching (LBE) technology used. The silicon substrate below the antenna is removed, resulting in a low-loss air cavity below the antenna. The transceiver link is tested with 15 cm separation, and an impressive 6-dB bandwidth of 35 GHz is obtained. A 25 Gbps wireless link is demonstrated by this transceiver with no equalization. One problem with direct conversion using no I/Q demodulation is that the demodulated signal's SNR is dependent on the phase difference between the Tx LO and Rx LO. A phase shifter is used in this test in case manual tuning is required to boost the SNR. Zhong et al. [56] have presented a 300 GHz QPSK transmitter for dielectric waveguide communication in a 65-nm CMOS process and an off-chip LO signal is used to drive on-chip frequency multipliers. The targeted data rate is 30 Gbps, which translates to around 20 GHz baseband bandwidth for QPSK assuming a roll-off factor of 0.3. Rodríguez-Vázquez et al. [57] have illustrated a 100-Gbps wireless communication link for a distance of 1 meter at 230 GHz. The I/Q mixer directly interfaces with an on-chip antenna to avoid bandwidth limitation introduced by LNA. An on-chip LO multiplier chain is used to convert the external 13.75–16 GHz LO to 220–256 GHz. The baseband bandwidth is around 14 GHz. This poses challenges as the spacing of LO spurs is comparable to this bandwidth. This leads to spurious modulation that overlaps with the desired signal. Nevertheless, 100 Gbps with an EVM of 17% is demonstrated. Takano et al. [58] have presented a 300 GHz 32-QAM and 128-QAM transmitter with 105-Gbps data rate in a 40 nm CMOS process. A heterodyne topology is used, and the LO frequencies for the two up-conversion stage are both set at 135 GHz. The IF frequency for the first stage is around 10 GHz, and high-pass filtering is used to suppress the LSB by approximately 10 dB. The single-balanced mixer is used in the first stage to intentionally leak the LO signal to the second stage. No on-chip antenna is used as this chip is intended to drive high-power THz devices such as traveling-wave tubes.

Recently, Headland et.al [59]. have developed external optics to enable broadband incoherent spectroscopy with terahertz-range CMOS cameras to reduce the cost and size of terahertz systems capable of real-time spectral analysis. A dispersive optic divides a terahertz beam into constituent colors, which are subsequently imposed upon the aforementioned terahertz camera at different angles. In comparison to pre-existing terahertz spectroscopy systems, the presented is low-cost, low power consumption, compact, and portable, enabling deployment in a broad variety of field environments. Further, a more-compact optical setup is realized, as the path of the beam is folded in a zig-zag fashion. A shorter propagation distance also reduces losses due to atmospheric absorption, which further increases the spectroscope's SNR. The reflective optics are fabricated using CNC milling of a solid metal. The THz-ICs have gained significant momentum [16] with remarkable demonstrations which have facilitated novel applications, such as plenoptic imaging [22]. However, it is challenging to develop single-chip electronic integrated circuit-based spectroscopy devices that exhibit a useful bandwidth except for the multi-mode antenna as demonstrated by Wu and Sengupta [60]. Further, it required significant calibration and processing to extract frequency information, which limits its scope of viable practical applications. The THz-ICs are a viable means to miniaturize and reduce the system cost as well as provide a pathway to mass production, which will reduce costs.

The scaling of silicon technology poses considerable challenges for THz-IC integration. First, advances in frequency performance are inherently linked to a reduction in device breakdown voltages. Second, the technology stack of foundry-level silicon technology presents an increasingly unfavorable environment for antenna and passive implementation in the THz band. Typically, several metal layers are embedded in a thin dielectric stack (BEOL) on top of a lossy silicon substrate. The BEOL fabrication enforces strict design rules on the density of metal structures, which escalate the modeling and design effort of on-chip antennas and passives. The design of on-chip antennas in silicon technology is particularly challenging given potential multimode propagation issues and the difficulties in achieving appropriately high directivity of the antenna system to provide sufficient link budget for practical applications. This motivated the wide use of backside radiating on-chip antennas with auxiliary silicon lenses.

The potential solutions to low antenna gain and radiation efficiency are as follows. (1) Massive multi-antenna transceiver. (2) Use of silicon lenses. (3) The array of the power amplifier and antennas incorporating spatial power combining. (4) Array of the synchronized signal source. (5) Harmonic generation and multiplication circuits. (6) PLL band synthesizer to increase tuning range and lower the close-in phase noise. Similar to the solution of antenna gain and radiation efficiency, the solutions of lack of amplification are as follows. (1) Maximum gain and output power condition for near $f_{\max}$ devices. (2) Over-neutralization for high-frequency gain improvement. The imaging setup was based on an optoelectronic terahertz time-domain spectroscopy system (TDS). A femtosecond pulse source was used to excite an optically gated transmitter. Through a delay path, the same source was exciting an optically gated receiver. The terahertz beam delivered by the transmitter was first collimated and then focused down to a focal point by a pair of lenses. At that point, an object was placed. Another pair of lenses were used to collect and focus the transmitted terahertz radiation through the object on the detector. The object was mechanically moved at the focal point of the imaging system. A two-dimensional image was created by recording the response of the detected signal. The first scanned objects were an integrated circuit in plastic packaging and a leaf with different water concentration levels. What made this experiment possible is the short time constant of the used detector in comparison with bolometers. A similar optical setup has been used to produce terahertz images with single output detectors. Polytetrafluoroethylene (PTFE) lenses were used to focus the beam of a modulated source to a focal point and then onto the detector silicon lens. While the object was moved in the vertical plane, the output signal of a single output detector was synchronously read and stored by a data acquisition system based on the lock-in technique. These signals were processed, and a terahertz image of the scanned object was produced. The dynamic range of the terahertz image would depend on the detector sensitivity and noise performance. In the configuration where a single detector is used, the chopping frequency of the pixel would define the modulation rate of the source for best SNR operation. Terahertz direct detectors based on silicon technologies are feasible and their performance is approaching the non-silicon-based devices as it has been demonstrated in the previous chapter. However, what arouses great interest in standard silicon implementations are the large-scale integration possibilities. Until now, non-silicon-based terahertz detectors have been more sensitive than the previously presented silicon solutions. However, their production is limited in number and external discrete readout components are necessary. Unlike that, standard silicon is a commercially available and reliable technology with high integration possibilities in a large scale. In other words, it is possible to integrate on the same chip the previously presented detector in a large number and make use of the technical possibilities to design a compatible on-chip analogue and digital circuits, with repeatability guaranty by the process for a large production volume. This is precisely what other terahertz technologies suffer from and limits their commercial development targeting hand-held devices for applications such as terahertz imaging.

11.4 Integrated Antenna Technology

The terahertz frequency regime of the spectrum has proven its potential for imaging applications with a good compromise between spatial resolution and penetration. The enabler of this development is the advancement of different technologies capable of generating or detecting terahertz radiation. The current technology for terahertz generation or detection is based on either pulsed time domain or frequency domain systems. The time-domain systems cover the broad range of terahertz spectra by the use of a laser pulse radiating into a photoconductive antenna. The sampling of the pulse is performed in the time domain and therefore the spectrum is recovered in amplitude and phase. There are different types of photoconductive switches available based in different technologies which produce significantly high bandwidth radiation; however, it is expensive and the power provided is still very low. These systems are commonly employed when it is required to perform imaging for processes that occur very fast in time. The frequency-domain systems can be characterized as that use microwave technology in the form of radiometers (commonly implemented using direct detectors), active heterodyne transceivers, or the photo-mixing of two lasers.

In this section, we focused on imaging systems that require non-cryogenic diodes as they can be potentially integrated with Monolithic Integrated Circuits (MIC) in order to perform commercial systems. Moreover, it will also concentrate on active systems in order to perform three-dimensional tomographic imaging. Considering the power budget restrictions and high losses that sources and receivers suffer at these frequencies, it is important to achieve a high level of integration among all the devices of the systems in order to avoid unnecessary additional losses. Due to the ability to integrate multifunction analogue as well as digital circuits on a single chip to improve performance and reduce the size of electronic devices, the MIC is referred as one of the key technological achievements. In the context of the antenna point of view in general, the antennas can be fabricated and integrated into the same fabrication process as the sensing or power converting devices that are used at their terminals [60]. Further, realizing specific impedance matching, frequency bandwidth, and high efficiencies remain also significant challenges. Based on the type of imaging system topology, the antenna configuration requirements can be broadly classified into two different areas such as:

Near-field imaging systems: These systems perform the imaging within the near-field region of the imaging system. The system is located at a distance closer than the Fraunhofer (the Fraunhofer distance is defined as 2D2/λ, where D is the main dimension of the imaging system antenna and λ is the wavelength) from the object. These near-field systems are designed with non-directive antennas because these systems require that the antenna itself does not perturb the response of the sample being imaged. The imaging resolution of these systems is not dictated by the antenna's dimensions but by the Nyquist sampling of the received field [61]. The near-field systems are widely used like microwave frequencies as reported in [62] and the antennas employed are planar antennas. When we scale these antennas to the terahertz frequencies, the substrates become too thin and fragile to fabricate or if they are not scaled, they suffer from high losses due to the trapped surface waves.

Far-field imaging systems: These systems perform the imaging within the far-field region (from the Fraunhofer distance to infinity on all the way). The optical systems employing reflectors or lenses are commonly used to achieve a proper resolution and high directivity is needed for the antennas that feed these optics. The conducting materials suffer from high losses in the terahertz regime of the spectrum; therefore, the phased arrays are rarely employed. The typically employed antennas for illuminating these optical systems are lenses or horns. The lenses suffer from multiple internal reflections because of the use of high dielectric materials and the horns represent a challenge to fabricate.

The image acquisition time for both types of systems is directly proportional to the number of receiving antennas. Therefore, the large arrays of antennas will be required in order to fulfil the incoming requirements of high-speed image acquisition of future terahertz imaging systems [63]. However, the large arrays increase considerably the difficulty and complexity of the overall process in terms of integration and fabrication of the antennas. These constraints are forcing the researchers to investigate novel and simplified forms of antennas, toward a large variety of new planar antennas (six designed to take advantage of wafer-scale integration) and toward new fabrication technologies, in order to provide highly integrated and high-efficiency antenna arrays.

11.4.1 State-of-the-Art Terahertz Integrated Arrays

This section introduces the recent advances of the current arrays of integrated antennas that can be found at THz frequencies. The arrays antennas can be divided into two main areas such as: (1) non-directive arrays and (2) directive arrays corresponding to near-field and far-field imaging systems, respectively.

Non-directive antenna arrays: The non-directive antenna arrays in which each of the antennas operate independently from the others are usually referred in the literature as antenna grids [64]. They are compounded of multiple non-directive antennas (planar slots or dipoles). In the submillimeter/terahertz wave band, these antenna grids are used to sample a plane wave and perform a certain active function on this area and this function can be a modulation, amplification, or a mixing of two fields [64]. These arrays do not contain a feeding network that combines the signal coming from all the antennas into a single port, but instead, an active component is integrated directly on each of the antenna output terminals [64]. This method avoids the high losses and fabrication problems associated with waveguide structures at a terahertz frequency. A maximum spacing between elements of $\lambda/2$ is required to accomplish the Nyquist sampling theory necessary to recover the image information from the target. In the microwave regime, these antennas are usually fabricated in thin substrates or surrounded by metallic pins in order to avoid the excitation of substrate modes. This integration is critical at high frequencies in order to avoid the high ohmic losses associated with transmission lines.

Directive antenna arrays: The directive antenna arrays found at submillimeter wave band are based on phased arrays, horns, and lenses. These antenna arrays are commonly used to illuminate an optical system like a reflector.

Phased arrays: The phased arrays are able to achieve high gain in the terahertz band only when the feeding network is made of a superconducting material. These superconducting lines avoid conductor losses, so are significant at these high frequencies. The superconducting effect occurs only below the superconducting frequency gap of the material and this gap frequency depends on the actual material. Moreover, these antennas can be fabricated with lithographic techniques and therefore easily fabricated in large arrays. In [65], an example of a phased array of a dual polarization slot antenna fabricated with superconducting lines is presented. The same microstrip made of a regular conductor would have a −5 dB loss for a length of just $\lambda/2$. However, this material requires cryogenic cooling in order to operate as a superconducting material. Because this thesis focuses on non-cryogenic imaging systems, these antennas will not be suitable for our purposes.

Horn arrays: While horn arrays are typically used at microwaves, there are only few examples at THz frequencies, especially above 300 GHz. This is because the fabrication of horn arrays is complex and requires high precision methods. The most common horn arrays are found at the lower end of the terahertz band and are typically fabricated using micromachining techniques. High-precision milling techniques have been proposed in [66, 67]. In order to reduce the cost and time of the fabrication of large arrays of horns using laser micro-fabrication, techniques based on photolithographic processes are a better option [68]. There are two different approaches that researchers are following: (1) One approach is to use selective etching silicon techniques to fabricate pyramidal horn antennas as proposed by [68]. It consists of using photolithographic processes to etch a pair of silicon wafers at 35 degrees (silicon can only be etched in a 90- or 35-degree angle due to the crystal structure). After applying a gold platted layer on the silicon, both wafers are assembled together. Lubecke in [68] shows an example of pyramidal antennas suitable for working at 585 GHz. However, horns fabricated with a 35-degree angle have low directivity and therefore they would not be suited for instance as a Focal Plane Array (FPA) where directive antennas are required. (2) Another approach is to stack thin gold-plated silicon wafers with tapered holes etched at a 90-degree angle. The tapered holes are obtained by applying a deep reactive ion etch (DRIE) to the silicon wafers, achieving vertical sidewalls. This technique is strongly limited by the tolerances of the fabrication process. This tolerance depends on the photolithographic process, the number of wafers employed, and the alignment process of those wafers. Consequently, the larger the directivity, the longer the horn and the larger the number of wafers. Since the tolerance problem is related to the excitation of higher order modes in the horn and this process is accumulative, the larger the number of the wafers, the tighter the tolerances need to be.

All in all, even by using photolithographic processes, there has not been developed an integrated and efficient method to fabricate arrays of horns. Simpler antenna designs made of a small number of wafers are required in order to reduce the impact of the alignment tolerances that occur when a high number of wafers are used.

Lens arrays: Integrated lens antennas are the most suited antennas for THz applications for their focusing capabilities and fabrication techniques for manufacturing. Even though at THz, the adoption of single lenses as a focusing element for a single detector is common, it is still hard to see them in large arrays. There are a few examples of lenses arrays fabricated at THz as in [69] where every lens is assembled manually (one by one). Several restrictions from the antenna point of view have been identified that currently limit the size and the electromagnetic performance of arrays of antennas at terahertz frequencies. The integrated antenna arrays for imaging applications are divided as follows.

11.4.2 Integrated Planar Antenna Arrays

This research has focused on the study of an integrated planar array called retina for a near-field imaging system. This retina consists of an array of antennas printed on an electrically thick substrate. The array antenna element is designed to cancel the surface waves that propagate within this substrate and also to be integrated with the rest of elements in the system. The imaging system is based on a near-field Modulated Scattering Technique (MST) system. This system has been characterized and general formulation and quality parameters have been defined. An assessment of different technologies available for the fabrication of the antenna arrays has been studied and different approaches have been followed.

11.4.3 Integrated Focal Plane Antenna Arrays

The goal is to design an array of a high number of antennas to be used as a FPA of a reflector and consistent with the silicon fabrication capabilities. The idea was to provide simple antenna designs made of a small number of wafers in order to reduce the impact of the alignment tolerances (especially in the case of horns) that could be integrated in an array of detectors fed by waveguides. One approach considered was to design and to fabricate an array of horn antennas by stacking gold plated silicon wafers with tapered holes. But in this case, the silicon is selectively etched with a 5-degree angle (instead of 35 or 90 degree). This etching allowed a reduction in the total number of wafers and therefore an improvement of the wafer alignment tolerances. This technique was still in an early stage of development and an improvement in this angle controllability was required. The idea was briefly presented in [70] but it has not been developed further. The approach followed in this thesis was the design and fabrication of an array of silicon lenses antennas. The whole antenna is designed to be consistent with two fabrication processes developed: one based on silicon laser micro-machining and another using silicon photolithographic micromachining. Silicon laser micromachining was already used in [71, 72] to fabricate hemispherical lenses. On the other hand, silicon photolithographic micro-machining is a technique used in the optics regime like in [73].

11.5 Planar Antenna Array for Near-Field Imaging

In this section, a planar antenna array for a specific near-field imaging system at terahertz frequency is presented. To sample the sufficient number of scattered field in space to obtain a sharp estimation of the existing scattered field distribution necessary to recover the currents induced inside the object produced by the incident or illuminating field in the near-field imaging system is a key concept. The well-known technique to sample field distributions is the modulated scattering technique (MST) which has been used [74, 75] to perform indirect measurements of the electric field distribution. It measures the field distribution by observing a scattered field of a probe placed at a certain position and its scattering properties are modulated during the time of the measurement to discriminate the field scattered by the probe from others. Initially, this modulation was generated by a mechanical rotation which limits the applicability of the technique; however, now a more commonly employed technique is the loading of the probe with some electronic device with two differentiated impedances by using a PIN diode [75] or a photodiode [76]. A single probe was employed to perform the measurements at an early stage; however, with the development of arrays of MST probes called retina reduced the measurement time, allowing the development of fast antenna radiation pattern measurement systems and developed the real-time microwave cameras used in material characterization [77] or tomography applications [78]. To move toward terahertz frequencies, we face different technological challenges in all the subsystems involved in the generation, transmission and detection of signals, making very difficult to attain a sufficiently high signal-to-noise ratio (SNR) for tomography imaging. This technique has been chosen because of its capabilities to perform high-speed imaging and employed to evaluate and study the design and fabrication of planar antenna arrays (retina) at terahertz frequencies. In general, the retina is a planar antenna array which is combination of non-directive antennas placed over a dielectric substrate and each antenna operates independently from the others. However, the antenna efficiency is the major existing problem at terahertz frequencies which is associated with this dielectric substrate. In the next section, the near-field imaging system and its key design parameters are discussed.

The efficiency of planar antennas associated with the surface-wave propagation within the substrate is studied; last, the tomographic imaging algorithm and its imaging capabilities are presented.

11.5.1 Near-Field Imaging System Design

A near-field imaging system based on MST has been designed as shown in Fig. 11.4 to evaluate the performance of these planar arrays. This near-field imaging technique consists of illuminating the target with an antenna which will be called illuminator (Port 1) and measure its scattered field along a probing plane also called retina (Port 2). Further, the retina re-radiates this field to another antenna that is

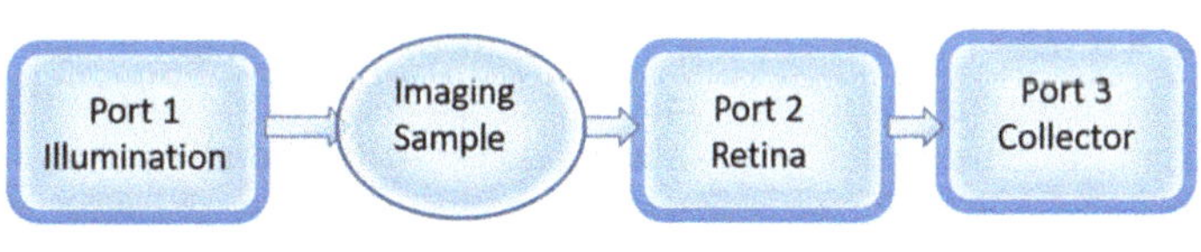

Fig. 11.4 Schematic of near-field imaging system

known as a collector (Port 3) which gather all the fields. A switching element is placed in each probe antenna port and provides two different load impedances to the probe antenna. The following two main advantages of this system are: (1) the obtainment of a multi-pixel image using a single high-frequency receiver which increases the acquisition time of the image and (2) the absence of microwave line distribution networks in the system which reduces the conductor losses which are very significant at terahertz frequencies. Further, the important parameters which are required to design an efficient near-field imaging system at terahertz frequency are presented. It is also discussed in detail in Chap. 9.

11.5.2 Retina Design

The design of the retina and other components employed in the system should maximize the parameter which is the difference of the transmission coefficient is the representative index of the MST system. The retina is a planar array of antennas which couples the radiation to both sides for the array (from the illuminator to the collector) and requires a front-to-back ratio close to one. The planar antennas at terahertz frequency suffer from significantly high losses like conductors as well as surface wave modes inside substrate. Therefore, these losses need to analyze and quantified for the different antennas as well as the number of modes propagated. Further, various switching technologies currently available at terahertz frequencies in order to study and analyze their capabilities and integration possibilities. This parameter can be dissected among three other quality parameters previously described:

Modulation efficiency: The difference in the reflection coefficient due to two states of the switching element impedances is known as the modulation efficiency and its main aim is to measure how well the probe antenna of retina couples to the switching element. At the terahertz frequencies, the impedance states provided by the available devices can be far from this condition and they cannot be freely chosen by the designer. In this case the optimization process is based in designing an antenna with a given impedance of the physical probe antenna which maximizes the modulation efficiency for the given switching element.

Surface wave efficiency: At the terahertz frequencies, the major problem of planar antenna arrays is associated with the dielectric substrate losses. The surface wave efficiency constitutes a power loss that falls within the overall difference of the transmission coefficients parameters. The surface wave efficiency which quantifies the amount of power loss due to the surface wave modes that propagate inside the substrate as a result of the lack of scalability of its thickness at millimeter and submillimeter wave frequencies. The antennas of the retina need to be carefully chosen in order to minimize the effect of substrate thickness in the overall system efficiency.

Coupling efficiency: This parameter describes the losses due the coupling between the illuminator-retina-collector. It depends on the matching of the illuminator, retina, and collector impedances and the scattering parameters. In several cases, the collector and illuminator are not able to be simulated within the system due to the large size of the retina (with high number of elements). In this case, the design of the retina can be adapted to the maximization of surface wave efficiency, modulation efficiency, and depending only on the impedance of the probe antenna and switching element.

11.6 Hybrid Electronic–Photonic Systems

In this section, we have emphasized on the developments in integrated technology across electronics and photonics, and the need for convergence between them to develop dependable and deployable THz systems for future applications. In particular, we focus on integrated systems, and the need for reconfigurability, adaptability, and scalability, beyond the classical metrics of sensitivity and efficiency.

Electronic approach: In recent years, there have been many advances in device fabrication technology for integrated circuits functioning in the THz range, including both silicon and III–V substrates. On the evaluation only by transistor-level metrics, the InP-based HEMTs and HBTs exhibit superior performance at high frequencies when compared with any silicon-based device [79]. The cut-off frequency ($f_{\max}$) (frequency at which the unilateral power gain of the device falls to unity) is a metric which is used to compare various technologies [80]. In another word, this metric fundamentally defines the maximum frequency at which the device can provide power amplification or establish an oscillatory behavior. The silicon-based technology such as CMOS offers a higher level of integration, but suffers from lower maximum frequency, typically in the range of 300–350 GHz [79]. However, due to limitations related to parasitic and contact resistances, this metric is not improved with size scaling [46]. This progress is extremely promising as it can simultaneously allow energy-efficient complex signal processing in CMOS with the THz operation in SiGe. However, the potential challenge is to scale these device-level experimental demonstrations to wafer-level processing with the tight process control required for high yield on complex integrated systems. In addition to this, others include noise figure, 1/f noise corner, breakdown voltages, current driving capability, and thermal stability. All these parameters affect the entire stack from block-level to system performance, impacting the quality (phase noise) of terahertz signal synthesis, the sensitivity and dynamic range of terahertz receivers and imagers, and the power generation and handing capability for terahertz sources.

Electronic–photonic hybrid approach: The propagation of terahertz signals is quite different from that of signals at another regime of the spectrum in several ways which offer both challenges and opportunities for the system design. In general, the terahertz signals are more directional and propagating as diffracting beams rather than omnidirectional broadcasts. In this way, the terahertz signals are more similar to the laser beams than that of the radiofrequency wireless links. Due to this high directionality, one can easily visualize the system configurations in which signal processing functions are separated from the source and/or detector subsystems. Various researchers have been pursuing the demonstration of specific device functions by fetching inspiration from the realm of optics and laser science. These devices are agnostic with respect to the particular choice of the radiation source and could therefore be integrated into a wide range of different system configurations in several cases. The versatility yield by this approach may offer significant advantages in the design of flexible and power-efficient architectures.

Future of THz integrated systems: The recent advances of terahertz integrated systems will be greatly impacted by future potential applications. The terahertz imaging and sensing capabilities particularly in the lower part of the spectrum (~0.1–0.3 THz) are promising for non-destructive quality control, 3D-imaging, radar, and gesture recognition in autonomous systems, vehicles, robotics, and industrial automation. The integrated circuit technology must play a major role for the demand of compact, efficient, high-performance sensing devices operable at low power and deployable on a large scale. We expect that the device performance in silicon, particularly SiGe-based devices, will continue to improve with maximum frequency reaching up to 1 THz in the near future. At higher frequencies beyond 1 THz, it is likely that the solution will be heterogeneous, with the front ends being realized in III–V technology and the rest of the system in silicon. This could be implemented with separate chips packaged in multichip modules, or with 3D-stack integrated circuits and III–V/silicon heterogeneous integration in which there have been promising developments in recent years. However, the technology of choice depends on the application, which at THz, can be particularly diverse spanning from the sensing, spectroscopy, and imaging to the wireless communication.

The foremost challenge to enable real-world applications in terahertz sensing, spectroscopy, and imaging is the need for efficient and widely reconfigurable chip-scale systems. Since a long time, it has illustrated promising capabilities, but its deployment has been limited due to the complexity of technology at hand. The two most prominent examples involve the relatively narrow absorption lines of rotational or vibrational excitations in gases, and the somewhat broader lines that result from lattice vibrations in crystalline solids, particularly in molecular crystals [3]. Consequently, one can envision using terahertz spectroscopy for identification of the chemical composition of powders [4] or plumes [5]. The absorption strength of various polar molecules peaks in the terahertz range, and numerous techniques have been explored recently to enhance the sensitivity and specificity of gas sensing [6]. The ability of a sensor to offer both the highly specific and sensitive detection relies critically on the tuning range of the system among other parameters with mid-infrared fingerprint spectroscopy. One of the key challenges for terahertz spectroscopic sensing is that the spectral lines of interest are often broader or more widely spaced than the tuning range of typical solid-state sources. New developments in broadly tunable solid-state sources would prove extremely valuable in this context. Some recent works have shown promising advancements in this regard.

The area of THz-IC technology has experienced significant development since the last decade. In the terahertz frequency range, these two techniques that is based on integrated electronic and integrated photonics have been combined and many emerging systems can be classified it as photonics-inspired or hybrid. Sengupta et al. [81] have reviewed the development of terahertz integrated electronic and hybrid electronic–photonic systems, and presented important functionalities for the communication, sensing, and imaging applications. Several advances in integrated systems have emerged including new architectures that are multifunctional and reconfigurable and break the trade-offs of classical approaches to electronic system design.

So far, the optical systems are bulky and expensive as compared with that of an integrated circuit. Moreover, the terahertz systems has not yet delivered such potential due to a lack of low-cost, portable, efficient technologies, even though THz regime of the spectrum is promising and can be used in a wide range of applications in sensing, imaging, and communication [1]. However, there has been a significant surge of progress in enabling integrated, compact, and efficient chip-scale terahertz technology, which could close the THz gap in meaningful ways in the last decade. This progress is a result of a cooperative effort stretching across a wide range of areas including solid-state and photonic devices [82], two-dimensional (2D) materials [83], heterogeneous integration [84], and packaging and system demonstrations [85]. Considerable work has been directed toward miniaturized THz technology demonstrated with quantum-cascade lasers [86], microbolometers [87], nanowires [86], novel plasmonic nanostructures [88], metamaterials [89], and ultrafast photoconductive semiconductor materials [90]. Significant recent effort has been dedicated to accomplishment techniques compatible with (or realized in) solid-state semiconductor technology (III–V and silicon based) that can operate at room temperature and can be manufactured at a low cost, exploiting economies of scale. This is a critical departure from the focus on the search for the "perfect" terahertz device and toward a more holistic approach for realizing new system-level properties that can enable a moderately efficient system, which is versatile and programmable. Programmability could include electronic reconfigurability of the wave front and polarization of the emitted terahertz fields, for applications in communication, radar and imaging, or dynamic spectral control of the radiated fields for spectroscopy and hyperspectral imaging. Such versatility is often required for advanced applications and is generally lacking in many current non-integrated terahertz platforms. If history is a guide, then the field of radiofrequency (RF) integrated circuits and systems provides a powerful lesson. Modern-day wireless infrastructure relies on the incredible complexity and reconfigurability of current RF integrated circuits, which would not have been possible without the level of integration supported by silicon integrated circuit technology. The field of terahertz technology is beginning to experience a similar revolution. Indeed, based on the advances in the field over the past decade, it can be argued that terahertz electronics is at an inflection point and the field is moving from bridging the THz gap to bridging the "technology" and "application" gap. Owing to its unique position in the spectrum, researchers across electronics and photonics have converged to address the technological challenges in the field. One result of this effort is how various technologies for terahertz generation have progressed over the past decade in their ability to generate signals in this range [91].

Although power is by no means the only important criterion, it is a useful metric for illustrating recent progress. On the electronics front, III–V-based semiconductor technologies such as InP heterojunction bipolar transistors (HBTs), high electron mobility transistors (HEMTs) and GaAs-based Schottky diodes are now capable of generating power levels at room temperature in the 100 μW–mW range [92], almost two to five times higher than a decade ago, particularly at frequencies beyond 1 THz. Silicon-based integrated technology, providing a platform for massive integration, has demonstrated complex phased array, imaging, and communication systems with output power reaching up to the 100 μW range at 1 THz. Photonic methods of terahertz generation have evolved as well, reaching tens of μW of continuous-wave power with photo mixer-based approaches [93] and mW power level in the 1–2 THz range with different frequency generation or parametric generation using nonlinear materials [94]. While plots of this nature are often used to show the existence (or absence) of technology solutions in the terahertz range, the requirements for practical terahertz systems are much more complex, ranging from signal generation (and its efficiency) and detection (and its sensitivity) to fast programmability. The specific issues include beam and frequency reconfiguration, integrability, scalability, size, form factor, and cost. Ultimately, the choice of a solid-state technology for a given application and frequency range has to take into consideration all these metrics, across various levels, from device to system.

In this chapter, we have focused on developments in the terahertz integrated technology across the electronics and photonics and emphasize the need for convergence between them to develop dependable and deployable terahertz systems for future applications. In general, the focus is on integrated systems and the need for reconfigurability, adaptability, and scalability, beyond the classical metrics of sensitivity and efficiency.

11.7 Tomography Imaging Techniques

To sample a sufficient number of scattered field values in space to recover the currents induced inside the object by the incident or illuminating field is the key aspect of the tomographic imaging system. The object that we want to image needs to be illuminated by a plane wave generated by a transmitter and then the scattered field by the object requires to be measured in an array of points. This multiport measurement can be achieved by either virtually generating it with the movement of one antenna, or by using a real array of multiplexed antennas. Even though with the first approach the coupling between antennas is not present and the receiving system is simpler, the second one is certainly advantageous for a real time system requirement. The imaging system proposed in this corresponds to a planar imaging scenario with a real array of antennas in order to perform high speed imaging. The reconstruction algorithm employed for the reconstruction is based on a technique called mono-focusing [95], which is used for systems of multiple receivers and a plane wave illuminator as a transmitter. This algorithm is based on the reconstruction of the local electrical properties of the object by means of synthesis of a receiver focusing operator. This focusing operator provides a compensation of the amplitude and the phase variations suffered by a wave due to the presence of the object. The reconstruction algorithm forms every image point of the local electrical properties of the object by the synthesis of one focus group (mono-focusing) of antennas (receivers) [96]. The received scattered fields are weighted by a focusing operator in order to be focused on a unique point. In order to predict the performance of the system employed, a canonical scenario has been simulated of the image reconstruction algorithm. It consists in the calculation of the theoretical scattered field distribution that a metallic cylinder scatters [26] over an array of Nr receiving antennas in an air medium. Once obtained this field, the image can be reconstructed with procedure previously explained. This scenario has been chosen as it is easily replicable in reality, and therefore the evaluation of the imaging capabilities of the array can be evaluated.

Frequency sampling criteria: The multiple frequencies can be combined to improve the quality of the image and allow the acquisition of a large number of spatial frequency components of the spectrum. A coherent combination of the different images at each frequency is used to enhance the features of interest of the individual images without having the destructive interference between images [95].

Space sampling criteria: In order to yield the multiple view, the system rotates around the scenario and makes a complete 360-degree rotation around the subject, obtaining a complete set of data from which images may be reconstructed covering all the views. A minimum number of views are required to avoid loss information [97] and guarantees.

11.8 Summary

Recently, the THz-IC design has emerged as an exciting area of research with breakthroughs in practical utility; however, the expected system cost and integration level of terahertz components are the frontiers which establish the bottleneck for adoption and commercialization of the terahertz technology. This chapter has reviewed the present status of THz-ICs targeted toward potential applications such as imaging/sensing and communications and highlights some of the fundamental trends which have been emerged with its design. With the advanced terahertz source and receiver components in both CMOS and SiGe HBT process, technologies are now available in the terahertz frequency range. The fundamental performance trends of different implementations, including oscillator or multiplier chain-based power generation, and heterodyne or direct power detection, have been reported in the literature. They demonstrate the technology capabilities, array scaling, and mixed-signal integration complexity which are being utilized to enable a new set of applications in the terahertz regime of the spectrum. The silicon-based THz-ICs have demonstrated the potential to leverage different terahertz imaging and sensing applications. The research in THz-ICs has enabled the cost-effective realization of terahertz applications such as tomographic imaging, multi-color imaging, high-resolution radar imaging, near-field imaging, and compressed sensing. Further, the power generation capability of THz-ICs and recent advances in THz-IC technology, which exploit the massive scalability and design space of silicon technology, are the major driving forces behind THz-IC advancement. Furthermore, a new generation of terahertz systems-on-chip (SoC) assures to extend the usefulness of the terahertz imaging systems. Thus, the THz-ICs should be an enabler for new imaging modalities, opening new applications as well as markets and will be seen as a compact, low-cost, and modest alternative to traditional terahertz equipment. Moreover, exploitation of integrated antenna arrays at terahertz frequencies for imaging/sensing and communication applications particularly on the technologies and the fabrication potentiality which can be used and are currently available. The current imaging systems require a large array of antennas in order to achieve the high-speed image acquisition which considerably increases difficulty and complexity to yield highly integrated and efficient antennas that is the recent demands most terahertz applications. The sources and receivers currently suffer the power budget restrictions and high losses with a high level of integration among all the devices and components of the systems and sub-systems at these frequencies. Thus, the antennas needed for these systems have to be integrated within the fabrication processes and technologies as the sensing and power converting devices are used at their terminals. Moreover, the terahertz regime of the spectrum has proven its potential for imaging and sensing applications with a good compromise between spatial resolution and penetration; however, it is facing many technological difficulties in all the subsystems involved in the signal generation, transmission, and detection. Further, the THz-IC designers are also confronted with an interdisciplinary set of challenges on the device, electromagnetic design, and packaging levels which could not overcome by the transfer of classic microwave and optical wave design techniques to the terahertz regime.

References

1. D.M. Mittleman, Twenty years of terahertz imaging: Invited. Opt. Express **26**(8), 9417–9431 (2018)
2. P.H. Siegel, Terahertz technology. IEEE Trans. Microw. Theory Tech. **50**(3), 910–928 (2002)
3. B. Fischer, M. Hofmann, H. Helm, G. Modjesch, P.U. Jepsen, Chemical recognition in terahertz time-domain spectroscopy and imaging. Semicond. Sci. Technol. **20**, S246–S253 (2005)
4. K. Kawase, Y. Ogawa, Y. Watanabe, H. Inoue, Non-destructive terahertz imaging of illicit drugs using spectral fingerprints. Opt. Express **11**, 2549–2554 (2003)
5. D.M. Mittleman, R.H. Jacobsen, R. Neelamani, R.G. Baraniuk, M.C. Nuss, Gas sensing using terahertz time-domain spectroscopy. Appl. Phys. B Lasers Opt. **67**, 379–390 (1998)
6. A. Sampaolo, P. Patimisco, M. Giglio, M.S. Vitiello, H.E. Beere, D.A. Ritchie, G. Scamarcio, F.K. Tittel, V. Spagnolo, Improved tuning fork for terahertz quartz-enhanced photoacoustic spectroscopy. Sensors **16**(4), 439/1–439/8 (2016)
7. C. Kulesa, Terahertz spectroscopy for astronomy: From comets to cosmology. IEEE Trans. Terahertz Sci. Tech. **1**(1), 232–240 (2011)
8. Z. Taylor, R. Singh, D. Bennett, P. Tewari, C. Kealey, N. Bajwa, M. Culjat, A. Stojadinovic, H. Lee, J.-P. Hubschman, E. Brown, W. Grundfest, THz medical imaging: In vivo hydration sensing. IEEE Trans. Terahertz Sci. Tech. **1**(1), 201–219 (2011)
9. F. Friederich, W. von Spiegel, M. Bauer, F. Meng, M. Thomson, S. Boppel, A. Lisauskas, B. Hils, V. Krozer, A. Keil, T. Loffler, R. Henneberger, A. Huhn, G. Spickermann, P. Bolivar, H. Roskos, THz active imaging systems with real-time capabilities. IEEE Trans. Terahertz Sci. Tech. **1**(1), 183–200 (2011)
10. K. Cooper, R. Dengler, N. Llombart, B. Thomas, G. Chattopadhyay, P. Siegel, THz imaging radar for standoff personnel screening. IEEE Trans. Terahertz Sci. Tech. **1**(1), 169–182 (2011)
11. M. Kemp, Explosives detection by terahertz spectroscopy-a bridge too far? IEEE Trans. Terahertz Sci. Tech. **1**(1), 282–292 (2011)
12. N. Shimizu, T. Ikari, K. Kikuchi, K. Matsuyama, A. Wakatsuki, S. Kohjiro, R. Fukasawa, Remote gas sensing in full-scale fire with sub-terahertz waves, in *Proceedings of the IEEE MTT-S International Microwave Symposium Digest*, (Baltimore, MD, USA, 5–10 Jun., 2011), pp. 1–4
13. Y. Yang, M. Mandehgar, D. Grischkowsky, Broadband THz pulse transmission through the atmosphere. IEEE Trans. Terahertz Sci. Tech. **1**(1), 264–273 (2011)
14. K. Ajito, Y. Ueno, THz chemical imaging for biological applications. IEEE Trans. Terahertz Sci. Tech. **1**(1), 293–300 (2011)
15. M. Tanaka, H. Hirori, M. Nagai, THz nonlinear spectroscopy of solids. IEEE Trans. Terahertz Sci. Tech. **1**(1), 301–312 (2011)
16. P. Hillger, J. Grzyb, R. Jain, U.R. Pfeiffer, Terahertz imaging and sensing applications with silicon-based technologies. IEEE Trans. Terahertz Sci. Tech. **9**(1), 1–19 (2019)
17. J. Grzyb, B. Heinemann, U.R. Pfeiffer, A 0.55 THz near-field sensor with a μm-range lateral resolution fully integrated in 130 nm SiGe BiCMOS. IEEE J. Solid State Circuits **51**(12), 3063–3077 (2016)
18. D. Wang, K. Schmalz, M.H. Eissa, J. Borngräber, M. Kucharski, M. Elkhouly, M. Ko, H.J. Ng, D. Kissinger, Integrated 240-GHz dielectric sensor with dc readout circuit in a 130-nm SiGe BiCMOS technology. IEEE Trans. Microw. Theory Techn. **66**(9), 4232–4241 (2018)
19. C. Wang, B. Perkins, Z. Wang, R. Han, Molecular detection for unconcentrated gas with PPM sensitivity using 220-to-320-GHz dual frequency-comb spectrometer in CMOS. IEEE Trans. Biomed. Circuits Syst. **12**(3), 709–721 (2018)
20. K. Statnikov, J. Grzyb, B. Heinemann, U.R. Pfeiffer, 160-GHz to 1-THz multi-color active imaging with a lens-coupled SiGe HBT chipset. IEEE Trans. THz Sci. Technol. **63**(2), 520–532 (2015)

21. N. Weimann, M. Hossain, V. Krozer, W. Heinrich, M. Lisker, A. Mai, B. Tillack, Tight focus toward the future: Tight material combination for millimeter-wave RF power applications: InP HBT SiGe BiCMOS heterogeneous wafer-level integration. IEEE Microw. Mag. **18**(2), 74–82 (2017)
22. R. Jain, J. Grzyb, U.R. Pfeiffer, Terahertz light-field imaging. IEEE Trans. THz Sci. Technol. **6**(5), 649–657 (2016)
23. P. Hillger, R. Jain, J. Grzyb, L. Mavarani, B. Heinemann, G.M. Grogan, P. Mounaix, T. Zimmer, U. Pfeiffer, A 128-pixel 0.56 THz sensing array for real-time near-field imaging in 0.13 μm SiGe BiCMOS, in *Proceedings of the IEEE International Solid-State Circuits Conference*, (San Francisco, CA, USA, 2018), pp. 418–420
24. X. Wu, K. Sengupta, On-chip THz spectroscope exploiting electromagnetic scattering with multi-port antenna. IEEE J. Solid State Circuits **51**(12), 3049–3062 (2016)
25. J. Grzyb, U. Pfeiffer, THz direct detector and heterodyne receiver arrays in silicon nanoscale technologies. J. Infrared Millim. THz Waves **36**(10), 998–1032 (2015)
26. C.A. Balanis, *Advanced engineering electromagnetics* (Wiley, Hoboken, NJ, USA, 2012)
27. K. Sengupta, D. Seo, L. Yang, A. Hajimiri, Silicon integrated 280 GHz imaging chipset with 4×4 SiGe receiver array and CMOS source. IEEE Trans. THz Sci. Technol. **5**(3), 427–437 (2015)
28. D. Filipovic, S. Gearhart, G. Rebeiz, Double-slot antennas on extended hemispherical and elliptical silicon dielectric lenses. IEEE Trans. THz Sci. Technol. **41**(10), 1738–1749 (1993)
29. E. Johnson, Physical limitations on frequency and power parameters of transistors, in *Proceedings of the IRE International Convention Record*, vol. 13, (New York, USA, 1965), pp. 27–34
30. U.R. Pfeiffer, Y. Zhao, J. Grzyb, R. Al Hadi, N. Sarmah, W. Förster, H. Rücker, B. Heinemann, A 0.53 THz reconfigurable source module with up to 1 mW radiated power for diffuse illumination in terahertz imaging applications. IEEE J. Solid State Circuits **49**(12), 2938–2950 (2014)
31. Y. Yang, O.D. Gurbuz, G.M. Rebeiz, An eight-element 370–410-GHz phased-array transmitter in 45-nm CMOS SOI with peak EIRP of 8–8.5 dBm. IEEE Trans. THz Sci. Tech. **64**(12), 4241–4249 (2016)
32. G. Karolyi, D. Gergelyi, P. Foldesy, Sub-THz sensor array with embedded signal processing in 90 nm CMOS technology. IEEE Sensors J. **14**(8), 2432–2441 (2014)
33. R. Al Hadi, H. Sherry, J. Grzyb, Y. Zhao, W. Forster, H.M. Keller, A. Cathelin, A. Kaiser, U.R. Pfeiffer, A 1 k-pixel video camera for 0.7–1.1 terahertz imaging applications in 65-nm CMOS. IEEE J. Solid State Circuits **47**(12), 2999–3012 (2012)
34. Z.-Y. Liu, L.-Y. Liu, J. Yang, N.-J. Wu, A CMOS fully integrated 860-GHz terahertz sensor. IEEE Trans. THz Sci. Tech. **7**(4), 455–465 (2017)
35. D. Yoon, J. Kim, J. Yun, M. Kaynak, B. Tillack, J.-S. Rieh, 300-GHz direct and heterodyne active imagers based on 0.13-μm SiGe HBT technology. IEEE Trans. THz Sci. Tech. **7**(5), 536–545 (2017)
36. R. Al Hadi, J. Grzyb, B. Heinemann, U.R. Pfeiffer, A terahertz detector array in a SiGe HBT technology. IEEE J. Solid State Circuits **48**(9), 2002–2010 (2013)
37. E. Seok, C. Cao, D. Shim, D.J. Arenas, D.B. Tanner, C.-M. Hung, K.O. Kenneth, A 410 GHz CMOS push-push oscillator with an on-chip patch antenna, in *Proceedings of the IEEE International Solid-State Circuits Conference*, (San Francisco, CA, USA, 2008), pp. 472–629
38. O. Momeni, E. Afshari, High power terahertz and millimeter-wave oscillator design: A systematic approach. IEEE J. Solid State Circuits **46**(3), 583–597 (2011)
39. H. Aghasi, A. Cathelin, E. Afshari, A 0.92-THz SiGe power radiator based on a nonlinear theory for harmonic generation. IEEE J. Solid State Circuits **52**(2), 406–422 (2017)
40. Y. Tousi, E. Afshari, A high-power and scalable 2-D phased array for terahertz CMOS integrated systems. IEEE J. Solid State Circuits **50**(2), 597–609 (2015)
41. R. Han, C. Jiang, A. Mostajeran, M. Emadi, H. Aghasi, H. Sherry, A. Cathelin, E. Afshari, A SiGe terahertz heterodyne imaging transmitter with 3.3 mW radiated power and fully-integrated phase-locked loop. IEEE J. Solid State Circuits **50**(12), 2935–2947 (2015)

42. N. Nguyen, R. Meyer, Start-up and frequency stability in high-frequency oscillators. IEEE J. Solid State Circuits **27**(5), 810–820 (1992)
43. F.F. Sizov, V.P. Reva, A.G. Golenkov, V.V. Zabudsky, Uncooled detectors challenges for THz/ sub-THz arrays imaging. J. Infrared Millim. THz Waves **32**(10), 1192–1206 (2011)
44. J. Marczewski, D. Coquillat, W. Knap, C. Kolacinski, P. Kopyt, K. Kucharski, J. Lusakowski, D. Obrebski, D. Tomaszewski, D. Yavorskiy, P. Zagrajek, THz detectors based on Si-CMOS technology field effect transistors–advantages, limitations and perspectives for THz imaging and spectroscopy. Opto Electron. Rev. **26**(4), 261–269 (2018)
45. H. Ruonan, Z. Yaming, K. Youngwan, K. Dae Yeon, S. Hisashi, A. Ehsan, K.O. Kenneth, Active terahertz imaging using Schottky diodes in CMOS: Array and 860-GHz pixel. IEEE J. Solid State Circuits **48**(10), 2296–2308 (2013)
46. M. Schroeter, T. Rosenbaum, P. Chevalier, B. Heinemann, S.P. Voinigescu, E. Preisler, J. Boeck, A. Mukherjee, SiGe HBT technology: Future trends and TCAD-based roadmap. Proc. IEEE **105**(6), 1068–1086 (2017)
47. K. Statnikov, J. Grzyb, B. Heinemann, U.R. Pfeiffer, 160-GHz to 1-THz multi-color active imaging with a lens-coupled SiGe HBT chip-set. IEEE Trans. Microw. Theory **63**(2), 520–532 (2015)
48. R. Lachner, Towards 0.7 terahertz silicon germanium heterojunction bipolar technology – the DOTSEVEN project. ECS Trans. **64**(6), 21–37 (2014)
49. F. Schuster, D. Coquillat, H. Videlier, M. Sakowicz, F. Teppe, L. Dussopt, B. Giffard, T. Skotnicki, W. Knap, Broadband terahertz imaging with highly sensitive silicon CMOS detectors. Opt. Express **19**(8), 7827/1–7827/6 (2011)
50. S. Boppel, A. Lisauskas, M. Mundt, D. Seliuta, L. Minkevicius, I. Kasalynas, G. Valusis, M. Mittendorff, S. Winnerl, V. Krozer, H.G. Roskos, CMOS integrated antenna-coupled field-effect transistors for the detection of radiation from 0.2 to 4.3 THz. IEEE Trans. Microw. Theory **60**(12), 3834–3842 (2012)
51. J. Park, S. Kang, S.V. Thyagarajan, E. Alon, A.M. Niknejad, A 260 GHz fully integrated CMOS transceiver for wireless chip-to-chip communication, in *Proceediongs of the Symposium on VLSI Circuits (VLSIC)*, (2012), pp. 48–49
52. J. Park, S. Kang, A.M. Niknejad, A 0.38 THz fully integrated transceiver utilizing a quadrature push-push harmonic circuitry in SiGe BiCMOS. IEEE J. Solid State Circuits **47**(10), 2344–2354 (2012)
53. M. Hitzler, S. Saulig, L. Boehm, W. Mayer, W. Winkler, N. Uddin, C. Waldschmidt, Ultracompact 160-GHz FMCW radar MMIC with fully integrated offset synthesizer. IEEE Trans. Microw. Theory Tech. **65**(5), 1682–1691 (2017)
54. A. Hamidipour, R. Feger, S. Poltschak, A. Stelzer, A 160-GHz system in the package for short-range mm-wave applications. Int. J. Microw. Wirel. Technol. **6**(3–4), 361–369 (2014)
55. Z. Wang, P.-Y. Chiang, P. Nazari, C.-C. Wang, Z. Chen, P. Heydari, A CMOS 210-GHz fundamental transceiver with OOK modulation. IEEE J. Solid State Circuits **49**(3), 564–580 (2014)
56. Q. Zhong, Z. Chen, N. Sharma, S. Kshattry, W. Choi, K.O. Kenneth, 300-GHz CMOS QPSK transmitter for 30-Gbps dielectric waveguide communication, in *Proceedings of the IEEE Custom Integrated Circuits Conference (CICC)*, (2018), pp. 1–4
57. P. Rodríguez-Vázquez, J. Grzyb, B. Heinemann, U.R. Pfeiffer, A 16-QAM 100-Gb/s 1-M wireless link with an EVM of 17% at 230 GHz in an SiGe technology. IEEE Microw. Wirel. Compon. Lett. **29**(4), 297–299 (2019)
58. K. Takano, S. Amakawa, K. Katayama, S. Hara, R. Dong, A. Kasamatsu, I. Hosako, et al., 17.9 A 105Gb/s 300GHz CMOS transmitter, in *Proceedings of the IEEE International Solid-State Circuits Conference (ISSCC)*, (2017), pp. 308–309
59. D. Headland, R. Zatta, P. Hillger, U.R. Pfeiffer, Terahertz spectroscopy using CMOS camera and dispersive optics. IEEE Trans. Terahertz Sci. Tech. **10**(5), 513–523 (2020)
60. X. Wu, K. Sengupta, Single-chip source-free terahertz spectroscope across 0.04–0.99 THz: Combining sub-wavelength near-field sensing and regression analysis. Opt. Express **26**(6), 7163–7175 (2018)

61. K.R. Jha, G. Singh, Terahertz Planar Antennas for Next Generation Communication. Springer: Switzerland, Jan. (2014)
62. M.A. Abou-Khousa, M.T. Ghasr, S. Kharkovsky, D. Pommerenke, R. Zoughi, Modulated elliptical slot antenna for electric field mapping and microwave imaging. IEEE Trans. Antennas Propag. **59**(3), 733–741 (2011)
63. K. Cooper, R.J. Dengler, N. Llombart, B. Thomas, G. Chattopadhyay, P.H. Siegel, THz imaging radar for standoff personnel screening. IEEE Trans. Terahertz Sci. Tech. **1**(1), 169–182 (2011)
64. A. Moussessian, M.C. Wanke, Y. Li, J.-C. Chiao, S.J. Allen, T.W. Crowe, D.B. Rutledge, A terahertz grid frequency doubler. IEEE Trans. Microw. Theory Tech. **46**(11), 1976–1981 (1998)
65. G. Chattopadhyay, C.-L. Kuo, P. Day, J.J. Bock, J. Zmuidzinas, A.E. Lange, Planar antenna arrays for cmb polarization detection, in *Proceedings of the 15th International Conference on Terahertz Electronics. IRMMW-THz. Joint 32nd International Conference on*, (2007), pp. 184–185
66. M.M. Kangas, M. Ansmann, B. Horgan, N. Lemaster, R. Leonardi, A. Levy, P. Lubin, J. Marvil, P. McCreary, T. Villela, A 31 pixel flared 100-GHz high-gain scalar corrugated nonbonded platelet antenna array. IEEE Antenn. Wirel. Prop. Lett **4**, 245–248 (2005)
67. J. Leech, B.K. Tan, G. Yassin, P. Kittara, S. Wangsuya, J. Treuttel, M. Henry, M.L. Oldfield, P.G. Huggard, Multiple flare-angle horn feeds for sub-mm astronomy and cosmic microwave background experiments. Astron. Astrophys. **532**, A61 (2011)
68. V.M. Lubecke, K. Mizuno, G.M. Rebeiz, Micromachining for terahertz applications. IEEE Trans. Microw. Theory Tech. **46**(11), 1821–1831 (1998)
69. A. Iacono, A. Freni, A. Neto, G. Gerini, In-line x-slot element focal plane array of kinetic inductance detectors, in *Proceedings of the 5th European Conference on Antennas and Propagation (EUCAP)*, (2011), pp. 3316–3320
70. N. Llombart, B. Thomas, M. Alonso, C. Lee, G. Chattopadhyay, L. Jofre, I. Mehdi, Silicon based antennas for thz integrated arrays, in *Proceedings of the 5th European Conference on Antennas and Propagation (EUCAP)*, (2011), pp. 3176–3179
71. T.H. Buttgenbach, An improved solution for integrated array optics in quasi-optical mm and submm receivers: The hybrid antenna. IEEE Trans. Microw. Theory Tech. **41**(10), 1750–1760 (1993)
72. J.C.S. Yates, J.J.A. Baselmans, A. Endo, R.M.J. Janssen, L. Ferrari, P. Diener, A.M. Baryshev, Photon noise limited radiation detection with lens-antenna coupled microwave kinetic inductance detectors. Appl. Phys. Lett. **99**(7), 073505/1–073505/3 (2011)
73. S.-K. Lee, M.-G. Kim, K.-W. Jo, S.-M. Shin, J.-H. Lee, A glass reflowed microlens array on a si substrate with rectangular through-holes. J. Opt. Pure Appl. Opt. **10**(4), 044003 (2008)
74. A. Cullen, J.C. Parr, A new perturbation method for measuring microwave fields in free-space. Proc. IEE-Part B Radio Electron. Eng. **102**(6), 836–844 (1955)
75. J.H. Richmond, A modulated scattering technique for measurement of field distributions. IRE Trans. Microw. Theory Tech. **3**(4), 13–15 (1955)
76. G. Hygate, J.F. Nye, Measuring microwave fields directly with an optically modulated scatterer. Meas. Sci. Technol. **1**(8), 703–709 (1990)
77. A. Joisel, K.J. Bois, A.D. Benally, J.-C. Bolomey, R. Zoughi, Embedded modulating dipole scattering for near-fifield microwave inspection of concrete: Preliminary investigations. Surf. Sens. Appl. **3752**, 208–214 (1999)
78. M.T. Ghasr, M.A. Abou-Khousa, S. Kharkovsky, R. Zoughi, D. Pommerenke, Portable real-time microwave camera at 24 GHz. IEEE Trans. Antennas Propag. **60**(2), 1114–1125 (2012)
79. S.P. Voinigescu, S. Shopov, J. Bateman, H. Farooq, J. Hoffman, K. Vasilakopoulos, Silicon millimeter-wave, terahertz, and high-speed fiber-optic device and benchmark circuit scaling through the 2030 ITRS horizon. Proc. IEEE **105**(6), 1087–1104 (2017)
80. G. Gonzalez, *Microwave Transistor Amplifers: Analysis and Design*, 2nd edn. (Prentice-Hall, Upper Saddle River, 1997)

81. K. Sengupta, T. Nagatsuma, D.M. Mittleman, Terahertz integrated electronic and hybrid electronic–photonic systems. Nat. Electron. **1**(12), 622–635 (2018)
82. J. Stake, A. Malko, T. Bryllert, J. Vukusic, Status and prospects of high-power heterostructure barrier varactor frequency multipliers. Proc. IEEE **105**(6), 1008–1019 (2017)
83. L. Ju, B. Geng, J. Horng, C. Girit, M. Martin, Z. Hao, H.A. Bechtel, et al., Graphene plasmonics for tunable terahertz metamaterials. Nat. Nanotechnol. **6**(10), 630–634 (2011)
84. T.E. Kazior, Beyond CMOS: Heterogeneous integration of III-V devices, RF MEMS and other dissimilar materials/devices with Si CMOS to create intelligent microsystems. Philos. Trans. Math. Phys. Eng. Sci. **372**, 20130105/1–15 (2014)
85. H.J. Song, Packages for terahertz electronics. Proc. IEEE **105**, 1121–1138 (2017)
86. K. Vijayraghavan, Y. Jiang, M. Jang, A. Jiang, K. Choutagunta, A. Vizbaras, F. Demmerle, G. Boehm, M.C. Amann, M.A. Belkin, Broadly tunable terahertz generation in mid-infrared quantum cascade lasers. Nat. Commun. **4**(1), 1–7 (2013)
87. K. Peng, P. Parkinson, F. Lan, Q. Gao, N. Jiang, Y.-N. Guo, F. Wang, et al., Single nanowire photoconductive terahertz detectors. Nano Lett. **15**(1), 206–210 (2015)
88. J. Gu, R. Singh, X. Liu, X. Zhang, Y. Ma, S. Zhang, S.A. Maier, et al., Active control of electromagnetically induced transparency analogue in terahertz metamaterials. Nat. Commun. **3**(1), 1–6 (2012)
89. N.K. Grady, J.E. Heyes, D.R. Chowdhury, Y. Zeng, M.T. Reiten, A.K. Azad, A.J. Taylor, D.A.R. Dalvit, H.-T. Chen, Terahertz metamaterials for linear polarization conversion and anomalous refraction. Science **340**(6138), 1304–1307 (2013)
90. H. Tanoto, J.H. Teng, Q.Y. Wu, M. Sun, Z.N. Chen, S.A. Maier, B. Wang, et al., Greatly enhanced continuous-wave terahertz emission by nano-electrodes in a photoconductive photomixer. Nat. Photonics **6**(2), 121–126 (2012)
91. A. Maestrini, I. Mehdi, J.V. Siles, R. Lin, C. Lee, G. Chattopadhyay, J. Pearson, P. Siegel, Frequency tunable electronic sources working at room temperature in the 1 to 3 THz band, in *Terahertz Emitters, Receivers, and Applications III*, vol. 8496, (International Society for Optics and Photonics, 2012), p. 84960F
92. I. Mehdi, J.V. Siles, C. Lee, E. Schlecht, THz diode technology: Status, prospects, and applications. Proc. IEEE **105**, 990–1007 (2017)
93. S.-H. Yang, R. Watts, X. Li, N. Wang, V. Cojocaru, J. O'Gorman, L.P. Barry, M. Jarrahi, Tunable terahertz wave generation through a bimodal laser diode and plasmonic photomixer. Opt. Express **23**(24), 31206–31215 (2015)
94. M. Scheller, J.M. Yarborough, J.V. Moloney, M. Fallahi, M. Koch, S.W. Koch, Room temperature continuous wave milliwatt terahertz source. Opt. Express **18**, 27112–27117 (2010)
95. L. Jofre, A. Broquetas, J. Romeu, S. Blanch, A.P. Toda, X. Fabregas, A. Cardama, UWB tomographic radar imaging of penetrable and impenetrable objects. Proc. IEEE **97**(2), 451–464 (2009)
96. J.M. Rius, C. Pichot, L. Jofre, J.C. Bolomey, N. Joachimowicz, A. Broquetas, M. Ferrando, Planar and cylindrical active microwave temperature imaging: Numerical simulations. IEEE Trans. Med. Imaging **11**(4), 457–469 (1992)
97. R. Harrington, *Time-Harmonic Electromagnetic Fields* (McGraw-Hill, McGraw-Hill Electrical and Electronic Engineering Series, 1961)

Index

I. Malhotra, G. Singh, *Terahertz Antenna Technology for Imaging and Sensing Applications*, https://doi.org/10.1007/978-3-030-68960-5

Lightning Source UK Ltd.
Milton Keynes UK
UKHW020736170522
403097UK00002B/13

9 783030 689629